T0253739

Standards

Infrastructures Series

edited by Geoffrey Bowker and Paul N. Edwards

Paul N. Edwards, *A Vast Machine: Computer Models, Climate Data, and the Politics of Global Warming*

Lawrence Busch, *Standards: Recipes for Reality*

Standards

Recipes for Reality

Lawrence Busch

The MIT Press
Cambridge, Massachusetts
London, England

First MIT Press paperback edition, 2013

© 2011 Massachusetts Institute of Technology

This book was set in Stone Sans and Stone Serif by Toppan Best-set Premedia Limited.

Library of Congress Cataloging-in-Publication Data

Busch, Lawrence.
Standards : recipes for reality / Lawrence Busch.
    p.   cm.—(Infrastructures)
Includes bibliographical references and index.
ISBN 978-0-262-01638-4 (hardcover : alk. paper), 978-0-262-52505-3 (pb.)
1. Standardization. 2. Standardization—Social aspects. I. Title.
HD62.B87   2011
389′.6—dc22

                                        2011007504

Contents

Prologue

Consider the following vignettes:

1. July 1847: American soldiers were on the outskirts of Mexico City. The Mexican general, Santa Anna, concentrated his forces in Churubusco, where there was a fortified bridge and Franciscan convent. American soldiers advanced slowly under heavy gunfire. In the convent, the Mexican soldiers and a group of Irish American deserters known as the San Patricios began to run low on ammunition. Santa Anna had another wagonload of ammunition brought to the convent. But the ammunition was of the wrong caliber for all the guns save those of the San Patricios. The Americans soon captured the convent and entered Mexico City (Nordstrom 2006).

2. California produce growers are caught between a rock and a hard place, or perhaps we should say between the birds and the bugs. New environmental programs include standards for providing a place for wildlife habitat. But food processors, worried about bacterial contamination of food that might make people sick as well as ruin their brand names, are encouraging a set of standards known as "clean farming techniques." These techniques include poisoning rodents and the removal of wetlands that harbor bacteria (Beretti and Stuart 2008; Martin 2006).

3. As part of its obligations under the Kyoto Protocol, itself part of the United Nations Framework Convention on Climate Change, the United Kingdom developed a rather substantial market for carbon emissions. However, at least initially, the UK government failed to establish standards for emissions trading. Thus, certain of the emissions credits purchased did not actually qualify under the Kyoto agreement. Other reductions would have happened regardless of the schemes, while still others were only temporary. Recently, the UK government had to step in to begin to certify that carbon offsetting schemes actually met the requirements of the

agreement (British Broadcasting Company 2007). Although the United States has yet to sign the agreement, the same problems appear to exist with respect to claims made by some U.S. firms (Elgin 2007; Green and Capell 2008).

4. In the late 1950s China embarked on the Great Leap Forward. Peasants were encouraged to build backyard steel furnaces and to produce industrial goods. Some years later in Guinea, the Chinese were building a small dam. A large number of dump trucks resembling the French Citroën 55 were imported. However, unlike the French trucks, over time, their chassis began sagging. This continued until the drive shaft was no longer connected to the rear axle. At that point the truck would grind to a halt and would usually be discarded at the side of the road. Discussions with mechanics revealed that engine parts from one truck did not necessarily fit others; each truck had apparently been made individually. In short, not only was the steel frame of the truck not of a sufficiently strong (standardized) alloy, the engine parts were not standardized. As a result of these deficiencies, the trucks remained serviceable for only a short time, after which they became piles of scrap metal.

5. The Panamax standard refers to the maximum size ship that can navigate the Panama Canal. It consists of a length of 294.13 meters, a beam (width) of 32.31 meters, and a draft of 12.04 meters (Autoridad del Canal de Panamá 2005). Since about 4 percent of world trade goes through the canal, many ships are built to precisely these dimensions. Bigger ships are more efficient but cannot fit through the canal. For example, the *MOL Encore* was built to meet those specifications (*Ship-Technology* 2008). It can carry thirteen rows of standard shipping containers on its deck, each 8 feet (2.438 meters) in width. Manufacturers of cardboard boxes designed to fit inside shipping containers without wasting any space must make them of such dimensions that the sum of the widths of the boxes is 7 feet 7 inches (2.3111 meters), as the interior width is slightly smaller than the exterior width. Products that do not fit within these dimensions must either be disassembled for shipping or shipped as unpackaged cargo requiring special handling at much higher rates.

6. Is Pluto a planet? Like the other eight planets, it goes around the Sun in an elliptical orbit. Indeed, dictionary definitions identify planets as large heavenly bodies that revolve around stars. But recently the General Assembly of the International Astronomical Union—the organization that sets standards for planets—determined that Pluto was not a planet after all. It was demoted to the category of "dwarf planet," a title it now shares with Ceres and an object called UB_{313}. Planets have been confined to our solar

system and to bodies that have sufficient mass to have cleared the neighborhood around their orbit (Sykes 2008). Not all astronomers are happy with this decision and hundreds have signed a petition to return to the older definition. Moreover, soon afterward, in March 2007, the New Mexico House of Representatives voted unanimously in favor of a resolution that Pluto is a planet (Holden 2007). That in turn raised another question: Just which earthly body has the final authority over the classification of heavenly bodies?

What unites all these vignettes is that, although they are about entirely different topics, they all involve standards. In fact, one can pick up the newspaper nearly every day and find stories about standards—standards for people, for the environment, for consumer products, for the welfare of animals, for accounting for public funds, for the acceptable stress of highway bridges, for health care, for education, for just about everything. Yet these standards are nearly always considered separately; their similarities and differences *as standards* are hardly ever discussed. In this book I aim to remedy that situation.

Acknowledgments

No volume of this complexity can be the work of one person. Like all authors, I have benefited from conversations and correspondence with numerous colleagues, acquaintances, friends, students, and critics. Paul Thompson provided comments on an early draft of the manuscript. William H. Friedland, Stefano Ponte, Elizabeth Ransom, Vladimir Shlapentokh, and Josh Woods carefully perused the entire manuscript and provided very helpful comments. In addition, the late Susan Leigh Star spent countless hours with me discussing a wide range of issues. Warren Samuels, Nicolas Mercuro, and Alan Schmid advised me on some fine points of economics. Kyle Powys White provided some useful comments on justice. Students in my graduate class on science and technology—Sana Ho, David Holt, Tazin Karim, Delanie Kellon, Xueshi Li, Allison Loconto, and Margaret Robinson—were subjected to the painful process of reading the manuscript as well. Martin Brett kindly replied to an unsolicited email and provided me with help in locating some medieval manuscripts on standards. Rhoda H. Kotzin helpfully pointed me toward the work of J. O. Urmson. Marietta Rice introduced me to the Families and Democracy Project.

Several short stays away from my office were particularly helpful. First, the Genomics Policy and Research Forum at the University of Edinburgh graciously provided me with a month of largely uninterrupted time during which I could discuss and work on the book. Similarly, the Centre for Business Relationships, Accountability, Sustainability and Society (BRASS) at the University of Cardiff provided me with two months of uninterrupted time.

I have split my time between Lancaster University and Michigan State University for the last several years. Hence, part of this work was supported by the Research Programme of the ESRC Genomics Network at Cesagen (ESRC Centre for Economic and Social Aspects of Genomics). The support of the Economic and Social Research Council is gratefully acknowledged. In addition to my Michigan State University colleagues, many of my new colleagues at Lancaster, especially Brian Wynne and John Law, have endured listening to my comments about standards and have raised important questions. Talbot Huey and Agnes Widder, among other librarians at Michigan State University and Lancaster University, went out of their way to help me find a variety of ephemera.

My wife, Karen Busch, tolerated my endless rantings on various aspects of standards over various meals, while driving from one place to another, and even in bed before (or instead of) sleeping. Perhaps as a result of that, our four grandchildren—Grace, Lilu, Niko, and Zoe—may eventually find this book to be of help in making sense of an increasingly complex world. To all of these institutions and people, the standard thanks and disclaimers apply. A heartfelt thanks to each and every one of them. Any errors are entirely mine.

Introduction

[W]hat I would like to show is . . . how a particular regime of truth, and therefore not an error, makes something that does not exist able to become something. It is not illusion since it is precisely a set of practices, real practices, which established it and thus imperiously marked it out in reality.
—Michel Foucault (2008, 19)

The hardcover edition of this book was typeset using a 9-point proportional font, bound in signatures of sheets of paper that meet the guidelines of the American Library Association, and should last for several hundred years. Normally readers pay little or no attention to these seemingly trivial facts. Nor do most philosophers, historians, or social scientists. We may marvel at the engineering of brick walls by the ancient Chinese, the clarity of sound of our new iPod, or the skill employed by a heart surgeon. Similarly, we may curse the complexity of the tax forms we complete each year and the speeding ticket we have just received. However, most of us lose interest rapidly at the mention of standards for bricks in the ancient world, for the earpieces in our iPods, for enforcement of traffic laws, or for the construction of tax forms. Yet without people concerned about and working on—indeed, obsessed with—these and myriad other standards, our modern world would be an impossibility. Yes, it is true: we have never been modern, as Bruno Latour (1993) asserts in the title of his book, but we have expended an astonishing amount of energy trying to be so. The modern project has not been successful to date and likely will not be in the future, but a project it has been nevertheless. Standards have been central to it.

Yet even as standards have been the subject of discussion in education, health care, information technology, and product quality circles, among others, only a handful of authors have examined just what standards are and what standards in these very diverse fields have in common. Do standards for scientific instruments, citizens, health care, athletes, horses, frozen peas, tax forms, and automobiles have anything in common? Is that connection a fundamental one that says something about the way in which we organize our world, or is it merely a semantic curiosity, perhaps a leftover from previous and now archaic meanings?

Every day as we go about our business we take for granted a vast array of standards, each of which has been and continues to be the subject of intense negotiation. Paradoxically, we see these standards (or rather we do not see them) as taken-for-granted aspects of our daily life. As the great champion of liberty Benjamin Constant ([1815] 1988, 75) observed nearly two centuries ago, "man adapts himself to those institutions that he finds already established, as he does to the laws of physics. He adjusts, in accordance with the very defects of such institutions, his interests, his speculations and his entire plan of life."

Put differently, we find ourselves in a position like that of Monsieur Jourdain in Molière's celebrated play, *Le bourgeois gentilhomme*, who did not know that everything he said was in fact prose. When we walk from one room to another inside our homes, or take a stroll in a city park, or ride the elevator to an upstairs office, or purchase a meal in a restaurant, or open a book to read its contents, or apply for a driver's license, or take an exam, or wash up after work—in each of these instances and many more we tacitly accept and adapt to a vast array of standards.

Perhaps because these standards are so taken for granted, they are rarely the subject of discussion in circles beyond those in which they are formulated. They are even more rarely the subject of discussion in the public square, in democratic institutions of government, or among friends. Indeed, standards are so taken for granted, so mundane, so ubiquitous, that they are extremely difficult to write about. They are usually noticed only when they fail to work. After all, history is largely the record of *events*, of things that are not routine, of changes that have occurred. In attempting to write this book I quickly discovered that what seemed a rather straightforward project involved inquiring into a vast number of subjects about which I can hardly claim any expertise. Yet, as I hope to make clear, standards shape not only the physical world around us but our social lives and even our very selves. Indeed, standards are the recipes by which we create realities.

The heart of the difficulty is that all standards invoke the linguistic categories we also use to organize the world. For example, a standard for a Valencia orange implies that the user of the standard knows the meaning of Valencia (a particular variety that was developed in California but is named for Valencia, Spain) and of orange (a type of citrus fruit). Of course, such a standard would not apply to a member of the House of Orange who had been named Valencia. While we could—and often do—make the definition more explicit, to do so we must invoke other categories that remain undefined.

A vast literature exists in linguistics on the nature of categories, and I was much tempted to dig into that in writing this book. However, while standards must necessarily invoke the categories that language provides, they are about much more than that. Standards are about the ways in which we order ourselves, other people, things, processes, numbers, and even language itself. To put it slightly differently, standards are where language and world meet; better still, the widespread use of standards and their virtual inescapability illustrates the interpenetration of language and world. Indeed, even as some standards are the subject of more or less formal definitions in words, others are physical objects. Thus, as a category of experience, standards span the material and the ideal, the positive and the normative, the factual and the ethical, the sacred and the profane.

This book can hardly claim to be the last word on the subject of standards. Indeed, as one colleague remarked in passing, it would take at least a dozen volumes to adequately address the subject. I suspect that the reader is even less likely to read a dozen volumes than I would be to write them.

Nor can it be said to be the first word, as others have also written about standards. What I hope can be said is that it is provocative, challenging, and perhaps offers some new insights into this much neglected subject.

Perhaps this work is best seen as that of someone who got his feet wet in the subject, and eventually waded in up to his waist. It is certainly not that of someone who has been swimming in deep waters far from shore. Given the vastness of the subject, if I had done that I would surely have lost my bearings and drowned. But by staying close to shore I may have missed some features of the ocean of standards that confronts us. I have little choice but to leave that to the readers to judge, and to future observers to correct. As the great Argentinean essayist Jorge Luis Borges (1964, 201) wrote, "every writer creates his own precursors." We don't have that luxury with our readers.

But, it may be argued, this subject has been addressed before by a vast array of scholars, scholars who did not use the term "standard" at all or did not use it in the way that I do in this book. Here I must concede the point. Thousands of social scientists have written about mores, norms, habits, customs, manners, laws, rules, practices, guidelines, social structures, and other recurring and even standardized features of the social landscape. Their work has paralleled that of the thousands of other authors—usually scientists and engineers—who have written about the characteristics of (or standards for) myriad objects. And thousands of philosophers have discussed what we mean when we hold certain values to be dear to us. Indeed, that is certainly the case, and anything that I say here must necessarily be said while standing on the shoulders of quite a few giants. But let me suggest some differences between what these scholars have done and what I shall attempt to do.

First, there is a tendency among scholars and the general public to assume the independence of these concepts and the phenomena to which they refer. This is reflected in the tendency of each discipline to study only one or a few of these standards. Hence, one can say with only a bit of exaggeration that anthropologists study only customs and traditions. Jurists study only laws and legal interpretations. Sociologists, despite the strong interest in law displayed by some of the founders, among them Émile Durkheim and Max Weber, relegate law to a minor subdiscipline and focus instead on social structure. Educators study educational standards, while health professionals study medical standards. Computer scientists study only standards for software and hardware. Food scientists study only standards for food products and processing equipment. In short, each field tends to reinvent this group of phenomena.

Moreover, many of these scholars distinguish sharply between the world of human interaction and the world of material objects. For the most part, those in the humanities and the social sciences study people, while those in the natural sciences and engineering study things. Occasionally they peer over the border between people and objects. They study the physiology of human bodies, or the impact of technologies on society. They study the economics of widget production or how we configure our bodies when we sit. But they hardly ever venture across the border. Borrowing a page from the science studies literature (especially Latour [1987, 2005]), I shall deliberately transgress these boundaries.

Second, I shall attempt to go beyond mere transgression, arguing that the boundary between humans and other material objects itself is (or at least has become) a barrier to our understanding and to needed actions. Indeed, the boundary, the wall, has been there for so long that we scarcely notice its artificiality anymore. Although it has become an unquestioned aspect of the landscape, its existence can and should be challenged.

Third, I argue for symmetry between standards for people and standards for things. Rather than assuming that they are and must be different, I begin by asserting their essential sameness, and see where that leads me.

Fourth, in constructing some of these new concepts and edifices, I do not attempt to build another wall. After all, tearing down one wall simply to move the boundary elsewhere and construct another, perhaps even more impenetrable wall there would ultimately be a delusional and self-defeating task. Instead, I attempt to show how these various categories overlap and interpenetrate each other, as well as how different perspectives may illuminate different aspects of a unified world of people and things.

Fifth, in this unified world, as a great deal of the science studies literature has shown, we no longer need to explain the actions of people solely by the actions of other people or the actions of things solely by the actions of other things. We will have at our disposal a new means for understanding how our world is put together, and perhaps for putting together better future worlds. We will see that the world is and must be made up of people and things. We will also see that standards are (better or poorer) recipes that we use to hold the world of people and things together.

My task here is made all the more urgent by the proliferation of people and things. The human population has now reached an unprecedented size. Living together with other humans and with a vast array of things is now an imperative. But our ability to remake the physical world threatens to lead to our very extinction, whether through nuclear annihilation or through human-induced global environmental change. We are in a position not unlike that of the sorcerer's apprentice, unable to stop the processes we have put in motion. But perhaps the asymmetrical treatment we provide to humans and nonhumans—the different means by which we create standards for humans and for things—is part of the problem.

Put differently, looking at the world of people and things, of material objects and thoughts, of ideals and realities, of social structures and human agencies, through the lens of standards provides us with one perspective, one vantage point, one means of understanding, one avenue of action—but surely not the only one. The world is far too vast for any one perspective, no matter how nuanced and insightful, to capture all of its infinite complexity. This is why in English and in several Romance languages we talk of doing research, or *recherche* or *ricerca*—literally of searching again.

But why should standards be needed at all? Why can't we simply agree on the characteristics of things, people, practices, and move on? Why do we need to endlessly negotiate over standards at all? Indeed, in a perfect world in which everyone made rational choices, in which perfect information was readily available, in which what went on in my head was precisely the same as what went on in everyone's head, we would not need standards. We would all be like Humpty Dumpty, who argues in Lewis Carroll's *Through the Looking Glass*, "When I use a word, it means just what I choose it to mean—neither more nor less." In such a world the meaning of all words, things, services, practices, measurements, and so on, would all be equally clear. Standards would be largely redundant since, as we shall see, standards always produce *partial and impermanent* orderings and never complete ones.

But of course, that is not the world we live in. We live in a messy world, one in which myriad standards, habits, traditions, and laws that we collectively make collide with each other and must be modified. Furthermore, all the choices we make in the course of our everyday lives must also necessarily be built on assumptions about the permanency of these structures. What we prefer to think of as a free choice is shot through with standards and regulations of multiple kinds.

Standards and Choices

While standards and choices clearly involve different forms of action, the one virtually unconscious and automated, the other conscious and goal-oriented, in practice, both are implicated in all situations. It is merely a question of which account should be in the foreground, which in the background. Following is a typical "choice" situation as described from a decision-making, rational choice, or neoclassical economics perspective:

Mary wants to buy breakfast cereal. She goes to the supermarket and finds a wide range of different cereals. She quickly decides that it is corn flakes that she wants, but she is undecided as to which brand. She compares the prices and determines that Brand Z is the most economical. Moreover, since she considers that corn flakes are a commodity, with little or no difference among brands, she decides to buy Brand Z. She picks the box off the shelf, takes it to the cashier, and pays for it.

Such a scenario is highly plausible, but the act of choosing it makes no sense unless a number of key conditions involving standards are realized:

1. Joe, the distributor's agent for corn flakes, has stocked the shelves in his usual manner such that the boxes do not come crashing down on Mary when she takes one off the shelf. He has put the corn flakes on a high shelf as the lower shelves are reserved for other cereals (usually with high sugar content) targeted at children.

2. The contents of the corn flakes boxes are of the same volume, weight, and quality for each box of the same size and brand because the machine

tenders at the Brand Z Corn Flakes Company are particularly careful in their monitoring of the machines that mix the ingredients, flake and bake the corn, and fill and seal the boxes.

3. Mary can read the words on the package and understand that it contains corn flakes and not dog food or detergent. (It should be recalled here that a significant portion of the world's population is illiterate.)

4. Mary is numerate and understands that $1.59 is less than $1.72, and that 13 oz. is less than 1 lb., 2 oz. (Recent studies of members of isolated societies strongly suggest that all but the simplest forms of numeracy are learned [Beller and Bender 2008; Gordon 2004; Wiese 2007]. All of us who memorized multiplication tables in elementary school know that advanced forms of numeracy are an acquired skill.)

5. The box that Mary grabs has an expiration date one month from now, while those farther back on the shelf have later expiration dates. This is so because Joe rotates the stock of corn flakes such that none of it is ever stale.

6. Mary and the cashier must use habitual gestures, words, and actions that lead to the successful completion of the joint project. For example, Mary must know that supermarkets are self-service stores in which the buyer is expected to remove the box of corn flakes from the shelf, carry it to the cashier, and place it on the conveyor belt. She must know that she will be expected to pay for the corn flakes. Moreover, she must pay with U.S. dollars (if in the United States), and not with some other currency or by bartering using the novel in her purse.

7. The project is interpreted by both Mary and the cashier such that they both act in a manner that brings the sale to fruition. This does not necessarily mean that they interpret it in the same way, only that they engage in actions that lead to the completion of the joint project. It may well be that the cashier detests her job and is ready to burst out into angry speech, but refrains from doing so out of fear of being embarrassed or possibly losing her job. Mary may well be hungry and may wish to sample some of the corn flakes, but she knows that custom (and perhaps law) prohibits her from doing so until she pays for the item.

For Mary to make the conscious decision to buy the corn flakes, then, many other persons must be engaged in making other decisions as well as engaged in habitual actions, and many standards must be maintained and supported. All of this is nearly invisible to Mary and the cashier unless the process breaks down somewhere. It is equally invisible (or taken for granted) to the economist, marketing researcher, or social scientist who describes the above scene in terms of decisions made by Mary. In fact, no matter how much we might wish to pin down every last detail leading to Mary's decision, we would find ourselves at a loss (although admittedly, the more spatially, temporally, and socially removed we got from the situation, the less consequence such knowledge would have for us).

The reverse is true as well. The following situation is described in terms of norms, structures, habits:

John is a dedicated runner. He is awakened by the alarm clock at 6:00 a.m. every morning, rain or shine, puts on his running outfit, and goes for a run. He knows the route like the back of his hand. He exits his home on Apple Street and runs to the corner of Maple, where he turns left. At each corner he glances both ways before crossing the usually deserted streets. He continues for eight quiet blocks until he reaches Walnut. There he turns left again and waves at a little boy who always seems to be sitting by the window of his home at this hour of the morning. The boy waves back. John continues eleven blocks to the somewhat busier Main Street. He makes another left onto Main Street and runs past a few pedestrians. He continues until he reaches Apple again, where he makes another left turn. Then he continues down Apple to his home. By 7:00 a.m. he is in the shower washing off the sweat. While running, John thinks about his family and all the things he will do that day—almost anything but running.

This plausible scenario glosses over myriad decisions that must be made for the habitual situation described to come into being:

1. Many years ago, someone or some group designed the street grid that includes each of the streets noted in the account. The group determined the names, the width of the streets, the locations of the sidewalks, and the type of pavement.

2. Someone designed the running shoes that John is wearing, competing against other designs deemed too costly, too flimsy, or as providing inadequate arch support. Through a complex distribution chain in which prices for various parts of the shoes and costs for transport were established, these shoes eventually came into John's possession.

3. Yet other persons designed and built the alarm clock that John uses, deliberating at length over its design, functionality, and ease of use, as well as ensuring that it met hundreds of safety and other standards.

4. The short piece of sidewalk at the corner of Maple and Jefferson was replaced last week after the city inspector, Jane, noted that it was tilted more than the maximum one inch and represented a pedestrian hazard. Moreover, since by law each of the property owners along John's path is required to keep his or her stretch of sidewalk clean on penalty of a fine, John's path is normally clear.

5. The traffic on Maple is slow because city workers have installed speed bumps in the middle of each block, discouraging many drivers from using

that street. This makes it unnecessary for John to pay too much attention to traffic.

Each of these decisions—a number sufficiently vast that we could never reconstruct them all—is taken for granted by John as he runs through his neighborhood, allowing him to devote his thoughts to other subjects. These decisions are invisible to John, so much so that he doesn't even notice the new coat of paint on the house at 435 Maple. Were that not the case, were it necessary for John to worry about where to put his foot each time it came down because the sidewalks were nonexistent or in terrible disrepair, whether his shoes would last through the entire hour, or whether passing vehicles might unexpectedly cross his path, he would likely not run in the first place.

In short, conscious, rational decisions *and* standards, habits, and routines are necessary in *both* situations. It is not a case of either/or but of both/and. In fact, some combination of decisions and standards is essential. If everything for us were to be standardized, habitual, routine, we might get along fine in the short run, but we would set ourselves up for a great fall. When the world changed, we might not notice until it was too late, likely resulting in catastrophe. But if everything were to require conscious decisions, we would be equally lost since we would not know where to begin. What would words mean? How would actions be understood by others? All forms of action would rapidly become impossible.[1]

It is also worth emphasizing that both conscious decisions and habitual actions take place in situations containing both people and things. The situation includes those persons and objects deemed relevant by Mary or John, based on their respective interpretations of that situation. Indeed, the situation is interpreted in large part by virtue of the relations between the persons and things in the situation and their relations to the environment surrounding it. Thus, Mary can tell who the cashier is by virtue of that person's garb and position next to the register and conveyor belt. Mary can choose the box of corn flakes she wants because of the orderly placement of the items for sale on the shelves and the signage pointing her to the cereals aisle.

At the same time, Mary's situation does not include the cans of tuna fish farther down the aisle or the actions of George, who is shopping for canned fish. Similarly, John's situation does not include the occasional vehicle that passes by or its occupants. Those persons and objects are part of the environment, the background, within which the situations are set, but they are irrelevant as they do not enter into either the judgments of the person or persons involved in the situation as to what type of situation it is or the choices or habitual actions undertaken by those persons.

Of course, all situations are tentative. It is possible that George, in a great hurry to get home, runs past Mary and bumps her, causing the corn flakes box to fall and break open. It is possible that in his hurry he pushes Mary aside just as she reaches the checkout line, causing the cashier to politely ask George to step aside and let Mary pass. Or perhaps when Mary gets to the register she discovers she has left her money at home and cannot pay for the corn flakes. We might even envision an earthquake occurring, one sufficiently violent to make Mary forget the corn flakes and leave the store by the nearest exit. We could, of course, extend this set of what-ifs indefinitely, and we could do the same for the story of John.

What do these two tales tell us? Judgments are required to define each case. Both John and Mary must judge that a given situation is of such and such a type, requiring such and such actions. In Mary's case the situation must be judged as one in which a choice must be made based on a calculation, while in John's case it must be judged as one requiring habitual action. (We, as observers, must also judge the situations to be of one or another type.) Moreover, if the party or parties to either situation suddenly found that the situation had changed, that new phenomena had entered into the situation, they would need to engage in a reevaluation, a new decision, or another judgment. Put differently, if the habits, standards, and rules that had previously held were found wanting—and in a world in which change is endemic this possibility must always eventually be the case—then they would need to be revised, modified, or replaced. Furthermore, what would on the surface appear to be matters purely or largely of individual psychology are intimately bound up with a larger temporal, spatial, and social world. Finally, what would appear to be matters solely about standards must necessarily include choices as well. But I have just argued that standards are paradoxically tentative, impermanent, and subject to revision. How could that be the case?

How Standards Make Metaphors Real

That standards are incomplete and tentative becomes apparent when one realizes they are *measured comparisons*. Standards always incorporate a metaphor or simile, either implicitly or explicitly. This is most obvious in the case of standards that are physical objects. An example is the weights used on a scale. They are used to *compare* the weight of the weight (a pound, for example) with the weight of some object for sale (a piece of meat). But the metaphor does not stop there. A *standard* weight can only be used (properly) by comparing other weights that are used in everyday commerce to it and reasoning metaphorically that this weight used in weighing meat is the same as this other (standard) weight. Here the meaning of "the same as" implies a caveat: it is "the same as" *for the purposes at hand*. The equivalence of the two weights is limited (as all metaphors are) by the purpose to which they will be put. Thus, we would not expect weights used to measure meat to be acceptable for weighing subatomic particles. Furthermore, the equivalence is based on the characteristics of interest in a given situation. For example, we might well have two weights that are said to be the same but that differ markedly in their shape and metallic composition, or two pieces of meat of the same weight, but one a nice juicy steak and the other mostly fat and gristle.

This same type of material metaphor is found in the many reference materials used to calibrate all sorts of scientific and industrial equipment. For example, some 1,300 test materials have been developed by the National Institute of Standards and Technology (figure 0.1).[2] Reference

Figure 0.1
Set of Hoke gage blocks. The blocks were used to calibrate precision equipment in machine shops. The faces of each block were plane and parallel. They were of correct nominal length within 0.000005 inch at 20°C (1919).
Source: National Institute of Standards and Technology.

materials of this sort are critical to the creation of modern industrial societies. Hence, by 1927 the former Soviet Union had begun to invest in the development of such standards (which was started under the czars) as a means of speeding up the process of industrialization (Shaevich 2001).[3] Today, more than 140 nations have their own national standards organization (Loconto and Busch 2010).

However, we can easily see that the metaphorical character of standards extends to other domains as well. I might try on a suit and declare that it is a "good fit." Though the language is far less precise, a metaphorical comparison is implicit here too: the suit is implicitly compared to other suits that I might try on, as well as to the (standard) shape of my body.

Nor need we stop at physical objects. When we say that Doris is a good fit for the job of head of marketing, we mean she can be compared favorably with the standard but abstract (and less precise) set of characteristics we would expect of a good head of marketing. These might include, for example, a degree in the field, experience in similar roles, and evidence of having increased sales in a previous position. In short, she measures up to the job.

Statisticians often use the term "goodness of fit" to evaluate whether a given statistical solution to a problem "explains" the data that have been collected and examined. The same approach applies, however, in all cases in which humans or nonhumans must be tested to determine their goodness of fit to a standard. (It should be noted that the tests and testing apparatus are not themselves part of the metaphor. Instead, they determine whether the metaphor applies. Put differently, they determine "what counts." Thus, for example, a balance might be used to determine whether a given weight is the same as the standard weight. The balance tests the weight to see if it measures up. Similarly, my body might be considered a test of the goodness of fit of a new suit. This contrasts with the case of the candidate for the position of head of marketing. She would be subjected to a number of tests, including, for example, completion of an MBA, documented experience in marketing, evidence of having increased sales in her current position, and so on. Here too the tests themselves are not part of the metaphor; however, they do allow us to determine whether the metaphor is appropriate.)

Through the metaphors that they make concrete, standards permit us to create complex sociotechnical networks. As standards are used, people and things are tested, and we determine what shall count. Those people and things that pass the tests or make the grade are drawn into various networks. In contrast, those persons and things that fail the tests do not count; they do not make the grade. They are *downgraded*. This book develops this theme, among others, by querying standards from a variety of perspectives. Below I provide a guide for traveling through the thicket of issues raised by standards.

Some Navigational Advice

As the title of this book suggests, standards are means by which we construct realities. They are means of *partially* ordering people and things so as to produce outcomes desired by someone. As such, they are part of the technical, political, social, economic, and ethical infrastructure that constitutes human societies. How this infrastructure is constructed, contested, modified, enforced, and abandoned is the overarching theme of this book. But standards themselves have a history, a technics, a politics, an economics, and an ethics, parts of which can be reconstructed. In delving into that history I have selected a central organizing theme for each chapter.

Standards mean many things to many people, and each of the myriad definitions of standards is still current today. As the title of chapter 1, "The Power of Standards," implies, standards are intimately associated with power. To highlight that power I carefully examine several standards while revealing some of their peculiar features. Although taken for granted, standards are hardly a unitary phenomenon. Examining the many contexts in which standards are invoked gives some clues to their power and helps distinguish them from norms, customs, traditions, and laws. Then I examine the importance of commensurability and symmetry between people and things in the formation and use of standards. I argue that all standards are of four types, although those types blend into each other at the margins. I also distinguish among standards, tests, and indicators, considering at length what is needed to construct a test. I note the metaphorical nature of standards and the measures incorporated in them, as well as the ways in which they are layered and path dependent. I examine how standards may be linked to different types of sanctions. I conclude by showing how standards are means by which we perform the world, how they are indeed recipes for realities.

Chapter 2, "Standardizing the World," begins by providing a brief (and necessarily selective) history of standards, culminating in the moral and technical project of standardization initiated with the rise of modern science. The last three centuries were marked by great standardizing projects for both people and things, as examples from many domains of social life attest. An important consequence of this standardizing movement was a tendency toward the standardization of cognition.

Chapter 3, "From Standardization to Standardized Differentiation," takes up the response to the project of standardization: the use of standards to differentiate. Beginning with the rise of non-price competition, interrelated changes in transport, communications, packaging, and the neoliberal reforms of the late twentieth century paved the way for standardized differentiation of both people and things. As a consequence, cognition has also been differentiated.

The worldwide explosion of standardized differentiation has been intertwined with demands for audits, for certifications and accreditations. After briefly examining the various forms of certification in chapter 4, "Certified, Accredited, Licensed, Approved," I consider why and how these processes have, seemingly paradoxically, both enhanced and undermined trust; how they have brought risk to the fore; and how technoscience has been implicated in the very processes of audit. The chapter concludes with a discussion of the violence that may be done by audits.

In chapter 5, "Standards, Ethics, and Justice," I suggest that standards are intertwined with ethics. Reflections of analytic philosophers on the nature of grading show that it necessarily involves ethical judgment. Standards may also be interrogated with respect to consequences, rights, and virtues. More pragmatically, I query how particular standards are justified and propose that different justifications are, and must be, employed in different social worlds. An extension of the concept of value chains to all values might be a helpful way of grappling with the ethical questions embedded in standards.

Standards also pose difficult challenges for democracy as they appear to require some form of technoscientific expertise if they are to be effective. This inevitably leads to the age-old debate about expertise and democracy, the subject of chapter 6, "Standards and Democracy." It seems that expertise, though necessary, it is not sufficient. New democratic means of governing our contemporary technoscientific society must be invented.

Finally, in the concluding chapter I take up the neoliberal challenge. We cannot afford to continue down the current path to standardized differentiation in light of the many crises facing the entire planet. I argue that failure to pay adequate attention to standards is tantamount to abandoning our notions of democracy, and I highlight twelve issues that must be considered if standards are to be fair, equitable, and effective.

That said, the book patently does not address a number of issues in the field of standards. For example, there is no thick description of the machinations of standards development organizations (for which see Salter 1988; Schmidt and Werle 1998), nor is there much on the relations between standards and innovation (for which see Blind 2004). I have written little about how or whether one should draw the boundary between voluntary standards and mandatory regulations, and I say nothing about the relative economic advantages and disadvantages of standards and standardization (for which see Egyedi 1995). Finally, while I more than occasionally stray into customs and traditions, habits and mores, on the one hand, and laws and regulations on the other, not doing so would force the analysis back into the very boxes I try to avoid. In contrast, I try to focus on those standards that are codified in language in one way or another or that are embodied in material objects.

1 The Power of Standards

[A]t least indirectly a vast amount of "private" activity affects the choices available to the people at large just as effectively as a governmental rule.
—Robert G. Dixon Jr. (1978, 10)

Any study of standards is complicated by the existence of numerous meanings of the term, which are often used virtually interchangeably. A few of these uses are most relevant to this book.

It appears that the term first came into general use in English at the Battle of the Standard in 1138. According to a contemporary observer, Richard of Hexham ([1138] 1988, 67), in this battle between the English and the Scots, the English

soon erected, in the centre of a frame which they brought, the mast of a ship, to which they gave the name of the Standard; whence those lines of Hugh Sotevagina, archdeacon of York:

Our gallant stand by all confest,
Be this the Standard's fight;
Where death or victory the test,
That proved the warriors' might.

On the top of this pole they hung a silver pix [i.e., box or vessel] containing the Host, and the banner of St. Peter the Apostle, and John of Beverley and Wilfrid of Ripon, confessors and bishops. . . . By this means . . . they might observe some certain and conspicuous rallying-point, by which they might rejoin their comrades, and where they would receive succour.[1]

In these few lines we can discern the outlines of some future uses of the term. Prominently, it is related to the notion of taking a stand, that is, of having a position, having standing, as well as being the best. It also establishes the link between a standard and a test—in this case, the "death or victory" that would prove or disprove "the warriors' might." Moreover, Richard of Hexham noted the legitimacy the standard had by linking it to the highest authorities of the day. It is the "king's standard," as well as a symbol of the Church. Likely this helped bolster the morale of the English soldiers, much as seeing it collapse would have been a blow to morale. Companies such as Standard Oil, American Standard, Standard Textile, The Evening Standard, the Standard Bank of South Africa, and thousands of others continue to use the term in this manner.

A second meaning brings us closer to its more common use today, as an exemplary measure or weight. In this sense it is similar to the object on a pole in that it refers to a physical object—for example, a standard weight such as a kilogram or a standard length such as a meter stick or standard time such as that measured by the atomic clock—and is in some sense the best, the most accurate and precise, for the purpose at hand. However, unlike the standard carried by a soldier or sailor, such standards are usually kept under lock and key in special facilities to guarantee their integrity. Indeed, many standards for kilograms are still kept in special environments to preserve their stability. Were they not kept in such a manner, they would likely change over time. At the same time, if they are to serve effectively as standards, they have to be occasionally removed from their protected environments so that they can be compared with copies, which can then be distributed across space and compared in turn with the weights or measures used in the profane world. Thus, as Joseph O'Connell (1993) has suggested, standards of this sort create universality by the circulation of particulars.

Furthermore, this type of standard is not confined to inert physical objects. Animals of all sorts are revered in this way. One of Napoleon's horses, Marengo, which carried its rider through the battle of the same name, has been immortalized in a painting, and its skeleton is on display at the National Army Museum in London. Similarly, prize bulls, pedigree dogs, and botanical and zoological specimens are kept in protected environments and occasionally compared with those in circulation.

In the same way, people who embody particular virtues or vices are frequently held out as exemplars, somewhat similar to the use of physical standards. In general, such persons are those who are long dead. While Shakespeare may have been correct in having Antony say that "The evil that men do lives after them; The good is oft interred with their bones," the reverse is true as well. In the American national pantheon, former presidents such as Abraham Lincoln are revered as courageous patriots. Douglas MacArthur is revered as a great general, Elizabeth Cady Stanton as a champion of women's rights, Martin Luther King Jr. as a champion of civil rights, and Babe Ruth as a great baseball player. Others are exemplars only insofar as they are reviled: hence, Benedict Arnold is seen (in the United States, but not in Britain!) as an exemplary traitor. Other nations have similar standards of greatness and revulsion.

In a few instances living persons are raised to this status; however, since living persons are fallible human beings—the bad has not yet been interred with *their* bones—their status as exemplars is far more tenuous. Indeed, in some societies, massive efforts must be made by the state apparatus to maintain such claims. And more often than not, that very effort undermines the claims; one need only look to Kim Jong-il, of North Korea and his designated successor, Kim Jong-un, to see the problems inherent in living exemplars.

In each of these instances, however, real living persons are compared with exemplars, in much the same way that weights used in ordinary everyday commerce are compared with those kept as standards. Of course, unlike weights, which can be compared with great precision, no similar balance exists to compare real persons with exemplars. Hence, reasonable people may disagree over what constitutes exemplary behavior.

A third meaning is more abstract and less precise. One may talk, for example, of someone who insists on a high standard of decorum, or who has a low standard of living. Unlike the highly precise physical standards, always embodied in *particular* objects (including human bodies), standards of this sort are necessarily far more ambiguous. They often involve both actions and an array of physical objects, themselves perhaps subject to a particular ordering. For example, a high standard of decorum would require both a set of actions by the person for whom the high standard is claimed and that person's employing an array of objects in a particular way. Furthermore, when the term is used in this manner, the moral character of all standards is far more evident.

A now dated but quite amusing standard of this type is the *Warrant of Precedence* used in India under the British Raj (see box 1.1). It provided a formalized standard for ranking every British subject in India. It was commonly used to specify the rights and duties associated with a particular rank. Not surprisingly, the queen was at the top of the list. Civilians with "less than four years' standing" were at the bottom. At formal gatherings such as banquets and state functions, it served as a standard of decorum.

To seat the governor-general next to a third-grade officer of the financial department was to commit a terrible faux pas, to show that one did not maintain the proper standards.

Box 1.1
The *Warrant of Precedence* (excerpt)

Precedence in India is regulated by a Royal Warrant dated the 6th of May 1871, a copy of which is subjoined.

VICTORIA, by the Grace of God, of the United Kingdom of Great Britain and Ireland, Queen, Defender of the Faith.

To all to whom these presents shall come, greeting.

Whereas it hath been represented unto Us that it is advisable to regulate the Rank and Precedence of persons holding appointments in the East Indies. In order to fix the same, and prevent all disputes, We do hereby declare that it is Our will and pleasure that the following Table be observed with respect to the Rank and Precedence of the persons hereinafter named, viz.:—

Governor-General and Viceroy of India. Governor of Madras. Governor of Bombay. President of the Council of the Governor-General. Lieutenant-Governor of Bengal. Lieutenant-Governor of North-West Provinces. Lieutenant-Governor of the Punjaub. Commander-in-Chief in India, when a Member of Council. Chief Justice of Bengal. Bishop of Calcutta, Metropolitan of India. . . .

FIRST CLASS

Civilians of 28 years' standing to rank with Major-Generals.

Advocate General, Calcutta. Residents at Foreign Courts and Residents at Aden, the Persian Gulf and Bagdad. Recorders of Moulmein and Rangoon. Advocates-General, Madras and Bombay. . . .

SECOND CLASS

Civilians of 20 years' standing ranking with Colonels.

Commissioners of Divisions. Directors of Public Instruction under Governments. Private Secretary to Viceroy. Military Secretary to Viceroy. Archdeacons of Madras and Bombay. Surveyor-General of India. . . . . .

THIRD CLASS

Civilians of 12 years' standing ranking with Lieutenant-Colonels.

Political Agents. Under-Secretaries to Government of India. Inspector-General of Education, Central Provinces, and Directors- General of Education, Oude, British Burmah, Berer and Mysore. Officers, 1st Grade. . . .

Box 1.1

(continued)

FOURTH CLASS

Civilians of 8 years' standing ranking with Majors.

Assistant Political Agents. Officers, 2nd Grade, Geological Survey. Officers, 3rd Grade, Education Department. Officers, 3rd Grade, Financial Department. Superintendents, 2nd Grade, Telegraph Department. Government Solicitors.

FIFTH CLASS

Civilians of 4 years' standing ranking with Captains.

Junior Chaplains. Officers, 4th Grade, Education Department.

SIXTH CLASS

Civilians of less than 4 years' standing to rank with Subalterns. . . .

Nothing in the foregoing rules to disturb the existing practice relating to precedence at Native Courts, or on occasions of intercourse with Natives, and the Governor-General in Council to be empowered to make rules for such occasions in case any dispute shall arise.

All ladies to take place according to the rank herein assigned to their respective husbands, with the exception of wives of Peers, and of ladies having precedence independently of their husbands, and who are not in rank below the daughters of Barons; such ladies to take place according to their several ranks, with reference to such precedence in England, immediately after the wives of Members of Council at the Presidencies in India.

Given at Our Court at Windsor, this sixth day of May, in the year of our Lord one thousand eight hundred and seventy-one, and in the thirty-fourth year of our Reign.

By Her Majesty's Command.

(Signed) ARGYLL.

(F. Dr.; W. A. L.)

Source: Encyclopaedia Britannica, 11th ed., 273–274 (Cambridge: Cambridge University Press, 1911).

A more specialized usage of the term is found in the notion of a gold or silver standard. Here the focus is on the guarantee of value associated with a particular coin or banknote. It is worth noting that this terminology has also crept into common usage as a synonym for "the best." Thus, some will claim that the Mayo Clinic is the gold standard for health care delivery, that Harvard University is the gold standard in education, and so on.

Another common use of the term standard is as a rule or norm (reflected in the term *norme* in French and similar terms in other Romance languages). This usage falls somewhere between the precision of the use of standard to indicate a physical object and the notion of high standards. But here too there is a certain ambiguity. Any rule or norm may reflect either an ideal to which one should strive or an average. For example, the ancient Greek poet Hesiod ([ca. 700 BC] 2007, II: 727–732) provided—with apparent solemnity, but rather amusingly to modern eyes—standards for urinating: "Do not stand upright facing the sun when you make water, but remember to do this when he has set towards his rising. And do not make water as you go, whether on the road or off the road, and do not uncover yourself: the nights belong to the blessed gods. A scrupulous man who has a wise heart sits down or goes to the wall of an enclosed court."

These ethical standards for male Greek citizens were more than likely honored in the breach, though clearly they were ideals to which it was claimed a self-respecting Greek male should strive. In contrast, since Karl Pearson initially described the bell-shaped distribution, statisticians have referred to plots of scores on tests as normal distributions. Therefore, on the one hand we may have a code of ethics, while on the other hand we might talk of the courses required for high school graduation. The former standard is a goal toward which one should strive, while the latter describes a typical or average high school graduate.

Furthermore, the use of the standard as a norm or average may be understood in a morally neutral or morally charged matter. Ever since Aristotle coined the notion of the golden mean, and likely before that, the average has sometimes been seen (quite paradoxically) as superior in some sense. Medical standards in particular are frequently subject to this type of interpretation. Being of normal height and weight and having a normal cholesterol level are often looked at in such a way as to identify the average as superior. Hence, being fat or thin may be seen neither as the result of natural variation nor as the result of a medical condition but as the result of a moral failing. Indeed, standards are still frequently used in this way in the social sciences and journalism to identify social conditions deemed deviant or pathological. As such, perhaps Garrison Keillor is right when he assures us that in the village of Lake Wobegon, all the children are above average.

Émile Durkheim on Norms

The term "norm" has been commonplace in the social sciences at least since the time of sociologist Émile Durkheim (1858–1917). Hence the question inevitably arises as to whether, and how, the notion of a norm differs from or is the same as that of a standard. While a definitive answer to this question is beyond the scope of this book, perhaps an initial attempt can be made to respond to it.

First, the notion of a norm as used by Durkheim is in some ways more limited than that of a standard. Norms apply to people and not to things. Standards apply to both, as well as to the interaction between people and things. They do not posit a world that is somehow purely social (or purely human) but rather a world in which both humans and nonhuman objects exist. A standard is also a more general term in that it may apply to many phenomena, ranging from that which is strongly prescribed to that which is neutrally perceived to that which is strongly proscribed. The term "norm" is usually used to refer to the two poles of this continuum but not to the middle.

Second, Durkheim's norms are somewhat mysterious. They seem to appear out of nowhere as part of the "collective conscience." In contrast, standards are always created by someone or some group. Hence, both Hesiod's standards for urination and those for a high school diploma are traceable to their originators. Standards may become anonymous, but they are always, at least in principle, traceable to their originators.

Third, unlike norms, which are assumed by virtually all those who study them to be all of the same "kind," standards are clearly of several kinds. Perhaps more important, those differences in kind matter. Different kinds of standards lead to different kinds of individual and social behavior, as well as to different kinds of organizational structures and systems of sanctions.

Fourth, Durkheim approaches the question, "How is society possible?," by positing a superorganic answer: norms are not the result of human interaction but rather are fetishized. By contrast, in the phenomenological or interpretive traditions, norms are emergent properties. There is little doubt that norms can and do sometimes take on a life of their own, becoming unquestioned. But Durkheim fetishizes norms by using them as explanations for human behavior. This is fallacious reasoning. It takes the form of using the explanandum as the explanans. That is to say, that which is in need of explaining, the norm, is used instead to explain certain forms of consistent behavior. This reasoning is essentially circular. Put differently, it posits that norm N is a thing that explains behavior A, while leaving norm N unexplained. It simply exists: as Durkheim suggests, sui generis. Of course, there may be norms that exist sui generis, but if this is the case, it needs to be shown rather than merely assumed.

Fifth, the notion of norm, as used from Durkheim to Parsons to contemporary sociology, unnecessarily assumes some high level of consensus. For example, Victor Nee and Paul Ingram (1998, 19) define norms as " standards of expected behavior that enjoy a high degree of consensus within a group or community." In contrast, I argue here that conformity to standards is far more the result of their taken-for-granted character than of any explicit or implied consent or consensus. In any case, consensus can hardly be taken for granted but must be demonstrated based on empirical research.

Finally, the notion of a standard is (or can be) more precise than that of a norm. Standards can be and usually are measured, tested, examined, revised. Norms, in contrast, are usually amorphous; they are rarely easily definable since they remain, as Durkheim claims, in the realm of the collective conscience. That is to say, for Durkheim norms are ideational phenomena that have material consequences. Standards are at once ideational and material. They span the ideal–material divide, or perhaps obliterate it.

Standards are the rules by which we are told we should live, and the range of possibilities presented to us when we make choices. Thus, standards are more than norms. Standards allow us to break away from the concept of norm, which has the unfortunate tendency to mean the average as well as to imply that breaking away from a given standard is necessarily deviant or pathological. At the same time, the precise character of standards can do violence to persons, a point I return to in chapter 4.

Yet another meaning of standards, and one closely related to the notion of average or normal, can be found in the notion of tolerance. In some sense one might think of this as a more precise casting of the standard as a norm or average. Particularly in engineering, it is common to define tolerances. A sheet of 1-cm-thick steel produced to a tolerance of ±5 mm is quite different from one produced to a tolerance of ±0.0001 mm. The sheet with a smaller tolerance (i.e., produced to a higher standard) will be more costly to produce than the one with a greater tolerance. Both, however, are produced to a standard, and whereas the latter can substitute for the former, the reverse is not usually true.

Moral and religious behaviors are subject to standards of tolerance as well. They are literally the limits of what behavior shall be tolerated. Thus, religious tolerance may be broadly conceived in many nations, but usually not broadly enough to include human sacrifice. U.S. religious tolerance excludes plural marriages; French religious tolerance does not include women or girls wearing head coverings or veils in public schoolrooms. Similarly, the range of behavior tolerated in an American college classroom is fairly broad but would not include coming to class nude. In the case of both people and things, tolerances are the maximum acceptable degree to which a thing or object may differ from some specified behavior without incurring some sort of negative sanction.

In sum, standards may imply that something is the best, or that it may be used as an exemplary measure or weight; or they may emphasize the moral character of someone or the superb qualities of something. Standards may also refer to rules or norms that embody the ideal or merely the average. Finally, standards may refer to tolerances permitted for both people and things. These various meanings are inextricably linked together. All say something about moral, political, economic, and technical authority.

Furthermore, all illustrate a point noted some years ago by Susan Leigh Star and James Griesemer (1989, 412): standards are often (perhaps always) boundary objects. Although they discuss scientific practices, their conceptual innovation applies to other situations as well. As they explain, "In conducting collective work, people coming together from different social worlds frequently have the experience of addressing an object that has a different meaning for each of them. Each social world has partial jurisdiction over the resources represented by that object, and mismatches caused by the overlap become problems for negotiation."

The soldiers who rallied round the king's standard, the moral character of a member of Parliament, the superb qualities of a diamond, the average cholesterol level in the blood, the tolerance for others with different religious beliefs—each of these things called a standard is a boundary object. They are places where persons with different histories, values, and desires are able to stabilize a set of practices that may well have different meanings to them. Hence, soldiers who believed that the king was God's representative on earth, as well as those who merely fought as they were told to do, could all identify that object on a pole as the king's standard. The diamond expert who knows that the specimen in front of her is nearly flawless and the groom who is about to buy it for his bride need not agree on the meanings attached to the standards for flawlessness. The scientist doing research on cholesterol, the medical technician who administers the blood test, and the person whose blood is drawn need not agree on the meaning of the cholesterol standard; they need only agree on the practices that bring it into use.

But wait! When I talk about standards am I merely engaging in a logical or linguistic fallacy caused by the various ways in which a single word is used? One might argue, for example, that standards for things such as automobile parts are largely unrelated to standards for health care or education. One might also distinguish voluntary standards produced by the private or nonprofit sector from mandatory regulations produced by government agencies, arguing that these are two very different things. But is this the case? I think not.

Two distinctions are generally offered as reasoning. First, some would distinguish standards for things from standards for people. Standards organizations currently make this distinction; in general, the myriad standards organizations are split between those organized around people (e.g., educational standards) and those organized around things (e.g., standards for bridges). But this distinction is a rather superficial one. We do not live in a world devoid of things but rather in one in which things must be taken seriously. Therefore, to the extent that we create standards for things, we implicitly create standards for humans.[2] Similarly, we cannot create standards for humans without creating standards for things.

In recent years some standards development organizations have begun to realize the difficulty of keeping standards for things and those for people apart. Hence, the International Organization for Standardization (ISO) has issued general management standards (ISO 9000) and environmental management standards (ISO 14000). These standards bridge the gap between people and things (see Brunsson and Jacobsson 2000; Loya and Boli 1999).

The second distinction often made is between private standards and public regulations. The former is said to give rise to voluntary rules and choices, the latter to mandatory rules and choices. But is this actually always the case? I think not, for several reasons.

First, private standards are often de facto mandatory (Olshan 1993). In some instances they must be followed in order to participate in a given market. In other cases deviation from the standards puts one at considerable risk for civil penalties in court.

Second, not all public regulations are mandatory. Many public regulations do not prohibit things outright, but encourage or discourage certain behaviors by positive (subsidies) or negative (taxes, fines) sanctions. In some instances, as Richard H. Thaler and Cass R. Sunstein (2008) argue, they provide a nudge in a given direction. In short, both private standards and public regulations make use of the entire range of positive and negative sanctions (although in general, public regulations tend to have stronger negative sanctions).

Third, it is argued that public regulation occurs through legislative voting mechanisms, while private standards are produced through consensus building. Here too the distinction appears overwrought. There are numerous examples of government regulations that have been developed through the use of legislative hearings, public and industry advisory committees, and the like. As well, many private standards are developed based on a consensus that exists largely among those who attend a given meeting. Both processes involve managed conflict, compromises, and iterative processes.

Fourth, it is sometimes argued that government regulation deals with a different set of issues than do private standards. But on closer examination, this distinction proves weak as well. When one compares standards and regulations across nations, one finds considerable variability in their scope. Moreover, it is not at all uncommon for privately produced standards to become the basis for both positive law (e.g., building codes) and case law (e.g., good medical practice). Lawyers tend to dismiss standards as outside the scope of law, but even they implicitly admit the blurred boundaries when they speak of standards as "soft law."[3]

Finally, it hardly needs noting that standards are nearly always designed to be within the scope permitted by law. In sum, private standards and public regulations are two similar and sometimes overlapping forms of governance, or of what Foucault (2007, 2008) called governmentality. They are two means of governing relations among us.

All this is not to say that public regulations and private standards are identical. Clearly, they are not. Redress of grievances varies considerably between the two. In general, one cannot be jailed for violating private standards. But we should not assume that public regulation and private standards are different; rather, we should demonstrate their differences (and similarities) through careful empirical examination.

Studying Standards

There are at least three ways in which standards might be studied. First, the very multiplicity of meanings of standards suggests that a phenomenological approach to their analysis might be useful. By this I mean an approach that carefully examines the multiple ways that standards are used, spoken of, employed, designed, put into common practice, and so forth. Such an approach will allow us to appreciate the enormous range of things to which standards are applied. It is the dominant approach used in this book.

A second approach is historical. I could attempt to uncover the historical development of multiple standards from antiquity to the present. Clearly, a comprehensive historical analysis would require multiple volumes. For example, merely describing the history of the standardization of coins and coinage over the last several thousand years would be a multivolume effort in itself. While such an effort would doubtless be of considerable historical value, it is largely beyond the scope of this work. That said, various historical references are incorporated into this work as needed.

Finally, one might focus on the fine technical details of some particular set of standards, noting the ever-advancing attempts to produce yet a better definition, greater precision, or more and better parameters for comparison. Some works of this type have been written, in particular for specialists in particular fields. While I will have occasion to draw on this type of analysis to illustrate certain points, in general, I avoid this degree of technical detail.

What is central to the analysis in this book is the intimate connection between standards and power. However much standards appear to be neutral, benign, merely technical, obscure, and removed from daily life, they are, I argue, largely an unrecognized but extremely important and growing source of social, political, and economic relations of power. Indeed, in our modern world standards are arguably the most important manifestation of power relations. Moreover, as Bruno Latour (2005, 64) notes, "power, like society, is the final result of a process and not a reservoir, a stock, or a capital that will automatically provide an explanation. Power and domination have to be produced, made-up, composed." In short, power is present only when it is performed or enacted.

However, this is not to suggest that standards have the kind of power we associate with a tyrant, lording it over his or her subjects with an iron fist. To the contrary, the power of standards lies in their very subtlety. It is because they are barely noticed, perhaps noticed only when their presuppositions are violated, that standards are powerful.

Let us briefly consider the ways in which standards are powerful. First, I offer a suggested definition of the kind of power of concern here: *the ability to set the rules that others must follow, or to set the range of categories from which they may choose.* When Caesar or a Ming Dynasty emperor or, more recently, Stalin or Mao was said to be powerful, it was precisely for this reason. While they might well have had the power to put a particular individual to death, or to take away that individual's property and offices and confine him or her to prison, or conversely, to reward a given (perhaps otherwise undeserving) person with honors or expensive gifts, that paled in comparison to their ability to set rules that others had to follow, to provide the categories among which other people might choose.

The public display of power takes place only on special occasions and in certain places. An army can march through the main square on holidays, but to do so every day would take away the awe the spectacle inspires. A public hanging might warn one's enemies of the consequences of certain behavior, but too many people dying or imprisoned might and often has sparked revolts and the overthrow of that powerful person or persons. In contrast, the setting of rules for others to follow is more subtle, often deflecting attention to the rule and away from the ruler. Nearly all of these rules involve the establishment of standards—standards for things as well as for people. Indeed, our very use of the term "ruler" both for someone who rules and for a measuring device reflects this dual character of power. And the use of the ruler as an object for disciplining students in the classrooms of yesterday strengthens that linkage.

Second, unlike the direct power often exercised by a ruler, standards display *anonymous* power. Even if we know who established them, standards take on a life of their own that extends beyond the authorities in both time and space. The premodern Chinese bureaucracy offers another relevant example here. For the highest-level positions, every several years an examination of some three days' duration was given in Nanjing in a special examination hall constructed for that purpose. Students spent years studying in the hopes of passing the examination and entering the civil service. Elaborate measures were taken to ensure that students were unable to cheat and that grading was truly anonymous. Success meant a career as a highly paid and privileged civil servant, perhaps even as a servant to the emperor himself. But only a small number of candidates actually passed the exams.

To our modern eyes the exams seem puzzling, as they often had little or nothing to do with the actual matters of administration but queried examinees on Confucian philosophy, Chinese literature, calligraphy, and history. When looked at from the point of view of standards, however, their purposes become obvious: the exams were designed to provide a means of upward mobility and to shift the center of authority from local landlords and merchants to fiercely loyal agents of the state.

Responsibility for passing or failing the exams was placed squarely on the shoulders of those who took them. As sociologist C. Wright Mills ([1976] 1956) suggested fifty years ago in a different context, those who failed the exams saw public issues (the small chance of passing and even the legitimacy of the examination system itself) as personal troubles (the failure to study sufficiently hard, to memorize the necessary texts, etc.). Put differently, the highly standardized exams were "naturalized." They were not subject to challenge but were seen as challenging those who took them. Moreover, they legitimated the centralized state precisely because of their apparent objectivity.

The history of the watt provides another instructive example. We know in considerable detail how the watt came to be measured as it is now. We know a great deal about the debates over how to standardize it, when it was first standardized—at the 1889 convention of the International Electrical Congress—and that it is now part of the international system of measures. We also know for whom it is named, James Watt. Of course, one is "free" to use some other measure to describe the electrical work produced by a 60-watt bulb—36.7 buschs, for example—but one does so at one's own risk! Since other measures are not the subject of international agreements, or widespread use and at least tacit agreement on their importance, they are not likely to be accepted.[4]

Moreover, once established, existing standards become "natural," and their very naturalness makes other potentially competing standards suspect. While occasionally one standard does replace another—hundreds of local measures were replaced by metric measures in France during the nineteenth century—making such a change is exceedingly difficult. Even in France, where metric measure was invented, it took nearly half a century for it to fully replace older measures (Adler 1995). And despite this, old ways persist even now. One need only consider the common use of the pound (*livre*) in France today to mean 500 grams, or the description and sale of nails in the United States not by their length or shank thickness but by their pennyweight!

Third, standards make things ready-to-hand or handy (*zuhanden*). The philosopher Martin Heidegger (1977) once suggested that a central characteristic of human–tool relations was their handiness. For Heidegger, things that were handy allowed us to perceive those aspects of the world that are revealed to us only through our transformation of it. For example, those characteristics of wood revealed to us by a hammer are quite different from those revealed by looking at a tree. Those revealed by a saw or by sandpaper are different as well.

In pre-industrial societies, handiness was produced only by direct experience. A carpenter knew precisely how to handle *his* hammer, how much force to apply at the tip, how to raise it and bring it down on the nail, by virtue of years of practice. This kind of handiness extended to the wood itself: a carpenter would know precisely how much force to apply to a given type of wood, how to shave off precisely the right amount so as to produce a smooth surface, and so on. Conversely, the hammer had to "respond" consistently to the carpenter. These aspects of the world could only be known by virtue of using a hammer, and only one that was handy. A hammer whose head fell off with each use could hardly be called handy. The same would have been true of all tools, and is still true today.

But the advent of standardized tools and especially machinery markedly changed the character of handiness. An example will help clarify what I mean. Some forty years ago I was working in Guinea and had a Chevrolet pickup truck at my disposal. In order to move a rather large object, I borrowed a Russian-made GAZ-51 truck. The young Guinean colleague with whom I was working was astonished to discover I was able to drive the Russian truck even though I had never been inside one in my life. What he did not understand was the handiness of the truck, or, to be more precise, the handiness of human–truck relations. By virtue of the standardized character of the interface between driver and vehicle (but not of the various parts of the vehicles), I was able to instantly understand what was necessary. Without thinking, I could depress the clutch, turn the key in the ignition, shift into first gear, depress the gas pedal while releasing the clutch, and drive off. The internal workings of the engine became a black box I had no need to inquire into unless the truck broke down.

The same would be true for a contemporary carpenter. Whereas in times past, carpenters made their own tools or had them custom-made for them, today they are likely to buy tools at a hardware store. If a carpenter's 16-oz. hammer breaks or is lost, he or she can purchase another one and immediately establish a relation of handiness with it. One consequence of the embedding of standards in everyday objects is that the objects acquire a kind of taken-for-grantedness that is not the case for nonstandardized objects. The hammer that was once handy for a particular carpenter is now handy for any carpenter. There might be dozens of hammers of the same brand and type in a shop, each seemingly identical to the others. From the vantage point of a carpenter, the hammer's origin is unknown and likely of no interest. Similarly, from the perspective of the producer of the hammer, the carpenter is anonymous and unknown. Moreover, the carpenter seeking to buy a 16.65-oz. hammer would look in vain, with no more chance of finding one than of finding a five-legged cow or an oak tree with three trunks.

As Heidegger noted, this handiness of everything marks a qualitative change in the relations humans have with the natural world. In our contemporary world everything is turned into "standing reserve." That is, everything becomes malleable, transformable, storable in different states. Hence, we can dam up great rivers and turn their force into electricity. Engineers can turn petroleum into gasoline to power our cars and into plastics to mold into a nearly infinite number of shapes. Managers, engineers, and scientists can turn everything—human beings included—into "resources," things to be drawn on when necessary to replenish declining stocks.[5]

The handiness of objects is matched by the handiness of organizations. The historian Lewis Mumford (1967) once described organizations as megamachines. He even went so far as to speculate that the machines we now employ were modeled on the human machines used to build monuments by ancient civilizations. Regardless of the validity of that claim, there is little doubt that Mumford was right in arguing for the machinelike character of organizations. Because the behavior of persons in organizations is standardized, organizations, like tools and machines, may become handy. I can put a stamp on a letter and drop it in the mailbox in East Lansing, Michigan, confident that in a week or so it will arrive in Germany. I can fill out a (standardized) form and be confident that a standardized driver's license will arrive in the mail several days later. I can even participate in an organization of which I am not a member by engaging in standardized behavior. For example, I can walk into the supermarket, put the items I wish to buy into a grocery cart, take them to the cashier, put them on the conveyor belt, slide my credit card through a reader, sign on the dotted line, take my bagged groceries, and leave the store. In short, I can participate in the store's handiness. I can consider it and all the people and things in it as if they were natural objects, as ordinary as a tree or a rock.

This very naturalness imbues standardized things and behavior with power. Here is what I mean by the power of standards: If someone tries to prevent me from entering a building by standing outside blocking the door, I am likely to be angry at that person. But if I find the door unopenable because of a snowdrift, I will only be frustrated and grumble about it. I will do the same if I find it to be locked and do not have a key. In short, the power of established standards is that they structure our expectations, because standards, like the world of nature, are seemingly "supposed" to be the way they are. What could be more powerful than something that is revealed as no less than a part of the natural world itself?[6]

Handiness can also be understood as predictability—predictability that allows us to get on with other projects. This aspect of handiness is empowering, in that the very taken-for-grantedness allows us to pursue other activities. Hence, the rules of the road allow me to feel in control, empowered when driving, since the behavior of other drivers is (usually) as expected. And conversely, when that handiness fails to materialize, when other persons or things do not perform as expected, then we are disempowered, as was Santa Anna in vignette 1 of the prologue.

Finally, the importance of power with respect to standards is reflected in the fact that the emergence of standards is almost invariably the result of conflict or disagreement. That conflict may occur (1) in the development of the standards, for example, within the confines of a technical committee designing a particular standard, where participants may disagree over the nature of the standard to be developed or where one or more participants may take part in order to block the creation of a new standard, or (2) by virtue of competition among supporters of several incompatible extant standards.

In general, it is fair to say that people do not normally have conflicts over things that do not matter to them. Conversely, when people engage in conflict, they care deeply about something. Although standards are often set by consensus, that consensus emerges only after considerable conflict and disagreement. This is the case because standards create winners and losers. The more that is at stake in these debates, the more rancorous people are likely to become. Recent debates over Blu-ray standards for digital video disks (DVDs) versus high-definition DVDs are a case in point. The companies involved invested vast sums of money in designs that were fundamentally incompatible. Widespread acceptance of one standard, Blue Ray, has led to the virtual abandonment of the other. Moreover, it is quite possible that, in part as a result of the protracted uncertainty posed by two competing and incompatible standards, both will lose, and a third standard, one for downloaded films, will capture the lion's share of the market (Gardiner 2008). Another example is when many actors stand to lose if a single standard is adopted and respond by deliberately creating conflicting standards. Cargill and Bolin (2007) argue that this is what has happened in parts of the IT industry.

Conflict is also related to precision, both in the case of standards for people and in the case of standards for things. A vague standard can be easily circumvented or even ignored when convenient, without fear of sanction. Thus, a standard of literacy is likely to be of little consequence when it consists of merely asking people if they can write, but it will be far more consequential if it requires taking a test demanding detailed knowledge of each paragraph of the U.S. Constitution. Indeed, tests of precisely that nature were used in the U.S. South in the last century to keep African Americans from voting. Similarly, the standard height of ceilings could vary considerably in a world in which walls were constructed of plaster and individually cut laths. But once standard building materials, such as (in the United States) 2 × 4s, and 4 foot × 8 foot sheets of plasterboard were made available, the variation in the height of ceilings was sharply reduced even as the speed at which wooden homes could be built increased. Furthermore, the need for skilled labor declined, waste was minimized, and costs were reduced, such that the use of nonstandard heights would result in considerable extra costs. This did not make construction of nonstandard buildings impossible, but it meant that they became considerably more difficult to build and costlier than standard buildings; indeed, as a result, having a new home with 9-foot ceilings has now become a point of prestige and status.

Furthermore, the *combination* of standardized building materials—that is, 2 × 4 studs each 8 feet in length, 4 foot × 8 foot sheets of plasterboard, and insulation designed to fit in the spaces between the studs—created a "system" of commensurable materials that could be coupled together. This in turn gave further impetus to a particular path of standardization.

Commensurability and Coupling

Standard home construction materials are an example of commensurable standards.[7] Like all standards, home construction standards may be either commensurable or incommensurable.[8] Hence, the 4 foot × 8 foot panel commonly used in the United States is easily translated into balloon frame construction on 16-inch centers. So is the fiberglass insulation typically manufactured to fit precisely in the space between the studs.

Consider the case of temperature. Degrees Celsius translates to degrees Fahrenheit using the well-known formula $F = 9/5 \, C + 32$. Hence, 32° F is exactly equal to 0° C. However, at many other temperatures, there is a very small margin of error. For example, 39° F is equal to 3.8888888 . . . 8° C. For most purposes, the difference produced by an infinitely repeating decimal is inconsequential, but in certain scientific experiments small errors might make a significant difference. Hence, it is fair to say that these two scales are nearly fully commensurable. Moreover, someone has to do the measuring; as such, achieving commensurability (or not) is an achievement and not a given.

In contrast, cuts of meat are only partially commensurable. Meat butch-ered according to British standards includes only some cuts that are similar to those produced by an American butcher; many cuts commonly avail-able in the United States are simply unavailable in Britain, and vice versa (figure 1.1).

Certain standards are fully incommensurable. They cannot be translated unless an intermediary is available. American television uses National Television Standards Council (NTSC) standards, while French television uses Phase Alternating Line (PAL); hence, one needs two television sets (or a television with dual circuitry and a special switch) to pick up both signals unless a converter is available. Furthermore, in some cases commensurabil-ity is deliberately blocked. Coca-Cola keeps its recipe (the standard for

Figure 1.1
Standard cuts of beef in the United States and the United Kingdom.
Source: Wikipedia commons.

making Coca-Cola) as a trade secret so that other brands of cola cannot duplicate its taste.

Money is often used as a standard of commensurability. Today it is often assumed that money is a kind of universal solvent that creates an equivalence among things or people. But as Richard Hadden (1994) suggests, this was not the case for the ancient Greeks. Aristotle insisted that monetary equivalence was only relevant for purposes of exchange. For him, two drachmas of shoes could be exchanged for two of wheat, but no other equivalence was established. One did not have the same worth as the other. In contrast, many centuries later, the philosopher Thomas Hobbes ([1651] 1991, 63) noted that "The *Value*, or WORTH of a man, is as of all other things, his Price; that is to say, so much as would be given for the use of his Power: and therefore is not absolute; but a thing dependant on the need and judgement of another." In modern times we have come to accept that the equivalence established through money goes far beyond exchange. Yet recent work by Michel Callon (1998b) and his colleagues challenges this position once again. Indeed, we can extend Callon and John Law's (2003) position by arguing that commensurabilities and incommensurabilities are the result of social action. Put differently, we can say that things must be *made* commensurable or incommensurable via a process of comparison. As Umberto Eco (1988, 512) notes (through a character in one of his novels), there are three rules for connecting things:

Rule One: Concepts are connected by analogy. There is no way to decide whether an analogy is good or bad, because everything to some degree is connected to everything else. . . .

Rule Two says that if *tout se tient* [everything holds] in the end, connecting works. . . .

Rule Three: The connections must not be original. They must have been made before, and the more often the better, by others. Only then do the crossings seem true, because they are so obvious.

Thus, standards designed to create commensurability may lead to incommensurability, and vice versa.

But commensurability is not merely a question of mathematical relationships; it is also sometimes a matter of what engineers refer to as coupling or interoperability. Some, but not all, standards are linked together by virtue of the need for physical interaction or connection or communication between the parts. A good example is provided by electrical plugs and sockets. They must be fully commensurable with each other. Put differently, plugs must be designed in such a manner that a normal adult can easily insert them into a socket. At the same time, the socket must be designed to accept that plug with relative ease, and must provide enough friction to hold it in place.

As anyone who has traveled around the world knows, electrical outlets vary from place to place. Thus the hapless foreigner may find herself with an appliance that, by virtue of its plug, cannot be connected to the electrical grid. This leaves only two other possibilities. First, one can cut off the plug, strip the wires, and insert the wires into the socket. Of course, this kind of jerry-built arrangement may not hold, and it increases the risk of electrical shock. Second, one can purchase an adapter that "translates" the shape of the plug on the appliance into that of the receptacle on the wall.

The problem is made more complex by the fact that the standard voltage for electricity at the socket varies from place to place. In the United States, 110 volts is the standard, while in much of the rest of the world 220 volts is more commonly used. (In still other places 100 and 240 volts are used.) Plugging a 110-volt appliance into a 220-volt socket will likely fry it. Conversely, plugging a 220-volt appliance into a 110-volt socket will likely result in sluggish or no performance.

Finally, electrical standards for alternating current also specify the number of cycles per second, with fifty or sixty cycles being the commonly accepted standards. For most electrical apparatus, the number of cycles is inconsequential. However, for digital clocks designed for sixty-cycle operation, use at fifty cycles will slow them down.

As international travel has become cheaper and more commonplace, a variety of solutions have been proposed and even implemented. Obviously, the most effective solution would be to develop a universal standard and replace all those sockets and all electrical equipment not conforming to the new standard. However, given the vast sums invested in the existing infrastructure, doing so is utterly impractical; there is considerable *path dependence* embedded in these standards. Therefore, second-best solutions are likely to remain commonplace for the foreseeable future. Among these are the use of adapters, as noted above; the development of adapters that can be used with a variety of sockets; the use of transformers to shift from 220-volt to 110-volt current; and the incorporation of sensors into appliances that automatically adjust the flow of current (as is now commonplace in laptop computers). Yet another approach, now found in many hotels and public buildings in China, is the redesign of wall sockets in new buildings to accommodate multiple forms of plugs (figure 1.2). Of course, each of these solutions requires conformity to several sets of standards. Similar issues of commensurability can be found in other networks, such as those for both mobile telephones and landlines (box 1.2).

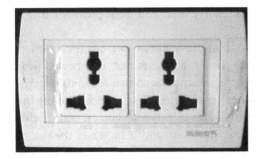

Figure 1.2
A socket capable of accepting a wide variety of plugs.
Source: Photograph by Neal Fortin. Used with permission.

Box 1.2
Digital mobile phone standards

One would have thought that the move to digital mobile phones would have been an opportunity to harmonize telephone standards globally. Unfortunately, this was not to be the case. In fact, during the first Bush administration telephone companies asked the government for guidance on the creation of a single standard for the United States. For ideological reasons the Bush administration declined to intervene (Russell 2006). In contrast, Europe rapidly developed an open standard known as GSM (Groupe Spécial Mobile) (Pelkmans 2001). As a result, there are several different standards for cell phones in the United States. Finally, through the development of intermediaries, American phones on different networks were made to communicate with each other, but most remain incompatible with those used in much of the rest of the world. Hence, the American traveler to Europe will likely need another phone. Furthermore, American telephone handset manufacturers found it necessary to make huge investments in new manufacturing equipment in order to produce telephones that met foreign standards, even as European manufacturers saw their fortunes soar as much of the rest of the world adopted the European standard.

However, commensurability of standards is important only when people and things come in contact with one another. In a world in which long-distance travel was rare, commensurability was of little consequence. Today, when nearly instantaneous communication with persons halfway around the world is commonplace, issues of commensurability are more urgent. How the world grapples with these issues over the next several decades will have considerable consequences for future generations.

Finally, it must be emphasized that commensurability is never complete. As Susan Leigh Star and Karen Ruhleder (1996, 132) suggest with respect to attempts to construct collaborative software for scientific research, despite heroic attempts at commensurability, "experimentation over time results in the emergence of a complex constellation of locally-tailored applications and repositories, combined with pockets of local knowledge and expertise." Errors of measurement, differences in local conditions, differences in the materials with which objects are constructed, and differences in local meanings attached to particular standards always limit commensurability. In some instances such minor incommensurabilities are of little or no consequence. The extra effort required to force a slightly nonstandard plug into an electrical socket will cause only momentary annoyance. In contrast, some standards that appear commensurable conceal incommensurabilities that are less easily overcome.

Lest the reader get a false impression, let me note that commensurability applies to humans as well as to things. The candidate for a professorship with a doctoral degree in engineering offered by MIT is not (fully) commensurable with a candidate with a degree from Fly-by-Night Mail Order University (although some persons may see the FBNMOU degree as better). Similarly, with the formation of the European Union, the partial commensurability of university degrees across the European continent has posed considerable difficulty for both prospective applicants for jobs and prospective employers. Just how does a French *doctorat d'état* stack up against a German *Habilitation*? Despite the Bologna Process, which seeks to harmonize different degrees, this remains a problem. But for many jobs where skills can be easily acquired in several hours, commensurability among candidates is the order of the day.

Symmetry

Bringing humans into the picture adds another feature to standards: their symmetry. That is to say, every standard for a person implies a standard for some thing or things, while every standard for a thing implies some standard for a person or persons. Moreover, every test is itself tested by those persons or things that are the object of the test. This is true for the metrification devices that produce the standards as well as for the standards themselves. Let us consider several examples.

Figure 1.3 reproduces a set of detailed standards for dress, a dress code of sorts, from the White Castle company around 1930–1949. All female counter staff workers were required to adhere to the dress code. It is important to note that these standards were neither solely for things nor solely

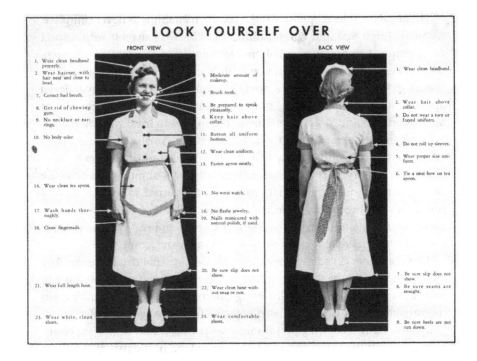

Figure 1.3

"Look Yourself Over" (used by employees ca. 1930–1949).

Source: Used with permission of White Castle Management Co. The White Castle image and material and the "WHITE CASTLE®" mark are the exclusive property of White Castle Management Co. and are used under license. No further use, reproduction, or distribution is allowed. Courtesy of the Ohio Historical Society.

for people but for people and things in a particular type of situation. The employee was requested to "look yourself over." This was not a matter of looking oneself over *tout nu*, or of looking over the uniform before putting it on, but of wearing the uniform in a particular approved manner. Herein one finds symmetry (1) between the person in the photograph and the person looking herself over, (2) between the uniform in the photograph and the one she is presumably wearing, and (3) between the person wearing a uniform in the photograph and the one wearing a uniform while looking at the photograph.

It may be objected that a case of this sort is merely an exception. But consider that the argument can be expanded thus: virtually all nations have standards for citizens. In general, citizens must be born in the nation in which they claim citizenship. Alternatively, they must apply and receive what is generally referred to as naturalization. Either way, standard things are involved. A citizen must produce a standard birth certificate demonstrating that she or he was indeed born in that nation. To be naturalized, a citizen often must pass a test demonstrating competence in various aspects of civics. That test itself must be standardized so that each applicant is treated in the same manner and citizens are easily sorted from noncitizens. But it should also be noted that the persons who take the test determine its validity. In short, they test the test.

The opposite situation arises with a quality control test for canned peas. Such a test might examine the quality of the peas themselves, the amount of foreign matter (e.g., other plant parts, dirt), whether the peas remain whole, their taste, size, and color, and whether they are fully cooked but not overcooked. While some of these processes are frequently automated, most require human oversight. They require that farmers exercise care in the growing and harvesting of the peas and that factory workers check innumerable gauges to ensure that the product is cooked at the right temperature for the proper amount of time. Still other workers must ensure that the complex processing equipment is functioning properly and is properly maintained. Finally, those persons charged with the quality control test must know how to sample the cans, how to cook the contents, and how to examine the contents for size, taste, and other attributes. Each of these workers must engage in standard actions, or, to put it slightly differently, each must adhere to a complex set of work standards.

However, the symmetry described here is not one of time, effort, or suffering. Many hours of discussion and debate are required to develop a test of civics to be used for citizenship applications, while the test itself likely is confined to a limited time. But the effort required by an applicant to pass the test is likely far greater than the effort required by the test developers in designing the questions. Moreover, the applicant may suffer greatly out of fear and anxiety, even as the test developers joke about the design of the questions. Similarly, the time and effort required to produce canned peas of a given quality are likely far greater than the time and effort required to see if they meet the standard.

Yet it is this symmetry that is essential to standards formation, implementation, and modification, that makes standards simultaneously technical and social devices. And it is this symmetry that introduces a moral element into standards. Clearly, looked at from one perspective, standards are technical details, perhaps essential in modern industrial societies, but mainly of concern to engineers and managers who need to produce and sell various products. For others, standards are bureaucratic devices designed, for example, to make it easier to keep records in a hospital. Yet looked at from another perspective, standards are a source of moral order. They are the raw materials out of which the social order is constructed. Not surprisingly, standards themselves come in several flavors. It is to that issue that I now turn.

Types of Standards

Not all standards are alike. For our purposes we can divide standards into four types, and we can see that each type can relate to either people or things (or perhaps more correctly, each type can relate to people *and* things). This typology is presented in table 1.1.

Table 1.1
A typology and examples of the use of standards for people and things

Type	People	Things
Olympics	Professional athletes, musicians, singers, dancers	Best wine of 1996, car of the year, best hotel in Paris
Filters	Citizen, Rotary Club member, student at a particular high school, medical school graduate	Safety of food, toys, and children's clothing; nation-states
Ranks	Associate professor, vice president, assistant director, general	Grain grades and standards, earthquake size
Divisions	Butcher, baker, candlestick maker, undergraduate major	Apple varieties, cloud types, types of crystals

Olympic Standards

"Olympic standards" are those for which there is one winner or just a few winners. The Olympic Games here serve as a metaphor for a wide range of standards whose purpose is to identify a single winner or a small group of winners in ranked order. In most instances professional and intercollegiate sports are measured in this way (e.g., baseball's World Series, the World Cup in soccer). Similarly, music has the Grammy awards, and actors compete for an Oscar. Virtually every Western nation has a national ballet award. And in the United States, every four years the nominee with the most (electoral) votes is elected president.

The situation just described for people also applies to things. Thus, the *Wine Spectator* chose the Château St. Jean Cabernet Sauvignon Sonoma County Cinq Cépages 1996 as its Wine of the Year 1999. The Renault Clio won the Car of the Year award in 2006. And each year horses compete for the elusive Triple Crown, an award given only when a single horse wins the Kentucky Derby, the Preakness Stakes, and the Belmont stakes. The award is so difficult to capture that in most years there is no winner.

What is central to each of these Olympic standards is that a single person or thing is identified as the best within a particular time or space. These standards nearly always reflect a winner-take-all approach (Frank and Cook 1995). They are *designed* in such a way as to produce a single winner that will be known as "the best." That means that, by definition, there are far more losers than winners. And the losers are usually quickly forgotten. (Quick: Who was Zachary Taylor's opponent in the U.S. presidential election of 1848?)

Filters

In contrast to Olympic standards, filters are standards that perform quite differently. The metaphor of the filter suggests the key aspect of this type of standard: some people or things can pass through the filter and thereby meet the standard, while others fail in this regard. Consider citizenship. While those persons born in the United States are automatically granted citizenship, becoming a naturalized citizen requires passing through filters that go far beyond birth. They include:

• continuous residence and physical presence in the United States;
• an ability to read, write, and speak English;
• a knowledge and understanding of U.S. history and government;
• good moral character;

• attachment to the principles of the U.S. Constitution; and
• a favorable disposition toward the United States. (U.S. Citizenship and Immigration Services 2006)

A variety of tests are used to ensure that prospective citizens meet these requirements. Similarly, persons who wish to become members of their local Rotary Club, enroll in a local high school, or graduate from medical school must measure up to a set of standards. In so doing they pass through a filter that marks them as a citizen, a student, a Rotary Club member, or a medical school graduate.

Filters are also used to sort among things. For example, in the UK and many other countries food must meet a set of safety standards in order to be sold. These standards can be quite complex, requiring strict limits on pesticide residues, bacterial contamination, types of packaging used, and time between the creation of the product and its sale. Safety standards are also to be found for the sale of toys (e.g., lack of sharp objects) and children's clothing (e.g., use of fire-retardant fabric). Even pets are required to pass through some sort of filter: most jurisdictions prohibit keeping wild animals as pets, permitting only domesticated animals (e.g., dogs, cats, certain birds).

Unlike Olympic standards, which identify a single winner as the best, filters are designed to eliminate the unacceptable. As is commonly noted when comparing medical doctors, in every class of medical students someone finishes last. But once medical school classes and other requirements are completed successfully, all graduates are able to hang a sign on their office door stating they are a medical doctor. Similarly, some foods easily pass through the food safety filter, while others just barely do so. But once they pass through the filter, they are considered to be safe. (If you find this unconvincing, then ask yourself when you last encountered a food product labeled "particularly safe" or saw a diploma on the wall in a medical office that noted the recipient had barely made it through medical school.)

In many if not most cases, filters lend support to the well-known saying made famous by John F. Kennedy but in use long before: "A rising tide lifts all boats." Put differently, filters may lead to generalized improvements. This was certainly the case when, after a multitude of boiler explosions resulting in considerable loss of life and property, laws were passed in the United States specifying certain features on boilers. Steamship owners complained that the costs of conforming to the prescribed standards would put them out of business. But such was not the case: steamship companies made the changes, and explosions declined rapidly (Burke 1966). In contrast, it was the failure to install standard filters that contributed significantly to the 1892 cholera epidemic, which was largely confined to Hamburg. In their zeal for laissez-faire, the merchants who governed Hamburg failed to install a drinking water filtration system of the sort that had become fairly common throughout much of Europe by that time. The epidemic was traced directly to the failure to introduce standard filters and filter standards (Evans 1987).

Just as a tide that rises too fast may capsize some ships, however, so a rapid change in a filter may have an exclusionary effect. That was certainly the case when soldered cans for food products were banned in the early twentieth century. Many can manufacturers found themselves out of business as they were unable to raise the necessary capital to switch to other forms of can production (Levenstein 1988).

Ranks

Ranks are a third form of standard. As the term implies, standards that rank persons or things put them in some sort of (usually) linear hierarchical order. An associate professor, therefore, occupies a rank higher than that of assistant professor and lower than that of professor. The standards for attaining the rank of associate professor are, at least in principle, greater or more difficult to meet than those required to obtain the rank of assistant professor but not so difficult as those required to attain the rank of professor. The same might be said for other intermediate ranks (see also box 1.1).

Things, too, are commonly ranked. In the United States most major cereals and oilseeds are ranked by a grading system as number 1, 2, or 3 and priced accordingly, with No. 1 being the highest grade. Most consumers never see lower-grade products, which are used as ingredients rather than sold directly in retail stores.[9] Moreover, what is true of things is also true of organizations. Large organizations such as hospitals and universities are often ranked based on a wide range of criteria.

Ranks differ from Olympic standards in that all of the objects or persons within a class receive a rank. While it might be argued that that is true of Olympic standards as well—someone or something must be in last place and in all the intermediate places—it is generally of little interest to those concerned. What is of interest is whether one did or did not get the medal. Indeed, in many Olympic competitions, only those persons or things near the top even receive a score; the rest are generally ignored, often to the chagrin of those interested in a particular outcome.

In contrast, where a person or thing stands in a ranking standard is usually of considerable importance, not only to that person or for that thing but to others as well. Furthermore, having a low rank is not necessarily stigmatized. The rank of assistant professor generally carries no stigma with it early in one's career, although if one has been an assistant professor for twenty-five years, then the rank does carry some stigma. But it is important to note that it is the particular individual rather than the rank itself that is stigmatized. Similarly, Hawaii was the last state to enter the United States, but its ranking is hardly a factor in state or national politics.

Ranks also differ from filters in that they have multiple hierarchical categories rather than the bimodal character of the filter. Grade A butter is ranked better than grade B butter and better than the product that was rejected for sale as butter because it didn't make the grade at all. The same distinction between ranks and filters applies to people.

Finally, ranks may be discrete or continuous. The professorial ranks are usually discrete, as are grades for butter. At the other extreme, the body temperatures of any group of persons form a continuous ranking. In principle, there is no limit to the number of places beyond the decimal point to which one might measure, although in practice few persons if any would care to know body temperature to, for example, twenty decimal places. It should be apparent from these examples that discrete ranks fade into continuous ones at the margin. Fine gradations can be used to create many ranks. However, they are more likely to be employed as a continuum or to be grouped into discrete ranks.

Ranks are standards that often confer rewards on people and things of higher rank, and sometimes penalties on those of lower rank. For example, grade A butter will likely command a higher price than grade B, and persons with more formal education will be likely to command greater incomes than those with less. Nor are the rewards associated with ranks confined to the monetary sphere. Those persons with higher ranks may be given the opportunity to make more important decisions (such as higher-ranking persons in the military, government, or corporations), while things of higher rank may be esteemed as art objects (e.g., excellence in industrial design). Of note, ranks are frequently contested. Some ranks are widely accepted and used, while others are subject to continued heated debate. But even the most well-established rankings can be challenged by those whose standpoint differs from the commonly accepted one.

Divisions

Finally, standards may take the form of divisions. Divisions are simply different categories that are unranked. Thus, apples may be divided into classes such as McIntosh, Granny Smith, Fuji, and so on.[10] Some persons may prefer one over another, but there is no widely agreed-upon ranking. The same may be said about the color of people's hair, or the residents of this or that city. Of course, at the edges, divisions slip into ranks. Occupational categories—the proverbial butcher, baker, and candlestick maker—are usually divisions, but there is little doubt that certain occupations rank higher than others on a prestige scale. Hence, judges are ranked higher than trash collectors, even though both occupations are necessary to any modern, urban society.

The widespread use of ever finer divisions takes the form of identity politics and product differentiation, respectively. On the one hand, persons who see themselves as parts of ever smaller groups are likely to believe that their identity is somehow unique, different. Based on that shared identity, they generally claim certain rights, sometimes including the right to keep others out. Somewhat paradoxically, identity politics simultaneously involves a proclamation of the uniqueness of individuals and social solidarity. Perhaps this is the case because proponents of identity politics desire to validate an identity that is found at a social location that is devalued, neglected, or ignored by others, and they must do it within the contemporary context of individualism and liberalism. On the other hand, product differentiation involves a proliferation of divisions among things such that they are claimed to be at least in some ways unique and different. Product differentiation is clearly and openly designed to sell the differentiated product without price competition from other similar products. In both instances, difference is then linked to an implied or demonstrably superior quality—at least superior for some group of persons. I return to this peculiar characteristic of divisions later in the chapter.

But it is also worth noting that divisions may be made (1) to distinguish among persons or objects, (2) because the divisions themselves serve some external purpose, or (3) to illustrate some sort of genetic relation among the persons or things so classified. Perhaps the most common form of division is that which divides based on some observable (with the naked eye or without instrumentation) properties of interest to those creating or using the divisions. Thus, apple varieties are based on differences in color, texture, flavor, and other attributes. Similarly, the divisions among cloud types (cumulus, nimbus, stratus, etc.) are based on certain observable features of clouds. Humans may be divided by nationality in a similar manner.

However, when teachers divide their classes into groups based on the first letter of family names (e.g., A–H, I–P, Q–Z) for the purposes of a class exercise, they are employing a standard of convenience. This is a convenient way to divide the class into roughly equal groups; the grouping is what is important, and not (usually) any property of the children themselves.

Similarly, when scientists randomly assign plants or animals to particular treatment groups, they do so to meet the expectations of their peers, and not to reveal any particular qualities of the divisions themselves.

Finally, some divisions may make a claim as to the *genesis* of a given set of persons or things. Thus, dividing living persons into children, parents, grandparents, and so on captures the kinship among the various group members. Similarly, all genetic and genealogical divisions—of organisms, soils, rocks, people—are designed to accentuate the historical relations among them.[11]

The Relationship among the Four Kinds of Standards

The four types of standards should be understood as fuzzy sets. That is, the categories are fuzzy at the edges. They overlap to some degree, rather than constituting fully discrete categories. This becomes particularly apparent when the four types are shown graphically, as in figure 1.4.

Across the *x*-axis are divisions or nominal categories. Such categories are here depicted as discrete, but they could be continuous as well. The key is that the categories are nominal: no ranking or hierarchy among them can be discerned. Although they are depicted of equal size in the figure, that need not be the case. Examples would be apple varieties and hair color. Height provides an interesting boundary case. Height is measurable on an

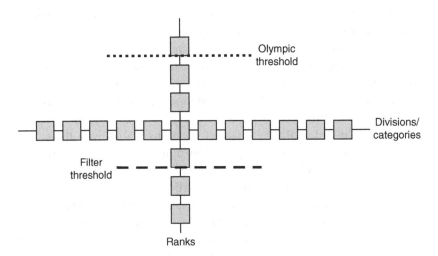

Figure 1.4
The relationships among four types of standards.

interval scale that one could use to order persons, and in some instances such ordering is called for, such as in determining the suitability of a player for the basketball team. But height may also be used as a nominal scale; an example here is the recording of height for medical studies. In addition, height might be used as a proxy for safety, as is commonly done on amusement park rides.

On the y-axis can be found various forms of ordinal categories. The gray boxes vertically aligned involve ranks. For example, military officers could be ranked in such a manner. As with nominal categories, ranks may be discrete or continuous. For example, apples might be ranked based on their brix level (sugar content), a continuous measure. George Lakoff and Mark Johnson (1980) argue that this kind of up–down spatial orientation emerges directly out of our human experience of the world and is then metaphorically applied to things like brix levels and military ranks.

The two other kinds of standards, Olympics and filters, may be seen as special binary cases of rankings. In Olympic ranking, only those entities ranking above some predefined upper level qualify, while in the case of filters, only those ranking above some predefined (usually) lower level qualify. Here again, the rankings may be continuous or discrete. For example, the Olympic runner who wins the gold medal must outrun all other runners in that particular race. The difference in running time between the winners of the gold and silver medals might be a fraction of a second. No matter. The exact difference in time between the first person and the second person to cross the finish line is not relevant to the determination as to who receives the gold medal. Much the same argument can be applied to filters.

An important finding of this exercise is that each of the four forms of standards may be tested via either quantitative or qualitative means. As David Miller (1965, 115) puts it, "we must first know what the ranking of items is on grounds other than the measure of their properties before we can use a measure of these properties as criteria for grading, and grading amounts to placing items in classes that have been ranked." For example, we must know that, ceteris paribus, taller basketball players have an advantage before we decide to measure their height to determine their suitability for the game. Similarly, we must know that redness is an indicator of ripeness before we measure the redness of tomatoes to determine their ripeness. In short, neither nominal nor ordinal standards necessarily imply measurements. People may be grouped by height with or without measuring them. Tomatoes may be grouped for color with or without the use of a colorimeter.

There are numerous ways, however, in which a given type of standard might blend or morph into another. The ordering of the letters of the alphabet would appear to be an example of divisions; each letter merely must be distinguished from the others in order to spell properly. But at the same time, the letters of the alphabet can be used to assign grades to students. In that case the letter A is ranked higher than the letter B. In addition, since telephone directories are organized alphabetically, there can be considerable advantage to businesses to have a name that begins with AAAAAA. Some companies deliberately invent names of this sort so as to appear first in the phone book (and presumably be selected first by a customer). Since only one company can be first, the standard becomes an Olympian one. Finally, the letters of the alphabet can be used as filters, as is commonly done at registration tables for professional meetings. Those with family names beginning with the letters A–M might be directed to one table while those with names beginning with N–Z might be directed to another. The same reasoning may be applied to search engines (box 1.3)

Box 1.3
Standards and search engines

Search engines on the Internet pose another ambiguous case. Typing in a search term will call up thousands of site addresses on the computer screen. In this instance the standard is embedded in the algorithm that is used to do the search. It matches (with more or less success) the term the user enters with sites on which that term appears. This would appear to be a standard that identifies divisions in the language introduced here. But the physical layout of a computer is such that some site must be first on the list, while another must be last. If the list is long, then only the first group of sites will appear on the first screen. Furthermore, since the algorithm is itself likely a trade secret, it may conceal a deliberate bias introduced by the owner of the search engine. For example, the XYZ site may be at the top of the list because its owner has paid a fee for that privilege. Doubtless other examples of concealed standards exist—standards that appear to be divisions but are in fact ranks. Indeed, Batya Friedman and Helen Nussenbaum (1996) argue that bias in computer systems can take three forms: preexisting bias with roots in institutions and prior practices, technical bias linked to constraints in the technology itself, and emergent bias that arises only in use. They argue that freedom from such bias should be among the criteria used to define good practice in programming. Doubtless, their argument could be extended to other fields of practice as well.

Finally, while standards may be classified into the four categories noted here, it should be understood that many, perhaps most, standards manifest each of the four categories even as one predominates. Box 1.4 provides examples of each.

Box 1.4
The fuzzy boundaries of types of standards

Although four distinct types of standards can be identified, in practice often several different types are linked together. One type dominates, while other types play a supporting role. For example, whereas there is only one gold medal winner each year for a given Olympic sport, there is a filter standard that determines who will be allowed to participate in the Olympics. There are divisions such that those competing in downhill skiing do not compete with those engaged in figure skating. The competition leading up to the selection of a final winner is itself a ranking exercise.

Something similar exists with respect to wheat grades and standards. Although this procedure focuses on *ranking* wheat (e.g., grades 1, 2, and 3), there is a filter for entry into the ranking in that the grain must be identified as wheat in order to be considered at all. Furthermore, wheat is divided into soft, hard, and durum, each with considerably different uses (for making cakes, breads, and pastas, respectively). Even the grades are somewhat porous: elevator operators commonly blend wheats in order to maximize their value. Therefore, a truckload of grade 3 wheat, the quality of which is just slightly below that needed for grade 2, might be blended with sufficient amounts of higher-quality grade 2 wheat to make it conform to the requirements of grade 2.

Yet another example of the multifaceted character of standards is the *filter* standard used to determine who receives a high school diploma. Here we find that students are commonly ranked—I myself graduated 526th in my class of 1089, hence barely in the top half—although this fact soon became largely irrelevant. In addition, there was also an Olympian standard: one person was chosen as the valedictorian. Furthermore, in some jurisdictions several different high school diplomas are offered depending on whether the student in question intends to go on to college, an office job, or a manual occupation; therefore, divisions can be identified as well.

Finally, apple varieties are an example of *divisions*, according to the terminology developed here. However, apples are also ranked based on size and quality, and they must pass through a filter defining them as apples and not, for example, Asian pears. Olympian standards for apples are less common, but doubtless there are those defined as the largest or the most perfectly shaped apples, or as produced on the highest-yielding tree. Such standards are commonly used at state fairs.

In short, despite the relative ease by which we may classify standards into four groups, these groups—standards for standards—are like all other standards. That is to say, they are fuzzy sets. They remain to some degree blurred at the edges no matter how much effort we put into distinguishing among them. These examples show both the limits and the situated character of standards. The very same categories, or things, or humans may be employed in different ways even though the categories used appear unchanged.

Standards, Tests, and Indicators

Standards, tests, and indicators are sometimes confused. Yet to understand standards, it is important to distinguish them from tests and indicators. A passport is an indicator of citizenship but it is not the test, which (usually) requires being born in or swearing allegiance to a given nation. Nor is it the standard, which is usually a legal document that specifies the requirements. The relationship among standard, test, and indicator is circular, but the circle is generally not a vicious one. For example, consider the case of the citizen in table 1.2. The standard is the law that proclaims that, for example, someone who is born in nation A or swears allegiance to that nation is, by definition, a citizen. The test consists of the determination by an officer of nation A where and when I was born or naturalized. A passport or identity card issued by the competent authority is an indicator that I am a citizen of country A. However, should it be necessary for me to prove my citizenship, I may well be asked to produce a passport or birth certificate, which serve as proxies for witnessing where I was born or naturalized.[12]

The same circular reasoning can be found with respect to nonhumans. For example, a legal document fixes the precise weight (or more precisely the mass) of a pound or kilogram based on a physical standard kept at the Bureau International des Poids et Mesures in Paris.[13] The test is performed

Table 1.2
Standards, tests, and indicators: Human and nonhuman examples

	Human	Nonhuman
Standard	Definition of citizen	Weight of package of beef
Test	Where born or naturalized	Scale
Indicator	Passport, identity card	Label or certificate

using a scale, itself certified to give an accurate weight within a given tolerance interval, based on a standard test. The label on the package provides an indicator of the weight of what is inside it. But in actual practice, if I buy a 16-oz. can of peas, I use the net weight printed on the label as a proxy for the actual weight of the product inside.

In some sense, then, each category tests the other, but only one is the focal point of our attention. Indeed, as Don Idhe (1979, 1983) suggests, to amplify one must also reduce. My automobile amplifies my speed, but it also reduces my awareness of the passing landscape. A test for college admission provides a numerical score but reduces the consideration of other criteria. The only exception to this rule is found in those instances in which things do not work as expected (or as they ought to work). On those occasions attention is shifted from the focal point to some other location, which in turn becomes the focal point. That other point is often one of repair (see Henke 2008).

Constructing Tests

Although it might seem that tests for various standards should be easy to create, in practice this process is both difficult and ongoing. Tests have several different features worthy of consideration here: precision, accuracy, time, cost, discrimination, reliability, and destructiveness.

Accuracy and precision Tests must be accurate and precise, although what constitutes accuracy and precision differs from standard to standard. An accurate test is one that provides a sufficiently error-free measurement for the purposes at hand. In contrast, a reliable test produces the same result each time it is used. To clarify the important distinction between these two terms, it is useful to consider two archers. The first archer shoots a number of arrows into a target and all land near the bull's-eye. In contrast, the second archer shoots the same number of arrows and all land farther from the bull's-eye but closer to each other. We would say that the first archer has higher accuracy while the second one has greater precision.

Now let us apply these notions to the weighing of meat on a butcher's scale. For this type of test, accuracy is far more important than precision. That is to say, we want the weight to be accurate within a given level of tolerance (say, one quarter of an ounce), but we are less concerned that every scale give precisely the same measurement. Indeed, a scale that was highly precise but consistently underweighed or overweighed the meat would be unsatisfactory for the customer or merchant, respectively. Of course, we would also expect that a scale used to weigh gold or diamonds would be both more accurate and precise than one used to weigh meat.

Time Another feature of relevance to the creation of tests is time. A butcher's scale will provide a test of weight fairly rapidly; no one would want to wait fifteen minutes to get a more precise reading.

In contrast, consider the case of a farmer delivering soybeans to a grain elevator. The soybeans need to be tested to see that they meet the standards required by the U.S. government as outlined in the *Grain Inspection Handbook* (U.S. Department of Agriculture 1997). Although the process of testing is complex and multifaceted, most farmers are able to have it completed within a relatively short period of time. Were the tests to take several days or even several months, they might be much more accurate and precise. But farmers would object: they want the testing done on the spot such that they can be paid immediately (and so that they can object and demand a retest).[14] The same is true for students taking exams. An examination that took ten days to administer might well be more accurate and precise, but most teachers would agree that the time needed to administer the test would take away from time that might be better spent learning.

Cost Tests are also constrained by cost. Some tests are relatively inexpensive, while others are extremely costly. The test required to determine how much the meat I am about to buy weighs is negligible in cost; indeed, the cost of a single use of the scale is nearly inconsequential when it is used thousands of times. In contrast, the flight test required for a new aircraft has a rather considerable cost. In medical testing it is common practice to begin with relatively inexpensive tests and to go on to more expensive ones only when initial results are inconclusive. Of course, if someone other than the person who makes the decision to test is paying for it, then cost may not enter into the equation at all. In paternity testing, cost may enter into the calculation only after the test has been completed. Furthermore, in some instances costs may be distributed in a manner that is unpredictable at the time of testing. Consider my test driving a new car. In most instances the cost of the test will be the cost of the gasoline, a trivial amount of wear and tear on the engine, and a small portion of the dealer's insurance policy. But if I am involved in a collision and it appears that I might have done it deliberately, questions will arise as to how the costs should be distributed: To me? To the dealer? To the insurance company? To the person whose vehicle I hit? Moreover, as with all other products of human endeavor, the technologies of testing are constantly changing; hence, the test that was extraordinarily expensive a few years ago might well be inexpensive today.

Discrimination   Discrimination is also an essential feature of tests. A test that everyone or everything passes can hardly be considered a test. The same applies to failure. In short, tests must distinguish in one of the four ways noted—Olympic, division, rank, or filter—among persons or things. Of course, discrimination is itself a term that can be used pejoratively (to discriminate against someone or something) or positively (to be discriminating about taste in persons or things). To reiterate points made earlier, testing determines who and what counts. It determines who and what makes the grade. Those persons (e.g., students) or things (e.g., grain) that do not make the grade are downgraded. Discrimination can extend to who or what is actually tested. It is not uncommon for some (usually very wealthy or politically connected) hospital patients to be given extra tests not administered to the general public. In contrast, elderly patients may be excluded from some tests on the grounds that they are not likely to live long enough to benefit from any change in treatment that might result from the test. Such positive or negative discrimination can be subtle or blatant. Whatever the case, the necessity for discrimination in determining who or what is tested and in testing itself imbues tests with an essential moral character. It also makes them subject to abuse. Finally, to return to our proverbial butcher, a scale that was unable to discriminate between 1.5 pounds and 2 pounds would be unacceptable to both the butcher and the customer, while there is hardly a need for one that can discern a difference of one-hundredth of an ounce.

Reliability   Tests must also be reliable. A test is reliable when it gives the same result each time it is administered, within some confidence interval. If a test is designed to time the speed of a computer chip, it should provide approximately the same result each time it is employed. In the same way, we would expect that a test for admission to a carpenters' guild would not vary significantly from day to day. As countless introductory statistics texts note, a broken clock is accurate twice a day, but it is not particularly reliable.

No test is always reliable. Tests are always run under a certain set of conditions. When those conditions change, then the reliability of the test will change as well. It is for this reason that most software is first tested in-house by the company or person who designed it. Then it is beta tested by volunteers eager to try it out. Finally, when it is broadly disseminated, bugs may be discovered that must be corrected in future releases.

Destructiveness   A final feature of tests is their destructiveness. For our butcher this is not an issue; scales are usually not destructive of the meat put on them. In contrast, canning factories commonly open every *n*th can to test the product inside. In so doing, the product must necessarily be destroyed.

The testing of passenger airplanes poses a problem in that it *must* be done in a nondestructive manner, given the expense involved. And clearly, tests of human beings are expected to be nondestructive for moral reasons. Killing or seriously injuring the subject of the test is normally seen as morally reprehensible; unfortunately, history provides us with all too many examples of people being subjected to one form or another of destructive testing, often without their knowledge. Indeed, in some instances, those who constructed the tests were apparently unaware of their injurious consequences. For example, educational tests biased toward middle-class white students often consigned poor or nonwhite students to educational dead-ends.

On a lighter note, Charlie Chaplin demonstrated the problem of destructive testing in his 1916 film, *Police* (figure 1.5). In that film Charlie bites into numerous apples, in each case deciding that the apple does not suit his taste, before the store owner demands that he pay for them. Then he unsuccessfully tries to divert the attention of the store owner by tasting a banana.

During World War II the issue of destructive testing took on more than normal importance. In particular, military ordnance of all kinds required

Figure 1.5
Charlie Chaplin on the destructive testing of apples.
Source: Charles Chaplin, *Police* (Chicago: Essanay Film Manufacturing Co., 1916).

testing and the extant statistical techniques required that fairly large samples be destroyed. To resolve the problem, Columbia University economist and statistician Abraham Wald (1902–1950) developed what came to be known as sequential analysis, which was initially proposed as potentially feasible by Milton Friedman (yes, the same one) and W. Allen Wallis. In March and April 1943, Wald "devised a sequential probability ratio test. This test established a decision protocol for each additional observation and signaled at what point the sampling procedure could be terminated with the researcher confident that the inspection was sufficient" (Klein 2000, 49). Today this approach is used in a wide range of applications where some destructiveness is necessary but destruction is costly.

To sum up, tests must be as precise, accurate, timely, inexpensive, reliable, discriminating, and nondestructive as necessary for the purposes at hand. Moreover, these concerns often force those who engage in testing to use an inferior test in order to get on with the business at hand. Indeed, in some instances a proxy measure must be developed because more direct measurement of the quality of interest is too costly, too slow, insufficiently discriminating, or overly destructive. Determining whether a given test is appropriate for a given situation requires making a judgment; as such, tests, like standards, are the subject of repeated negotiation. Such judgments are necessarily social, economic, ethical, political, and sometimes scientific, all at the same time. Moreover, tests often have sanctions that follow from them, a point I take up later in this chapter. But first let us examine the issue of measurement more carefully.

Metaphors and Measurements

Earlier I noted that standards always involve some sort of comparison. As illustrated in table 1.3, both humans and nonhumans may be measured by (or against) other humans or nonhumans. Moreover, within each category we may distinguish each of the four types of standards noted above. Thus, for example, in a 100-yard dash, humans are tested against other humans in an Olympian competition; if the race is properly conducted (i.e., it is conducted in a standardized fashion), there is a clear winner—the person who runs faster than all the others in the race. This contrasts with figure skating, which is also an Olympian competition but one in which the skaters are each judged individually based on tests for a set of standards (usually written, and with so many points assigned for skating skill, performance, choreography, etc.), and then the scores are compared to determine the winner. Of note, figure skating is also impossible without skates,

Table 1.3

Measuring humans and objects

		Test of	
		Human	Object
Measured by	Human	O: 100-yard dash, wrestling R: Lining up by height F: Trial by jury D: Human phenotypes (ectomorph, endomorph), sex, race	O: Best wine of 2006 R: Wine tastings, coffee cuppings F: Deodorant tests; trying on clothing (this is too small, or I am too large) D: Apple varieties
	Object	O: Figure skating R: Human weight, size of feet F: Height test for police officers D: Occupational aptitude test	O: Dog race R: Weight of paper (30-lb. bond) F: Meter stick to measure another meter stick, most rheological tests, tests of machines D: Paint color charts

Note: O = Olympic; R = rankings; F = filters; D = divisions.

while (at least in principle, and likely in practice in ancient Greece and elsewhere), a 100-yard dash requires no special equipment.

Similarly, we can find standards where objects are tested by humans and where humans are tested by objects. Furthermore, there are cases of objects being tested through their performance against another object. Finally, there are hybrid forms: an auto race is a hybrid form, as is a horse race. In both instances it is the *combination* that is tested: both the driver and the car, or both the rider and the horse, that win the race are praised.

It is important to note that these categories should not be reified (although they often are). In some sense, humans are implicated in all kinds of standards, including those in which objects are used to measure other objects. After all, it is humans who develop standards. Despite a number of attempts to develop standards by using "nature" as a measure (e.g., the initial basis on which the meter was developed), it appears that we are the authors—wittingly or unwittingly—of all standards. Similarly, things are implicated in virtually all standards. The naked runners of the 100-yard dash, even if unshoon, must have a starting and ending point for their race. Hence, all standards implicate both humans and objects, although both are not always the focus of our attention in the process of measurement.

Layers of Standardization

Standardization involves not merely the standardization of a given person or object but multiple standardizations, each of which reinforces the others in a network of nested relationships (Star and Lampland 2009).[15] Consider the example of a college aptitude test by which students will be ranked. First, there must exist a set of (at least two) categories that define who is eligible to take the test (e.g., high school students are eligible; nonstudents are not). Second, there must be a measuring device used to determine whether a given student meets the standard. Third, the device itself must be standardized in such a manner that at least some students qualify, that is, pass the test. It must also be reliable; that is, it must yield results that are "the same" within a given tolerance level. Next, there must be a set of standard procedures for administering the test and for reading and interpreting the results. This set might include the date and time of the test, the time allotted, the placement of seats in the room, the number of test monitors, the type of materials that may be brought into the room with the examinee, and so on. Finally, there must be some indicator, usually a certificate, that states that a given person has indeed scored a certain score on the test. That indicator itself must be standardized, such that others seeing it can recognize it for what it is. Only when all of these requirements are met (and probably others that are far less obvious) can we talk of someone having received a certain score on the test. Failure to adhere to these requirements can, and often does, call into question the validity of the test.

The same is true of standardized things. Consider the case of a test of the quality of cooking oil. First, only certain categories of objects can qualify as cooking oil. Thus, canola, soy, olive, and corn oils may be accepted as part of the class, but motor oil and spaghetti would not be. Second, there must be at least one measuring device that determines whether the oil meets the standard. For example, a machine that measures the clarity of the oil might be used. That test must be reliable in that repeated sampling of the same batch should give the same results (within some tolerance limits) and in that it must differentiate clarity from one batch of oil to another. Next, there must be a set of procedures used for administering the test and for reading and interpreting the test results. These procedural requirements might include a procedure for sampling, a required sample size, and the use of a standard light source to conduct the test. Finally, there must be a written indicator that the test has been administered and a given result has been achieved. This indication might be in the form of a logo or label that notes that the test has been passed, or one that provides the results to the buyer.

Thus, standardization requires many layers of standards, tests, criteria, and documentation. Only when all are found together can one make the claim that a given standard is "the standard."

Path Dependence

The combination of layering, coupling, and commensurability leads to what economists refer to as path dependence. Path dependence is perhaps best explained by looking at negative examples, that is, at difficulties encountered in introducing new standards. The creation of a new standard or the radical alteration of an old one often requires considerable investment, even as it threatens to destroy the value of existing infrastructure.[16] For example, in both France and India, the introduction of metric measure was fraught with difficulty (box 1.5).

Box 1.5
Metric measure in France and India

Even in France, where it was initially developed, the metric system was not fully accepted for more than half a century. Even then, the initial idea of ten-day weeks, ten-month years, and ten-hour workdays was never taken seriously. Why is this the case?

In eighteenth-century France, weights and measures were standardized only for each province or city, but time was standardized as twenty-four-hour days and seven-day weeks across Europe even if it was not always accurate. Moreover, weights and measures could be relatively easily replaced village by village and province by province. After all, everyone was already used to the idea that each locale had its own measures, and merchants had developed means to translate from one system to another when engaged in trade. But since time was standardized across all of France, replacing the standard with a new one would have required that everyone change within a very short period of time. Otherwise, simply knowing what day or time it was would have become quite a complex affair. Furthermore, timepieces were more complex to produce than weights and measures. Every clock and watch in the nation would have to have been altered or replaced at considerable cost. Similarly, calendrical reform would have required adjusting to new work schedules, and new dates for public and religious holidays.

Box 1.5
(continued)

> The India case is similar. As early as 1801 the British attempted to stan-
> dardize weights and measures, but the measure failed because of the difficul-
> ties of enforcement. In 1868 a proposal was made to convert India to the
> metric system, but it failed to gain sufficient political support (Verman and
> Kaul 1970). Only after independence in 1956 did India move rapidly toward
> the metric system. More than 150 different local systems of measure were
> replaced, a herculean task. The post office alone required 1.6 million new
> weights (Venkatachalam 1970).

In short, well-established standards encourage investments of time, money, organization, and material resources that meet the standard. They incur what economists call sunk costs. But standards also encourage the creation of complex organizational networks to which people become quite committed. As a result, some standards create path dependence: they make it costly (in terms of money, skill, organization, and social networks) to shift to an alternative development path since future actions are contingent on those taken in the past.

But it is also worth emphasizing that not all standards create path dependence for society as a whole. Some standards appear to create little or no path dependence. What are the criteria under which standards create path dependence? This is not an easy question to answer, but perhaps some of those criteria can be identified.

Reversibility Perhaps the most obvious criterion for path dependence is irreversibility, or more precisely the effort required to reverse a given standard or to replace it by another one that is entirely different. For example, once gasoline engines became the dominant means of propulsion of automobiles, it became quite difficult to change the situation. Indeed, virtually all efforts at innovation have been limited to hybrid vehicles or alternative means of producing gasoline (e.g., blending with ethanol). This is the case not merely because designing alternative vehicles is expensive and new technologies are difficult to develop but because the existing organizational infrastructure for servicing, repairing, and fueling automobiles would have to be rapidly replaced were one to shift to hydrogen, fuel cells, or electricity (Roberts 2004).[17] A similar situation can be found with respect to the school year (box 1.6).

Box 1.6
The school year

Initially, the school year was built around the need for farm labor in the summer, to plant and harvest the crops. In industrial nations that task is now accomplished largely by machinery. However, despite some regional and national differences, summer remains a time when schools are closed. Critics view this as a waste of infrastructure and note how much more students could learn if they were to go to school year round, with perhaps a few weeks off here and there for short vacations. Universities have attempted to make better use of their facilities by adopting a quarter or trimester system such that the same number of students would be in class all year. These attempts have largely failed.

However, there are considerable social and financial investments in the development of infrastructure surrounding the nine- to ten-month school year. Many teachers have summer nonschool jobs that are a source of additional income and an opportunity to get away from the classroom. Families plan to vacation together during the summer; in France, the entire nation goes on vacation in August! A wide range of seasonal tourism destinations has developed around the existing school year, including beachfront and other seasonal resorts and summer camps (often with unheated buildings). Seasonality has been established in related activities such as the annual ritual of purchasing school supplies and clothing at the end of summer. And many (newer) school buildings have been built with no provision for any form of summer ventilation were classes to be held.

Abandonment Standards differ in the degree to which they can be abandoned in favor of some alternative. For example, when I was a child, it was common for American men to remove their hats on entering any public building. I was taught to do that early in elementary school. Today, that standard is only honored in the breach. There was little or no path dependence since the effort required to stop removing one's hat was minimal. To help matters, men's hats—once the fedora was the mark of the well-dressed man in the United States—have to some degree gone out of fashion (although baseball caps are worn in some circles). Likely it was easier simply not to replace a worn-out hat than to buy a new one. Abandonment applies to standards for things as well. The shift from post-and-beam to balloon frame construction of wooden buildings also provides a case in point (box 1.7).

Type of standard It should also be apparent that path dependence is somewhat different for each of the four types of standards, Olympian, division,

Box 1.7
From post-and-beam to balloon frame construction

Post-and-beam construction was used for millennia to construct wooden buildings. It required heavy timbers that were mortised into place at each joint—a cumbersome task requiring a great deal of skill and labor. In the third decade of the nineteenth century the balloon frame was invented, either by Augustine D. Taylor or by George Snow, or perhaps both (Cronon [1991] credits Taylor; Sprague [1981] credits Snow). The balloon frame replaced the heavy posts mortised in place with lightweight dimensional lumber in the form of standard 2 × 4s that could be precut to standard lengths. As a result, far less wood was needed to construct a building of a given size. Moreover, 2 × 4s could be nailed in place, thereby requiring less skilled labor. Given this combination of advantages, post-and-beam construction with mortises and tenons for joints has essentially been abandoned. Path dependence was minimal since learning the new skills was relatively easy, lumber could be cut to whatever lengths the buyer specified, and few buildings required that the two systems of construction commingle; indeed, more lumber was suitable for the smaller sizes needed for 2 × 4s than for the older form of construction.

Today, in the United States and many other nations, the venerable 2 × 4 remains a key element in wooden construction. Changes have been made at the edges—literally, as the finished size has gradually diminished to the current 1½ inches by 3½ inches in the United States—but the underlying standard remains largely unchanged. Abandonment of the standards for dimensional lumber would likely come about only if a radically new means of home construction that supplanted wooden houses became widespread. Even then, given the enormous investment in wooden houses, it would take many years before the standard was abandoned.

ranking, and filtering. Divisions are perhaps the least path dependent as it is relatively easy to add a new category to existing divisions without (immediately) destroying those that already exist. Thus, the recently developed Pink Lady apple simply adds another category of apple to the long list that already exists. Of course, the history of apple growing, processing, and eating shows that some varieties virtually vanished from use as others took center stage.

Olympian standards present a more complex picture. In one sense they are not particularly path dependent since the number of winners (who are likely to be most protective of the existing standard) is always small. The development of an alternative Olympian standard based on different criteria does not necessarily undermine those that already exist. Thus, we can and do have multiple music competitions, multiple beauty pageants, multiple wine awards, and so on. On the other hand, many Olympian standards are based on a particular set of technologies, and changing those technologies changes the nature of the competition. Thus, the ball used by Babe Ruth to drive sixty home runs in one season is considerably different from the one used today. In contrast, professional golfers have been prohibited from using the latest club and ball technologies for fear that the entire competition would be undermined.

In contrast, both filter and ranking standards appear to be far more path dependent. For example, the (in)famous QWERTY keyboard—a filter standard—remains *the* standard for computer keyboards long after the reasons for its invention became irrelevant. Filter standards might be raised or lowered slightly with relative ease—even the QWERTY standard has been changed at the edges as new keys were added and some keys rearranged—but they are difficult to change in their entirety. They often have massive infrastructures built around them.[18]

Much the same applies to ranking standards. The tests used to determine ranks may change, thereby changing the scores received by some persons or objects, but the ranking itself is often locked in by a complex infrastructure built around it. For example, the standards used to determine the oil content of canola seeds remain largely unchanged for more than half a century, but the means by which oil content is measured has changed as the technologies of measurement have improved. Similarly, the questions on the Scholastic Aptitude Test have changed over the years, but the use of this test in college admissions—for those institutions that still use it— remains much as it was at its inception.

Coupling and interoperability Standards for things that are coupled or must be interoperable create strong path dependence and therefore are quite difficult to change. For example, railway gauge standards were first developed when George Stephenson, a British engineer, used the gauge of a mining railway (4 feet, 8.5 inches) in 1830 to construct the first steam-powered railroad in the world, the Liverpool and Manchester Railway.[19] Unfortunately, since most engineers at the time did not anticipate a future when railways would replace inland water transport, they paid little attention to Stephenson's de facto standard. As a result, a plethora of gauges were developed. Changing track gauges not only required considerable time, money, and energy, it also required a change in the placement of wheels on the trains themselves. The United States had as many as nine different gauges in place until the 1880s. Only by virtue of the federal government determining in 1862 that the transcontinental railroad would have the gauge already most widely used did the nation move toward uniformity—and that was conditioned by the fact that most other gauges in use at the time were broader, making it possible to put an extra rail down the existing track bed (Schmidt and Werle 1998). Even today, several countries—among them Australia and Argentina—have multiple track gauges in use. As Spain has been incorporated into the European Union it has been forced to adopt, gradually and at considerable cost, the gauge used by France and other nations to its north (Puffert 2002).

This problem is multiplied many times in the case of information technologies. Standards for computers are perhaps best considered as a complex web of interlocking standards where each depends on the existence of others. Hence, a desktop computer must conform to a wide range of standards. These include standards for (1) all sorts of physical couplings, to (2) the software it runs, to (3) the standards for the bits of data it produces or analyzes, to (4) the "user interface," including the keyboard, mouse, and screen. Of course, it is possible for someone to build a machine that lacks these interoperabilities and couplings, but such a machine would be difficult to use and would necessarily be disconnected from the IT networks that now span the globe. For this reason, perhaps no industry is as obsessed with standards as is the IT industry (Cargill 1997).

Decentralized decision making Finally, path dependence is partially the result of decentralized decision making in the creation of standards. Put differently, when many persons must agree in order to change the standard in question, then changing it becomes considerably more difficult. This is the case in the IT industry, and it is linked to the myriad interoperable standards. In contrast, a centralized national or international standards-setting body with enforcement powers can more easily change (or even replace) a given standard. This is the case since there are winners and losers every time a standard is changed or replaced. Those likely to lose are more often able to block change in the existing standard when they must be consulted about it.

An obvious example here is the QWERTY keyboard described above. It has remained the standard because manufacturers, typists, companies purchasing keyboard operated equipment, and so forth have been unwilling to take the risk involved in making the change to another, perhaps more efficient, design. Moreover, no organization—such as a National Keyboard Agency—exists to enforce such a decision. Paradoxically, precisely because of the lack of any national or other authority, the QWERTY keyboard maintains its grip on us. In contrast, U.S. liquor manufacturers switched

Table 1.4
A continuum of sanctions

	Sanction	Agency	Examples (people, things)	Typical standard types
	Required	Law	Driver training; net weight on food label	Filters
	Prescribed	Private standards	Corporate ethics code, consistent product quality	Olympic, filters, rankings, divisions
	Noted		Expertise; ice cream flavors	Divisions
	Proscribed	Private standards	Obnoxious behavior in the workplace; inconsistent product quality	Olympic, filters, rankings, divisions
	Prohibited	Law	Housing discrimination based on race; toxic chemicals in food products	Filters

←Negative sanctions to positive sanctions→

from using English measure for liquor bottles—a fifth was one-fifth of a gallon—to using metric measure (e.g., 750 mL) because a government agency required them to do so. While some manufacturers balked, they had no choice in the matter. Similarly, seat belts became the standard only when governments began to require their installation and use.

Standards and Sanctions

Standards also differ in terms of the sanctions that do or do not apply to them, as illustrated in the continuum in table 1.4. As the table makes clear, at both ends of the continuum sanctions are inscribed in the form of laws. (In Europe, standards written into law are known as regulations.) Conversely, at the center of the continuum we find alternatives that have no sanctions applied to them. Those standards that are prescribed or proscribed have sanctions associated with them, but these sanctions do not carry the force of law. These are what lawyers refer to as precatory standards; they are advisory rather than mandatory. Included here are myriad standards for humans and nonhumans, but generally state agencies are neither the source of these standards nor of their enforcement. Instead, it is usually private actors and non-state organizations that are the promulgators and enforcers of these standards. Indeed, lawyers call this type of sanction "soft law" since it is usually not found in any legal code but is sometimes enforceable through contract or tort law. For example, if I fail to meet the standards specified in a contract, you may take me to court for contract violation. Similarly, if you fail to live up to "good practice" or even "common practice" in an industry, and I suffer damages, I can usually sue under tort law.

It is also worthy of note that different types of standards are associated with different points on the continuum of sanctions. Those standards enforced by law are usually filters, while those that merely note but do not sanction differences are usually divisions. In contrast, those prescribed or proscribed are more likely to be of several types.

The continuum of sanctions described in table 1.4 suggests that as one moves from the center to either end of the continuum, sanctions should increase, either positively or negatively. However, in the everyday world that is generally not the case. Since the myriad standards for both people and things are determined seriatim, the punishment does not always fit the crime and the reward does not always fit the achievement. In Victor Hugo's celebrated novel, *Les misérables*, the French peasant Jean Valjean is imprisoned for nineteen years for stealing a loaf of bread. While the example is fictional, there are doubtless numerous nonfiction examples of punishments inappropriate to the circumstances. Similarly, we all know people who have been well rewarded for meeting dubious standards.

This is not to suggest, however, that the determination of appropriate sanctions is fixed. To the contrary, sanctions—both positive and negative—tend to change over time. From 1920 to 1933 the production and consumption of alcohol for drinking were prohibited in the United States. Although the sanctions against those who continued to drink were only spottily enforced, during that period sanctions were far more severe than either before or after. In contrast, sanctions against smoking were nonexistent in the 1950s. Initially they were introduced as proscriptions, but they have become quite commonplace today in the form of laws barring smoking in certain public places.

It should also be noted that sanctions (irrespective of their severity) may be embedded in material objects. Smoke detectors in airplane lavatories materialize the rules against smoking on airplanes. Doors that can only be opened from one side materialize the rules about the flow of traffic. Barriers in front of highly secured government buildings block the passage of vehicles. Placement of the produce counters near the entrance of many supermarkets ensures that even those persons who arrive not intending to buy produce will have to walk past those counters. In each of these instances the standard has been given material form—a form that is far more rigid than a sign or a piece of paper, thereby sanctioning certain forms of human behavior. That said, just as laws and punishments do not eliminate certain undesirable behavior or ensure that people always behave in the desired way, materializations do not ensure compliance. With some difficulty, the smoke detector in the lavatory might be disconnected, the produce counter bypassed, or the one-way door opened from the "wrong" side.

The Objectivity of Standards

As a result of the ubiquity of standards, their path dependence, their handiness, their anonymity, and perhaps especially their materializations, standards take on objective status. Or is it the objectivity of standards that makes them ubiquitous, path dependent, handy, material, and anonymous? Let me suggest a third alternative: standards are a means by which we construct objective reality, a means by which we create an objective world. That said, standards clearly vary in their objectivity. Some standards are more objective than others. Standards for threads on bolts are more objective than those for impressionist art. But to say this is to beg the question, since it avoids an explanation of objectivity itself.

There are several equally useful (often in different situations) notions of objectivity. Let us examine several of these and see how standards stack up. First, some speak of objectivity as the ability to measure things precisely and accurately. From this vantage point, metrological standards would be among the most objective. But numerous other standards for persons and things would also fit this criterion. Of course, critics might point out that precision and accuracy may well be misplaced; hence, a precise standard for shaking hands would be regarded as at best rather strange.

Second, the creation of objective standards may be understood as the avoidance of human subjectivity, its excision from a given situation. From this vantage point, an objective standard would be one that was measured not by humans but by machines. For example, a color sorter of the type used to sort fruits and nuts at high speeds incorporates this type of objectivity into the standards it employs (although, alas, a human being must establish the standard in the first place).

Third, some see objectivity as something that emerges out of a community of practitioners. From this perspective, most standards are objective, as they nearly always arise from the concerns of such a community. Indeed, standards development organizations (SDOs) strive very hard to create consensus within a community of practitioners before releasing a standard.

Fourth, some see objectivity as conformity with certain natural processes. Certain kinds of standard measures fit this definition quite well. For example, the length of a year is defined in reference to a natural phenomenon over which we have no control. The originators of the metric system saw the meter as a natural measure: It would be one ten-millionth of the distance of a meridian line drawn from the North Pole through Paris to the equator. (In point of fact, the measure was slightly short, given the imprecision of measurement at that time.) Similar claims have been made for measures of humans. For example, so-called IQ tests claimed to measure innate natural intelligence with more or less accuracy.

Yet another approach to objectivity is what Theodore Porter (1995) calls "mechanical objectivity." By this he means the use of (usually) numerical techniques and procedures to reduce judgment to such a degree that it is unnoticeable. Such mechanical objectivity is manifested in standards when they are designed to be quantitatively precise, thereby limiting the discretion of those who use them and making them visible and obvious, as well as apparently unaffected by personal bias, to all who see them. As Porter suggests, this type of objectivity is usually found embedded in those standards that can be easily interpreted by the general public—in bureaucracies, in applied sciences and engineering, and in certain aspects of medicine. This is the case since the practitioners in these fields are the most likely to be challenged by laypersons; hence, the employment of mechanical objectivity serves as a barrier to unwanted criticism.

One might make the case that standards should strive for the strong objectivity advocated by Sandra Harding (1991). That is to say, they should incorporate a wide range of viewpoints of persons who find themselves in different situations with respect to the standards in question. The commonly used consensus process in SDOs claims to do just that, to let whoever is likely to be affected by the standard in question participate in the committee designing it. But in practice, standards committees tend to be dominated by producer advocates. Users are rarely represented. Other affected persons usually are informed about the standard only long after it is produced.

Arguably, the concept of robustness used in evolutionary biology is the obverse of Harding's strong objectivity. For evolutionary biologists (e.g., Hermisson et al. 2003) and for philosophers such as William Wimsatt (2007), robustness refers to "the invariance of phenotypes in the face of perturbation." In short, where Harding argues for strong objectivity as emerging from the diverse standpoints of knowers, proponents of robustness emphasize the stability (or lack thereof) of that which is to be known. Likely, both are necessary for standards to become objective or robust.

In short, it appears that standards produce objectivity (or robustness) through their (more or less successful) spread across time and space. As Porter (1995, 15) puts it, "what appears as universal validity is in practice a triumph of social cloning." In some instances the very act of employing standards has unavoidable consequences. For example, some telephone survey centers prohibit laughter on the part of interviewers. Others allow it. Prohibitions are based on the desire to avoid biased responses. However, at the same time, successful interviews require a certain degree of rapport with the interviewee—rapport that may be created by joining in the laughter. One study suggests that either way, the framing of the entire interview is inevitably affected (Lavin and Maynard 2001).

But make no mistake about it: this is not to say that standards produce a world that is somehow unreal, fake, or bogus. Indeed, standards are real in ways not unlike the reality of the building in which you are likely now sitting.[20] The building is the result of a great deal of hard work. It resembles countless other buildings for which similar standards—for a vast array of materials employed in construction, from concrete to brick to wood, to the electrical and plumbing fittings, as well as for the people employed in the actual construction, including everyone from hod carriers to architects—were employed. Moreover, it orders the behavior of the humans (and the rats, cockroaches, ants, mice, and other nocturnal creatures) who inhabit it. And it will last for some period of time, but not forever. Such is the objectivity, the robustness, and the reality of standards.

But the literal reality of standards embedded in a building does not exhaust the reality-creating character of standards. The standards embedded in the computer I use, in the government to which I pay taxes, in the faculty handbook at the university where I work—each of these sets of standards takes on objective status. And, like the building in which I sit, each will remain stable, robust, objective, for a period of time, but not forever.

In time, the standards for both buildings and computers will change. Perhaps we will no longer use concrete for buildings; perhaps universities will become obsolete. But so long as they are in general use, so long as they continue to function in a manner tolerable to their users, the standards that allowed their construction and their ubiquity will remain taken for granted, natural, objective, real.

That said, there are times and places where standards do produce a world that is unreal, fake, bogus. A case in point: the *Manchester Evening*

News recently reported an incident in which the driver of a truck had followed his highly standardized GPS without wavering and had managed to get his truck wedged between several buildings in a small rural community (Bell 2008). Moreover, it appears that similar events had occurred sufficiently frequently that signs had been erected along the road warning drivers to ignore their GPS units and to use another route. Several villagers had tried in vain to flag down the errant driver. All this to no avail. The GPS unit and the standards supporting it were reality for that driver until his truck came to a halt and the world of road signs, narrow roads, and sharp turns intruded on it.

The Signs on the Wall

We can see many aspects of standards illustrated in practice in a very small space by hopping on an airplane. Indeed, on a flight from Detroit to San Francisco I wrote down the many standards of a written or symbolic form that could be identified in an aircraft lavatory.[21] There were quite a few, and they illustrate quite well how ubiquitous and taken for granted standards are, and how variable the sanctions associated with them are as well. Of course, many—perhaps most—of the standards associated with the lavatory were not available in written or symbolic form to me, such as those for the physical shape of the toilet and sink, the threads for faucet parts, the placement of the hot water tap on the left and the cold water tap on the right, and so forth. However, I will leave those invisible standards aside here and discuss only those that someone intended to bring to the attention of those on the plane (table 1.5).

Nearly all of the standards mentioned or implied by the indicators (signs and symbols) in table 1.5 are filter standards: either the person or the object makes it through the filter or does not. The only exception is the faucet, which provides the occupant with a choice of water temperatures (a division). Although there are no ranks displayed inside the lavatory, there is a sign outside that reserves some toilets for first-class passengers only (a ranking). There are no Olympic standards to be found.

Path dependencies are also implied. The general form of the toilet goes back to ancient Rome and perhaps earlier; as such, most people who fly are very likely to understand what to do on entering the lavatory. Similarly, the flush mechanism, although employing suction and chemicals on the airplane (in contrast to terrestrial toilets, which employ water alone), has a handle or button to press that is familiar to those who have used toilets before. The   sign, widely used today, would be familiar to most passengers.

Table 1.5
Indicators of standards noted in an airplane lavatory

Waste container must be installed (requirement)

Please lock door (written and symbol, a prescription)

Flush (a symbolic notation, a prescription)

Pull down for baby changing table (a symbolic notation, a prescription)

Push to turn on water (a symbolic notation, a prescription)

Turn knob to left for hot, to right for cold water (a symbolic notation, a prescription)

Towel disposal—push (a prescription)

As a courtesy to the next passenger may we suggest that you use your towel to wipe off the wash-basin. Thank you (a prescription)

Waste disposal—Push (a prescription)

Close door immediately when not in use (a prescription)

Return to seat (symbol, a prescription)

Needle disposal (a symbolic notation, informational)

No drinking water (symbol, informational)

No stowage inside (a written proscription)

Flush toilet—please do not put objects such as paper towels, diapers, or airsickness bags in toilet (a written proscription)

No smoking in lavatory (a written proscription, perhaps a prohibition)

No cigarette disposal (a written proscription, perhaps a prohibition)

Warning—Federal law provides for a penalty of up to $2000 for tampering with the smoke detector installed in this lavatory (a written prohibition and symbol)

Each of these physical and linguistic conventions would be difficult to change radically without causing confusion. Furthermore, given the modular construction of aircraft toilets, it would be difficult to make major changes to the standards for any one large part of it without a full redesign.

There are tests, too, that are either manifest or implied. For many, they are built into the physical environment of the lavatory. Pushing the flush knob causes the toilet to flush; pushing on the faucet causes water to spray out. Hence, in each instance the standard is tested by its use. However, in the case of smoking, the detector serves as the test, indicating the result by an alarm. However, most of the other prescriptions, information, and proscriptions either have no test associated with them or very weak ones. Given the lack of clear sanctions, this is perhaps unsurprising.

Symmetry is displayed in virtually every standard and indicator, in that each is concerned with the interaction between humans and nonhumans. Similarly, coupling or interoperability is implied. For example, the designers incorporated into the lavatory certain (necessary) assumptions about the typical human who would use it with respect to size, weight, mobility, and ability to understand English and certain symbols.

Each of these written or symbolic statements implies a host of layered standards—for human behavior, for the specific technology, and even for the wording of the statement or design of the symbol. Moreover, some of these standards have significant sanctions associated with them. These sanctions obviously include the prohibition against tampering with the smoke detector, and probably include the prohibition against smoking while in the toilet, but they also include some far less obvious positive and negative sanctions. For example, if you push the handle, you will be able to wash your hands. If you flush the toilet, you will not have to smell the unpleasant odor.

Even the best attempts at developing standards for lavatories and for the persons who use them will fail to capture the range of human bodies and cultural differences in defecating. For example, since the major airplane manufacturers are located in Western nations, the toilets are the type on which one sits. Were the major companies Japanese, they might have been equipped with squat toilets of the sort more common in Japan. Similarly, the heights of the toilet and sink presuppose a standard height (or height range). Both dwarfs and children might have some difficulty in reaching the sink. And passengers who are not fully ambulatory undoubtedly experience difficulty entering and exiting the lavatory itself.

Conclusion

Standards are recipes for reality, or perhaps for realities. Like recipes for foods, they may be well- or ill-conceived, the subject of careful analysis or of a slapdash throwing together of ingredients, and they may result in a tasty dish or one that is barely palatable. Moreover, like recipes, they implicate both people and things. Even when using the same recipe, the master chef and the novice may well produce quite different things. Some recipes must be followed extremely carefully if the expected results are to be achieved, while others can be easily modified.

What Heidegger (1977) noted about technology is equally true of standards. They allow us to "enframe" aspects of the world so as to make them appear as "standing reserve," as something that can be called on when needed to do our bidding. Heidegger noted the inherent danger in this, but as he also pointed out, the danger holds within it the potential for finding a way out. Standards (and technologies) are dangerous because they are so easily naturalized, because in following them we amplify certain aspects of the world while reducing others, and we are thereby over-whelmed by their (and our) power (Idhe 1979). The driver who blindly follows a GPS unit is a case in point.

But standards are also means by which we perform the world. The very act of repetition, of following the recipe (whether faithfully or not), creates a reality that is ordered, regular, and stable. But what if we take the idea of performativity seriously? Shakespeare said that "all the world's a stage." His players were enchanted; they were "merely players." But if we perform the world through standards (among other things), then we can perform not only better or worse but *differently*. We can ask a whole new range of questions about performing: Who is performing? What is being performed? Is this a performance that will liberate us? Enslave us? Who is resisting or counterperforming? How could reality be performed differently?

Johan Huizinga (1950) showed just how important this is. As he suggested, culture is play; it is an ongoing performance. Children may play with great seriousness or with frivolity. But once we leave childhood behind, we tend to perform with great solemnity. Some football fans take the game so seriously that they get into brawls with fans of other teams. The military play their war games in geographic areas known as theaters. Terrorists rehearse their deadly games as well. How much more serious about playing can one get?

The approach I am adopting here is akin to that taken by Annemarie Mol (2002). She argues that we "enact" the world through technical practices. We do not merely get to *know* previously hidden realities through our techniques and technologies, we *enact*, construct, perform, create realities through our practices. In her ethnography of arteriosclerosis treatment in a hospital, she argues that the realities felt by the patient as pain, performed by the surgeon as an artery in need of replacement, and viewed on a slide by the pathologist are not the same realities. As she explains, "The practices of enacting clinical arteriosclerosis and pathological arteriosclerosis exclude one another. The first requires a patient who complains about pain in his legs. And the second requires a cross-section of an artery visible under the microscope" (Mol 2002, 35). Indeed, a number of different realities now must be made commensurable, a task requiring considerable effort and not always successful. Only by making these diverse practices commensurable do they merge into a single phenomenon: arteriosclerosis.

Standards enter into this enactment as they provide the scripts, the playbooks, the librettos, the recipes that guide practice. The standards for the patient's pain may include the ability to walk without limping, as well as where the patient places the pain on a numerical scale. The standards for the surgeon include a wide range of practices summed up as good surgical practice. The standards for the pathologist include those for obtaining a tissue sample, for producing and properly reading a slide under a microscope. Each of these standards is a recipe for quite different things; when realized in practice, they must be cobbled together to form a coherent whole, in much the same way that the various dishes of a meal, each prepared according to a different recipe, may be arranged together to form a meal. "The singularity of objects, so often presupposed, turns out to be an accomplishment" (Mol 2002, 119). Standards, or perhaps networks of standards, may help or hinder in the production of that accomplishment.

In short, in attempting to understand standards we are quickly forced to the conclusion that a topic that is all about measuring, standardization, precision, and order is in fact quite messy. Indeed, were the world precise in its form, were physical laws and human laws always without exception and precisely measurable, there would be no need for standards at all. But it is the very messiness and only partially ordered character of the world— and of the standards themselves—that seems to lead us inexorably toward the use of standards of all kinds. In the next chapter I examine various projects to standardize the world; the chapter following that examines some of the responses: the use of standards to differentiate.

2 Standardizing the World

The great book of Man-the-Machine was written simultaneously on two registers: the anatomico-metaphysical register, of which Descartes wrote the first pages and which the physicians and philosophers continued, and the technico-political register, which was constituted by a whole set of regulations and by empirical and calculated methods relating to the army, the school and the hospital, for controlling or correcting the operations of the body.
—Michel Foucault (1977, 136)

Nothing is more remarkable, and yet less unexpected, than the perfect identity of things manufactured by the same tool.
—Charles Babbage (1835, 66)

The origins of standards are buried in the very distant human past, "in the dreamtime long ago." Standards likely originated in the division of labor, the specialization in the performance of various tasks that can be found in even the earliest human societies. The division of labor in turn is almost undoubtedly of biological origin, found in the sexual division of labor necessary to reproduction but extended by humans to a range of other tasks. These standards were and are often marked by rites of passage, by ritual ceremonies that mark transitions of one sort or another in life. These rituals themselves became standardized forms of behavior, easily recognized as such by members of a given human group. The very specialization implied in the division of labor meant that standards were essential—for the behavior of those in various roles and occupations (hunter, gatherer, pastoralist, arrow- or spear-maker, healer), as well as for the various articles and implements necessary to engage in those occupations.

As expressed in language, standards appear to be linked to four domains of great concern to our ancestors, and still of great concern to us today: counting, shape (later formalized as geometry), weight, and time. These four domains allow us to distribute persons and things in myriad ways. They are so central to our daily existence that we hardly notice their continued presence in everything we do. Some examples of their uses follow.

Counting Does what you say count? Can you be counted on? Does your work count? Will you be held accountable? Or are you of no account? Are you even?[1] Or uneven? Will you get even? Or are you the odd one out? Are you paired? Or impaired? Can you recount the story? Should I discount your words?

Shape Are you in good shape? Have you shaped up? Are you right and true? Are you an upstanding citizen? Are you valid? Or an invalid? Are you a square? Do you square off against others? Do you always get a square meal? Are you on the straight and narrow? Can you give me a straight answer? Do you always do the right thing? Or are you rather obtuse? Do you go straight ahead? Is your reasoning linear? Or circular? Do you produce vicious circles?

Weight Do you weigh in on the issues? Do things weigh on your mind? Do you throw your weight around? Do you pull your weight? Are you a heavy? Or a lightweight? Or merely excess weight? Do you do things in a heavy-handed way? Or do you have a light touch?

Time Are you on time? Are you short on time? Do you give people a hard time? Can you stand the test of time? Isn't it high time? Are you pressed for time? Do you live by the clock? Do you spend quality time with your children? Are you living on borrowed time? Do you have time on your hands? Has your time come?

Note that nearly every one of these expressions implies a standard against which one is tested. Each suggests a *distribution* of persons and things. Someone who can stand the test of time, who really counts, who is always in shape, whose words are weighty is someone to be taken seriously. One may make a similar case for things. While the metaphors listed above will be familiar to those raised in most contemporary Western nations, other cultures, other languages, other historical epochs each have their own metaphors for time, space, number, and weight, and each implies standards. For example, native Australians mark space through song, the length of time required to sing the song serving as an adequate measure of the space to be traversed. Yet as solely verbal expressions, these standards could hardly be (nor did they need to be) very precise.

The advent of written language brought standards to the fore in a new way. Written language appears to have had its origins not so much in the recounting of great exploits or in the retelling of epic poems but in the rather prosaic need to account for things and people. Indeed, many early languages were simply pictures of commonly traded products along with some system of numbering that allowed accountants to keep track of them.[2] The symbols used for both pictographs and numbers had to be sufficiently standardized that they could serve not only as an aide-mémoire for the writer but also could be similarly recognized by others who might learn to read them. Eventually, the written languages we know today emerged either as alphabetic or as character-based (or, as in Japanese, both).

But such alphabetic written languages demanded a kind of multiple standardization, at least in respect to nouns. To standardize its meaning, the word "apple" (1) must be spelled (more or less) in the same manner each time it is used, irrespective of the writer. Thus, one may spell apple (especially when used in a sentence: e.g., "I ate an apple"), aple, appple, appel, and so on, and others may still understand the word. Moreover, (2) each letter must be formed in (more or less) the same way each time it is formed by *any* writer. For example, apple, *apple*, **apple**, apple, *apple*, and **apple** are all easily recognized by persons literate in English as the same word.[3] Of importance, writing in a manner that is sufficiently standardized that it can be recognized by others is a learned skill, requiring—as anyone who spent hours practicing penmanship in elementary school knows—a considerable amount of work. In addition, (3) the written word must refer to the same class of fruits each time it is used. Thus, in the sentence "I ate an apple," one conversant in English would not understand this to mean either that I ate either a computer by that name or that I ate some other fruit. That said, as with all other standards, there is a certain degree of irreducible ambiguity. For example, certain Asian pears could easily be mistaken for apples, as could related fruits (crabapples, quince). Furthermore, (4) standardization of the word "apple" is assured by its *situated* use. Only when it is used contextually in a sentence or phrase can its stabilization as a standard be assured. Conversely, (5) by limiting the range of acceptable letters in the alphabet, certain words, sounds, and the like become difficult to express. For example, uniform resource locators (URLs) used for addresses on the Internet employ the ASCII character set. It in turn was developed with English language usage in mind. It makes no provision for diacritical marks. Thus, the Swedish town of Hörby has had to make do with the address www.horby.se. This would hardly be problematic were it not for the fact that Horby translates roughly as "adulterer village" (Pargman and Palme 2009). Finally, (6) none of this rules out the creation of neologisms: until the Apple computer was invented and named, use of the word "apple" to refer to a computer would have been either strange or meaningless. Now it is accepted and widely understood.

Alphabetic languages also permit arbitrary orderings that, as we shall see, are widespread in society. Thus, we have standardized the English alphabet as abcdefghijklmnopqrstuvwxyz. Norwegians add three additional letters at the end: æ, ø, and å. Other alphabetic languages have different standardized sequences of letters. The ordering of the letters is entirely arbitrary, but it allows us to file away a vast array of things and retrieve them at a later date.

Much as written language helped to stabilize and standardize a wide variety of things, so has the recording of temporal measures. Forecasting the seasons was of particular concern to the ancients, as their decisions to plant and harvest depended on an intimate understanding of this temporal quality of existence. But merely noting the changes was insufficient. To predict the changes, they had to be inserted into a calendrical grid; in other words, a grid had to be developed.

Hence, the ancient Egyptian priests worked hard to predict the annual flooding of the Nile Valley. The builders of Stonehenge and similar monumental "machines" around the world focused on seasonal predictions. Indeed, these standards *created* sharp breaks of seasonality within the continuity of time. That is to say, they marked off a given period to be labeled as a season. They made what was continuous discrete. Precise time-telling remained a long way off.

It is likely that many formal standards for things as we know them today came into being only with the creation of monetary exchange. In systems of barter, the parties to the exchange each inspect the object that the other wishes to exchange and, upon reaching agreement that the exchange is satisfactory to both parties, conduct the trade. Although rules of thumb might well exist that provided a rough equivalence based on relative value, bartering was often difficult as well as long and drawn out, especially given the incommensurability of the objects being exchanged. Finding a suitable collection of objects to produce an acceptable level of gain for each party is not easy. If I have a bed that I wish to exchange and you have shoes, it is unlikely that I would wish to exchange the bed for ten pairs of shoes.

Money changes all that by introducing a concrete abstraction (Sohn-Rethel 1978). I can sell you the bed for a given sum of money and use the money to buy something else. I can even hold on to the money until I decide what to buy. But for this sequence of transactions to be possible, there must be a standard for the money (today it is in the form of coins or bills), as well as for any weights, measures, or scales involved in the transaction. Without those standards, a money society is virtually impossible.

Furthermore, money is specifically designed *not* to be used for anything other than its exchange value (cf. Zelizer 1998). For this reason its value is usually printed or struck on it. Of course, money may be kept for its symbolic value: rare coins may be valued and collected for their rarity, or a $100 bill may be used to light a cigar, demonstrating the wealth of the person doing it. Money may also be employed for its use value: an ordinary coin may be used to stabilize a table. However, doing so removes the money from its designed and designated purpose—as pure standardized exchange value. For much the same reason, a crumpled and soiled bill and a worn coin retain their original value; in contrast, an automobile with a large dent in the fender will be worth less than one of the same model in excellent condition.

The world's great religions, each of which arose in the context of a civilization with monetary exchange, treat questions of standards with considerable gravity. Moreover, they treat standards as obvious, fixed, even deified (although doubtless someone determined what those standards should be). Consider, for example, the following lines from the Bible:[4]

Just balances, just weights, a just ephah [~21 dry liters], and a just hin [~3.8 liquid liters], shall ye have . . . (Leviticus 19:36)

But thou shalt have a perfect and just weight, a perfect and just measure shalt thou have: that thy days may be lengthened in the land which the LORD thy God giveth thee. (Deuteronomy 25:15)

A just weight and balance are the LORD's: all the weights of the bag are his work. (Proverbs 16:11)

Or consider the following passage from the Koran:

Allah indeed has appointed a measure for everything. (65.3)

Similarly, in the Hindu holy text, the Rig Veda (Hymn 154), one finds the following passage:

I WILL declare the mighty deeds of Visnu, of him who measured out the earthly regions,

Who propped the highest place of congregation, thrice setting down his footstep, widely striding.

And the great Chinese philosopher Confucius tells us in *The Doctrine of the Mean*:

To no one but the Son of Heaven does it belong to order ceremonies, to fix the measures, and to determine the written characters.

Each of these texts, in somewhat different yet essentially similar ways, links weights and measures to the proper behavior of persons and to the divine. One can speculate as to whether this indicates a comparison between divine perfection and earthly imperfection or whether it provides a means of emphasizing the importance of such standards. Perhaps it does both. That said, linking standards with the divine in no way substitutes for exhortations to adhere to standards and the implied punishments for those who do not.

From among many such exhortations in the Bible:

A false balance is abomination to the LORD: but a just weight is his delight. (Proverbs 11:1)

Ye shall do no unrighteousness in judgment, in meteyard [yardstick], in weight, or in measure. (Leviticus 19:35)

From the Koran:

O my people! serve Allah, you have no god other than Him; clear proof indeed has come to you from your Lord, therefore give full measure and weight and do not diminish to men their things, and do not make mischief in the land after its reform; this is better for you if you are believers. (7.85)

And

O my people! give full measure and weight fairly, and defraud not men their things, and do not act corruptly in the land, making mischief. (11.85)

Confucius extends the metaphor far beyond weights and measures and directly links it to all human behavior:

What a man dislikes in his superiors, let him not display in the treatment of his inferiors; what he dislikes in inferiors, let him not display in the service of his superiors; what he hates in those who are before him, let him not therewith precede those who are behind him; what he hates in those who are behind him, let him not bestow on the left; what he hates to receive on the left, let him not bestow on the right: this is what is called "The principle with which, as with a measuring square, to regulate one's conduct."

In short, each of these ancient revered texts links standards with divinity as well as with justice, and emphasizes the moral duty involved in upholding standards, both in the literal sense of providing true weights and measures and in the figurative sense of measuring up.

A Perfect and Just Measure Shalt Thou Have

Over the past several centuries we can discern first a standardizing trend and somewhat later a differentiating trend, corresponding to universalist and particularist claims for standards. The remainder of this chapter considers the trend toward standardization. Two important caveats should be kept in mind. First, the project of standardization was just that—a project. Like all projects, it was never realized in its entirety. Moreover, the very attempt to create a standardized world produced reactions that were certainly unexpected by proponents of the project. These reactions included resistance to standardization, using standardization to ends not intended by those who designed the standards, and getting tasks accomplished despite the standards. Second, the following story of the project of standardization is hardly the sole story of the last several hundred years that could be written. Others could have and have written very different stories focusing on different objects of interest. Thus, what follows is the story of an only partially successful project, and it is hardly the only story that can be told.

While standardization can be traced back to the origins of human civilization, it was given an enormous boost by the grand universalizing project known as the Enlightenment. The Enlightenment challenged the older particularisms of the past by laying claim to universal reason—reason that itself required the correct (standardized!) method—and careful attention to the empirical world.[5] Proponents of the Enlightenment were convinced that reason combined with empirical observation would allow us both to reveal the underlying workings of nature (by reading the Book of Nature as written by God) and to transform society so as to make it more rational and universal. Put differently, "The notion that predictability, accountability, and objectivity will follow uniformity belongs to the Enlightenment master narratives promising progress through increased rationality in control" (Timmermans and Berg 2003, 8). As a result, the seventeenth and especially the eighteenth and nineteenth centuries were a time of a great project of standardization, notably in the Western world. Despite their vast differences in perspective, a wide range of groups pressed for standardization in virtually every aspect of social life.

It is important to understand that the Enlightenment was not merely a philosophical movement but a set of profound changes in material practices. In the seventeenth century Thomas Hobbes and Robert Boyle debated the value of experimental science (Shapin and Schaffer 1985). Hobbes argued that the only true knowledge was that obtained from mathematics.

He noted that mathematical objects—squares, triangles, cubes—do not exist in our everyday world but are instead ideal objects. Moreover, Hobbes noted that the relations between them are valid because they are implicit—and sometimes explicit—in their definition. Hence, the sum of the interior angles of a triangle equals 180 degrees. Were I to find a three-sided plane object the interior angles of which summed to 179 degrees, it would simply not be a triangle.

In contrast, Boyle believed that experiment would provide true knowledge. He demonstrated that when the air was removed from between two hemispheres, the power of teams of horses could not pull the hemispheres apart. But Hobbes asked how one could know that the air was actually removed. How could one be sure that the seal between the two hemispheres was not faulty? He concluded that any experiment would fail to produce true knowledge as there would always be room for error—error always absent from the ideal objects of mathematics.

Contemporary science combines the insights of Hobbes and Boyle in a synthesis that is at once mathematical and empirical. On the one hand, it uses the tools of mathematics to introduce precision, and it uses statistics to produce standards of acceptable error. On the other hand, modern science requires that research be accomplished using highly standardized objects. Put differently, science creates its objects of study such that they conform, as much as possible, to the assumptions of probability statistics. Let's explore this issue a bit deeper.

Those persons usually credited with the creation of modern science generally believed that both the perfect knowledge of the sort described by Hobbes and the experimental knowledge of the sort proposed by Boyle were possible. For example, in 1637 René Descartes (1901, 161) proposed four rules of method:

The FIRST was never to accept anything for true which I did not clearly know to be such; that is to say, carefully to avoid precipitancy and prejudice, and to comprise nothing more in my judgment than what was presented to my mind so clearly and distinctly as to exclude all ground of doubt.

The SECOND, to divide each of the difficulties under examination into as many parts as possible, and as might be necessary for their adequate solution.

The THIRD, to conduct my thoughts in such order that, by commencing with objects the simplest and easiest to know, I might ascend by little and little, and, as it were, step by step, to the knowledge of the more complex; assigning in thought a certain order even to those objects which in their own nature do not stand in a relation of antecedence and sequence.

At the LAST, in every case to make enumerations so complete, and reviews so general, that I might be assured that nothing was omitted.

What is striking about these rules is how much the first three remain the standards to which virtually every scientific discipline adheres.[6] Most scientists work particularly hard to rule out alternative hypotheses and to separate their personal biases from their work.[7] But at the same time, as Karin Knorr Cetina (1981, 42) wrote some years ago, "Scientific papers are not designed to promote an understanding of alternatives, but to foster the impression that what has been done is all that could be done." And, should that be insufficient, most papers end with an escape clause: "More research is necessary."

Nor have scientists abandoned the project of dividing the world into smaller and smaller pieces. In biology, what started as studies of whole organisms has shifted to a fascination with the molecular. Indeed, for some biologists it is as if all behavior can be explained by the combination of various genes. Similarly, in engineering and the physical sciences nanotechnology is the current focus of interest.

The third rule follows logically from the second. At least in principle, most proponents of modern science believe that knowledge gained about the microlevel will help illuminate the larger world. Even critics must agree that this approach, usually defined as reductionism, has been quite productive. However, in recent years it may have reached its limits, heralded by the rise of complexity studies.

The last rule, however, was shown to be quite limiting. As dozens of philosophy textbooks have noted, complete enumerations are virtually impossible to achieve; the possibility that one might find a blue bear in nature always remains, even if the likelihood is extremely small. The apparent solution to this dilemma has been to abandon the notion of complete enumeration and to replace it with statistical sampling. Sampling is now the preferred method for gathering data on subatomic particles, blood components, preferences for various ice cream flavors, and opinions about world events.

Sampling has brought with it an ever-increasing array of standardized statistical techniques by which it is claimed (usually correctly) that analyses conducted with a properly identified sample apply to some larger population of things, people, or events. Different disciplines tend to favor different statistical techniques. Most biologists prefer to use analysis of variance. Most economists prefer some form of regression analysis. Nonparametric statistics are currently in vogue in sociology. In each instance a standardized test of statistical significance is employed, and the value of

0.05 (a probability of ≤5 percent that the findings of the analysis are due to chance) is defined as a standard cutoff point beyond which one cannot have sufficient confidence in the validity of the results obtained.

The practitioners of science understood the rules of method as a means to change the world, and change it they did. The invention of money may have introduced concrete abstractions into the practice of exchange, but it limited its use to exchange. In contrast, modern technoscience extended concrete abstractions into all aspects of social life. The theories of modern science, embedded in myriad scientific instruments and machines, came to surround us and to create a theory-laden and therefore standardized second nature. As the French philosopher of science Gaston Bachelard ([1934] 1984, 12–13) put it, "Scientific observation is always polemical; it either confirms or denies a prior thesis, a preexisting model, an observational protocol. It shows as it demonstrates; it establishes a hierarchy of appearances; it transcends the immediate; it reconstructs first its own models and then reality."

The history of modern science is the history of the creation of ever more precise, ever more standardized, ever more predictable experiments, machines, and experiments that require machines. Indeed, one could say that modern science allowed the deciphering of nature by ciphering it; it allowed one to make sense of nature in large part by turning it into numerical form.

But one may still ask why it is that scientific knowledge appears so universal in character. Why didn't French scientists develop different machines, different experiments, than their British or German or Swedish or Italian counterparts? The answer to this question has been suggested by Robert Wuthnow (1987). He argues that, despite the political fragmentation of seventeenth-century Europe, the practice of natural philosophy, and what would later be called science, was from its very inception similar across the continent. Scientific communication was initially in the language of scholarship, Latin, and only much later in the vernacular. The natural philosophers, later known as scientists, were hired by wealthy members of the aristocracy, many of whom were interrelated by marriage. Scientists themselves were from similar class origins. Nearly all male, and similarly educated, they looked to each other both for recognition and for validation.[8] When European settler colonies were created, they too became part of the network. For example, Benjamin Franklin corresponded regularly with continental mathematicians, scientists, and philosophers.

Moreover, what was true for the fledgling scientific community was equally the case for the instrument makers. From the very inception of modern (instrumental) science, success in replicating experiments required that one have access to the same or similar instruments. Initially, the craftsmen with the greatest skills in producing these instruments were almost always linked to trade (Daumas [1953] 1972). After all, it was ships that required the most sophisticated instruments of the seventeenth and eighteenth centuries. However, soon specialist scientific instrument makers began to emerge across Europe (Price 1957). Without these skilled craftsmen, the scientific achievements of those times—and today—would have been impossible. As one instrument maker opined,

Mathematical Instruments are the means by which those noble sciences, geometry and philosophy, are render'd useful in the affairs of life. By their assistance an abstracted and unprofitable speculation, is made beneficial in a thousand instances: in a word, they enable us to connect Theory with Practice, and so turn what was only bare contemplation, into the most substantial uses. (George Adams in 1746, plagiarizing from a 1723 work of Edmund Stone, quoted by Brown 1979, 86)

In so doing, the instrument makers helped to produce a standardized, controllable nature.

As a consequence, scientists tended to work on the same problems everywhere throughout the continent and later around the world. They were all part of what Diana Crane (1972) would later refer to as "invisible colleges." These early scientists worked largely on problems of concern to the enlightened despots of the time: the myriad problems of military technology—the trajectories of cannon balls, the design of faster ships, navigational aids—as well as of the construction of prosperous modern industrial states that would fill the kings' coffers with taxes. That meant that the three key problems of the day were of constant concern: theoretical understandings of water, wind, and heat (what would later be called mechanics) could, at least in principle and often in practice, be rapidly put to work to power machinery, mine the earth, protect against foreign invasion, or support projects of geographic expansion. Moreover, just as cooks and (al)chemists wrote down recipes for the preparation of food and medicine, respectively, scientists began to write down recipes for the use of water, wind, and heat to solve these practical problems. The combination of increasingly standard written texts describing the procedures to be used (eventually recorded in what were to become scientific journals) and standard instruments permitted scientists to replicate and improve on each other's work, to sort out the effective recipes from those that were ineffective, and thereby to make (more or less) standardized realities (see Goody 1977).

With a few standardized instruments, a relatively limited number of projects capable of scientific analysis, and a firm belief in both method and the power of mathematics, it is little wonder that scientists concluded that every problem had a single, universal answer. It required little further analysis to become convinced that there was one best, standardized way to perform all tasks, to build all objects, even to conduct one's life.

The unity of science movement at the beginning of the twentieth century took the standardization of knowledge through science yet further. Many of its proponents sought to limit knowledge to that which could be known through the use of the scientific method. Furthermore, its proponents attempted to establish a hierarchy of the sciences, with physics at the top, in the belief that all other sciences could, given sufficient time, effort, and financial support, be reduced to physics. To many of its proponents, science would replace older forms of knowledge and become the sole path to truth.[9]

To understand just how important standards are to science, consider what must be done to create the ordinary laboratory rat, used in the past and today in thousands of experiments in scientific laboratories around the world. Table 2.1 was compiled from a careful reading of the first manual on the creation of laboratory rats, written in 1931. What its authors, Milton J. Greenman and F. Louise Duhring, realized was that without standard rats, raised under standard conditions, fed standard feed, and allowed standard exercise, the validity and replicability of laboratory experiments were in doubt. As they put it, "It became more and more evident that clean, healthy, albino rats were essential for accurate research and that their production was a serious, difficult, and worth-while task" (Greenman and Duhring 1931, 3). Their careful, step-by-step approach to organizing the rats' entire lives helped make comparable the thousands of experiments that came later.

In short, the laboratory rat was *designed* to be precisely the opposite of a rat found in a natural setting. It was from a selected group of physiologically and genetically similar albino rats that were "gentled," fed a special abundant and varied diet, reared under highly hygienic conditions, observed frequently for any sign of undesired behavior, given plenty of opportunity to exercise, weighed regularly, and otherwise carefully monitored. Even details such as when mating should occur, what menus and recipes were to be used for feeding them, how cages should be stacked, and at what angle the drinking bottle should lean (10 degrees toward the rear of the cage), and what kinds of persons should be considered as caretakers were specified. The result was the production of "name-brand rats

Table 2.1
Selected requirements for standardization of albino laboratory rats

Adequate ventilation, sunlight, and temperature controls should be available.

Openings should be screened against flies.

Protection against contact with wild mice or rats should be available.

Building should be made of brick, stone, concrete, steel, and glass.

Water outlets should be available.

Electricity for light and power should be available.

Large colonies should have several rooms.

Cages should be hung from the ceiling or bolted to the walls, improving ease of cleaning floors.

Cleaning and sterilizing room for cages is necessary.

A sterilizer to sterilize all items brought in contact with colony is necessary.

A kitchen for food preparation should be available.

Food trucks should be used for distributing food.

Fiber boxes should be used to collect waste and dirt.

A separate room should be used for record-keeping.

The storeroom should be vermin-free.

Strangers should be prohibited from entering the cage room.

Cage should have dark space where rat can seek shelter

Bedding material provided for rats; pine or poplar shavings preferred.

Cages should be collapsible for ease of handling and sterilization.

Each cage should have a one-quart water bottle.

Water bottles should lean out and to the rear at a 10-degree angle.

Galvanized wire cloth screening should be used as a door.

Doors should not be hinged to cages so they can be easily removed.

Cages should be constructed so they can sit on top of each other in groups of four or five.

Provision should be made for exercise, e.g., with a treadmill or turntable.

Exercise should be recorded with the recording mechanism (Rotary Ratchet Counter) that counts revolutions in both directions.

Modified steel buckets should be used as carriers to transfer mice from one location to another.

Rats in experiments should be of the same age, weight, litter, and sex.

Caretakers should not be changed during an experiment.

Diet should be varied, including cereals, roots, tubers, red meats, etc.

Rats should not be picked up by the tail.

Rats should mate between 110 and 220 days of age.

Detailed records should be kept on each rat.

Source: Adapted from Greenman and Duhring (1931).

such as Osborne-Mendel, Long-Evans, and Sprague-Dawley" (Clause 1993, 331). Only such rats would henceforth be acceptable in laboratory experiments.[10]

In more recent years we have witnessed the development of specialized laboratory mice, developed by selecting for certain genetic traits. Perhaps best known is the OncoMouse, developed and patented by Harvard University biologists. It is designed to be susceptible to various cancers, allowing it to be "sacrificed" for our benefit. The OncoMouse, as Donna Haraway (1997, 102) notes, was bred to be an "inhabitant of the nature of no nature."

What is true of laboratory rats is true of thousands of other scientific materials, laboratory animals, cell samples, and scientific instruments (Clarke and Fujimura 1992). However, not all have been equally successfully standardized or equally widely used. Fruit flies (*Drosophila melanogaster*) are widely used in genetic studies because of their short reproductive cycle and comparative ease of breeding; in contrast, flatworms (Planaria), commonly used a century ago, are rarely employed today (Mitman and Fausto-Sterling 1992).

Moreover, even today, in fields in which standardized materials do not exist or are difficult to create, research may well be impeded. For example, drug-resistant malaria has become a major concern worldwide, yet at the moment, "There is no single, universally accepted, standardized protocol for in vitro assays. Different research laboratories take into account different factors, each of which can profoundly influence the level of drug response" (Basco 2007, 3). As a result, test results are incomparable, and improved treatment of the disease is hampered.

But science was hardly the only sphere of human endeavor in which standardization was enthusiastically embraced. It was embraced in virtually every aspect, every locus of human existence (although with variable effects). Time would become standardized. Colonies were created and standardized. Social movements would bring standardized benefits to humanity. Medicine, agriculture, education, families, fashions, factories, communications, management, the military, and knowledge itself would be made regular, uniform, homogeneous, standardized. It is to these multiple standardizations that I now turn.

A Time for Standards

The ancients were concerned with measuring the seasons. For telling time, looking at the position of the sun in the sky or the stars at night was usually

sufficient. Until modern times, very few people really cared if it was precisely 11:30 a.m. or 6:00 p.m. The invention of the astrolabe, the sundial,
and other time-telling instruments allowed greater precision in time-telling
for those who desired it, but hardly greater accuracy. After all, day length
is related to season and has greater seasonal variation as one moves away
from the equator. Even the development of the mechanical clock was of
little initial consequence since most clocks were not particularly accurate
and had to be adjusted constantly. Watches were even less reliable.

Two concerns spurred the quest for standard time: the need for precise
and accurate measures of longitude and the advent of the railroads. Both
were part of the projects of colonization and industrialization. The problem
of longitude had been pondered by sailors since ancient times. Knowing
the exact longitude was of relatively little consequence for plying the Medi-
terranean since the sea itself was not that large. Similarly, ships could sail
along the shores of continents without any great concern with longitude.
But when attempting to sail across oceans, the lack of longitudinal measurements meant one could be blown far off course and not even know it.

Galileo had proposed that by making multiple observations of the
moons of Jupiter one could calculate longitude. However, while this
method worked on land, watching the heavens through a telescope while
aboard an oceangoing vessel—even on a calm sea on a clear night—proved
nearly impossible (Drake 1978). Other astronomers tried variations on
Galileo's approach, to no avail.

Another strategy used at the time employed two clocks. One would be
set to local time at the beginning of a voyage and taken aboard ship. A
second clock would also be taken aboard ship and would be set to local
time at that location. The problem, however, was that no one had ever
built a clock that had such precision. Most had to be reset quite frequently.
In 1714 the British Parliament set up the Board of Longitude, the first
research-and-development body in the history of the world. It offered a
prize for the first person to develop a satisfactory means of measuring
longitude. After a lifetime of effort, an English clockmaker, John Harrison,
eventually developed the marine chronometer, a timepiece of astonishing
precision that made possible relatively straightforward measurement of
longitude (Sobel 1995). Although his own ambitions were quite limited,
Harrison had simultaneously advanced the building of a standardized
British Empire and all those aspects of our modern world that depend on
standardized means of time-telling.

However, beyond telling longitude, there was little practical use for his
expensive instrument. That required the advent of railroads, the owners

and operators of which soon created a demand for standard time. The problem was both simple and appalling: On single track lines, trains would set out from both ends and collide head-on somewhere along the line. Such collisions often led to a loss of both life and property. There were several solutions to this problem: The first involved allowing trains to go in only one direction on the track at a time. This had the disadvantage of leaving the track idle most of the time. A second solution involved building a second parallel track, but this involved considerable capital costs, costs often unjustified by the volume of freight and passengers. A third solution was to calculate the times that the trains left the stations and estimate their average speed, and to ensure thereby that one of the trains would be on a siding when the other came by.

But this last solution could be implemented only if the time was accurately calculated and standardized at both the starting and ending points. This was virtually never the case since local time was usually set by a clock in the main square of each town, a clock that was in turn set by examining the place of the sun in the sky. Individuals would then set their clocks and watches by the town clock. While this worked fine in a given town, it also meant that every town had its own unique time. Even with Harrison's chronometers, which provided great precision, time was not sufficiently accurately measured to prevent collisions.

The railroads also tried to create their own standard times. In the United Kingdom, an industry body, the Railway Clearinghouse, recommended in 1847 that Greenwich Mean Time (then called London Time) be used throughout the rail network as soon as it was permitted by the Post Office, the agency in charge of such things. By 1855 most clocks in Britain were set to GMT, but the legal system did not catch up until the Definition of Time Act of 1880 (Weller and Bawden 2005). In contrast, Cronon (1991, 79) notes that in the United States, which lacked an industry body such as existed in Britain, by the late nineteenth century, "Railroads around the country set their clocks by no fewer than fifty-three different standards— and thereby created a deadly risk for everyone who rode them." But this hardly resolved the problem.

The solution—much disputed at the time (see, e.g., Schivelbusch 1978; Zerubavel 1982)—was the creation in 1883 of what was initially known as "railroad time," that is, standard time zones within which all clocks and watches could be calibrated in the same way. Doing so created a huge demand for chronometers such that each conductor, station master, and engineer on each train could have one to ensure against collisions. Simultaneously, it increased the demand for accurate standardized timepieces among the general public, who wanted to be able to catch the 8:49 a.m. train to Edinburgh or arrive in New York in time to meet a client at 6:00 p.m.

Today, atomic clocks produce accurate and precise readings of standardized time. Thus, in much of the world we no longer look to the sky to determine the time but instead depend on an atomic clock—indirectly through the circulation of other clocks that are calibrated based on that one—that is accurate to within 30 billionths of a second per year (see National Institute of Standards and Technology 2007). Without this complex yet nearly invisible infrastructure of time, the standardized global transport networks of railroads, ships, trucks, buses, and airplanes, global supply chains that depend on just-in-time shipping, and hundreds of television and radio networks would hardly be able to function.

Yet even as clock time dominates the global economy, even as it is largely taken for granted by those of us enmeshed in that economy, the project of standardized time remains incomplete: "Children and the elderly, the unemployed, carers the world over and subsistence farmers of the majority world inhabit the shadowlands of un- and undervalued time. Women dwell there in unequal numbers. Their time does not register on the radar of commodified time. Their work is not accorded value in the capitalist scheme of things. Rather, it is rendered invisible" (Adam 2006, 124).

Thus, as successful as the enactment of standardization of time appears, it conceals those persons and things who resist or who are unable to conform to it. But perhaps no institution has been as obsessed with standardization as the modern military. And military standardization, too, has always met resistance.

The Military

The notion of the citizen soldier, now about two hundred years old, brought with it a wave of standardization. If large armies were to be well organized, they needed to know who they were and who the enemy was. The very word "uniform" reinforces the importance of uniformity, standardization, within the military.

Weapon standardization was a far more complex task than the creation of standard uniforms. Standard knives and swords were relatively easy to produce, since their standardization required no interchangeable parts. But standardization of rifles and artillery was a more drawn-out and difficult process. A standardized rifle had to have standard parts, such that a broken

part could be easily replaced on the battlefield by a soldier having no training in metalworking.

The first standardized gun parts were presented to an astonished crowd
in Paris at the Hôtel des Invalides by Honoré Blanc in 1790. They were
"part of a larger program by these military engineers to improve the effectiveness of the French army after the humiliations of the Seven Years' War"
(Adler 1997, 277).

On the one hand, there were technical problems to be overcome in
producing weapons with interchangeable parts. Resolving those technical
problems required the development of precision machine tools (to allow
production of parts that were "the same"), standard gauges against which
the parts could be checked for conformity, standard measures (so that
artisans in different locations could produce to the same specifications),
and standard projective techniques for mechanical drawings (figure 2.1).
Whereas perspectival drawings required considerable judgment in determining from which perspective an object was to be drawn, projective
drawings produced the view from nowhere (or perhaps everywhere). "Projective drawings achieve this effect, in part, by reducing the representation
of objects (and their decoding) to a set of formal rules. The goal is to limit
the discretion of both the person drawing the plan *and* the person interpreting it" (Adler 1998, 514, emphasis in original).

On the other hand, the artisans who produced such weapons vigorously
resisted the shifting of skilled work from their shops to the drafting boards
of engineers. Prior to the 1760s, craft guild members had considerable
freedom to define their own measurements, develop their own designs, set
their own work pace, and choose their own clientele. Thus, the imposition
of the "objectivity" associated with mechanical drawings and standardization was the subject of considerable resistance. Nevertheless, by the mid-
nineteenth century, most rifles and artillery were standardized.

Furthermore, the military needed standardized food products. Supplying an army, especially one bent on conquest, is a difficult task at best.
One needed to know the minimum nutritional requirements for each
soldier. One needed foods that could be carried long distances without
spoiling. Not surprisingly, for this reason, the process of canning owes its
origins not to the food-processing industry but to a bounty paid by Napoleon to whoever could create an inexpensive, effective means of preserving
fresh foods. Nicolas Appert claimed that prize in 1809.

Finally, the military needed to have disciplined, standardized soldiers.
This was achieved by the introduction of standard uniforms, haircuts,
lodgings, drills, units of infantry and cavalry, marching, saluting, and other

Figure 2.1

Examples of perspective (St. Mark's Square) and projective (Cochran House) drawings.

Sources: Reginald Blomfield (1912, plate following p. 60), Wooster Bard Field (1922, plate 26).

behavior both on and off the battlefield. The modern military machine was born.

However, as with standardized time, military standardization has been resisted. From conscientious objectors to draft dodgers to military deserters, from soldiers who painted unauthorized names on their airplanes to sailors who filled ships' cannons with black market goods, there have always been those who have resisted military standardization.

Colonization

Perhaps the most ambitious (and ultimately horribly flawed) attempt at standardization occurred as part of the spread of Western empires around the world starting in the late fifteenth century. It was the Spanish and Portuguese who established the first rules of empire. First, trade was to be with the "mother country," with a clear and often brutally enforced division of labor: raw materials would be produced in the colonies, while finished goods would be produced in the mother country. Moreover, within the boundaries of the colonies, (1) the laws of the mother country would largely prevail (though often in caricatured form), (2) the local populations would be subjects of the appropriate European crown, (3) metropolitan art, literature, and architecture would replace the indigenous styles, and (4) if at all possible, the locals would be converted to Christianity (sometimes by missionaries, sometimes by force, and sometimes by missionaries using force).

During this period first the Spanish and Portuguese, then the British, French, Dutch, German, Italian, and Belgian, empires were established. What Benjamin Constant ([1815] 1988, 74) wrote, in his critique of Napoleon's conquest of Europe, summed up the link between standardization and conquest: "Applied to all the parts of an empire, this principle [of uniformity] must necessarily apply also to all those countries that this empire may conquer. It is therefore the immediate and inseparable consequence of the spirit of conquest."

Of particular import was that even as these nations fought over territories, even as they scrambled to claim those parts of the world that were "unclaimed" for themselves, they agreed on the necessity of colonies for prosperity and the dangers of leaving as much as a single acre unclaimed. By the dawn of the twentieth century, nearly the entire world had been colonized. Among the few remaining places, beachheads had been established by Western powers all along the Chinese coast. Only a handful of exceptions remained: Thailand, Ethiopia, and the subordinate nations consisting of "repatriated" slaves in West Africa. Japan, threatened by colonization, determined to adopt Western social organization and technologies and to create its own colonies. Parts of China, Taiwan, and Korea became part of the Japanese Empire. And the United States, while only marginally a colonial power (in the Philippines, for example), simultaneously kept out meddlesome European powers and tried to reshape most of Latin America in its image. The belief that the colonies were inseparable from the mother countries was taken quite seriously well into the twentieth century. The French fought a protracted war in Algeria in the delusional belief that it was part of greater France.

With colonialism came standardization. Western legal systems were introduced and imposed, often (partially) replacing locally developed law. During colonial times such laws, and their corresponding conceptions of justice, held sway; even today they remain, in a sometimes tense relationship with indigenous notions of law and justice (Moore 1992). In vast areas Western notions of private property were introduced. Communal forms of ownership, as well as legal arrangements that treated land tenure as inalienable, were abolished and replaced by Western laws governing real property. Similarly, Western notions of (usually written) contracts were introduced, replacing older traditional relationships. Entire systems of police, courts, military, and other forms of law enforcement were imposed on the colonies. And with the law came the entire governmental structure, imported from the "mother country."

Nor was standardization limited to legal issues. The introduction and widespread use of Western-style clothing, Western ideas—even those of the radical opposition—Western languages, and manufactured goods all contributed to the attempt to standardize, to mimic, the metropolitan nations.

Education was organized along Western lines as well. Asian and African students in the schools of what were then French colonies famously began their history lessons with *"nos ancêtres les gaulois"* (our ancestors, the Gauls). Colonial subjects, it was argued, through education would become French citizens (although of course only a small fraction of them would actually attain that status [Ha 2003]). Schools in the British colonies taught students the history of medieval Britain but ignored the history of India or Africa. Textbooks were carefully written to illustrate the virtues of British culture. As one observer writes, "For many colonial officers, the colonial school textbook was no less than a weapon in the larger armory of colonial rule, albeit one that was lengthy, ponderous, and dry" (Talwalker 2005, 1). Perhaps Charles Trevelyan best expressed the approach of the colonial authorities in 1838 when he wrote of elementary schools, "Even on the narrowest view of national interest, a million [pounds sterling] could not be better invested. It would ensure the moral and intellectual emancipation of the people of India, and would render them at once attached to our rule and worthy of our alliance" (205).

Nor was the project of standardization limited to social organization. Nature was also to be remade. Western scientific expeditions discovered new kinds of plants and animals, and spread those deemed economically viable around the colonies. At the same time, Western animals—cattle, horses, sheep, goats, and pigs—were introduced to the Americas. Those that escaped into the wild soon became the dominant species on much of the two continents, helping to remake nature in a manner more suitable to colonists' tastes. In New Zealand, with its fragile island ecosystem, introduced animals and plants virtually replaced the existing flora and fauna with European imports (Crosby 1986). By 1800 more than 1,600 botanical gardens had been established to support work in the new field of economic botany (Brockway 1979). Plants were usually shipped from colonial gardens to those at Kew (Great Britain), Amsterdam, Paris, and other European capitals, where they were cataloged and evaluated in terms of their potential economic value for the empire. Then, those deemed valuable were "acclimatized" and shipped to gardens in the colonies to establish plantations. Stimulants such as sugar, tea, coffee, cocoa, and tobacco were particularly valuable as plantation crops, as were industrial crops such as rubber, indigo, cotton, and oil palm. The results of this vast exchange of plants and animals were manifold: the same standard plant and animal species (and often specific varieties) could be found in numerous locations. Plantation economies in one part of the world resembled those in other places. A standard system of botanical classification had been created and its use was extended around the world. (Indeed, the creation of ethnobotanies for each of the world's cultures had been suggested as early as 1821 [de Candolle and Sprengel 1821] but was ignored by the scientific community in favor of the unitary one initially developed by Linnaeus.)

In short, attempts were made to standardize both the social organization and the natural environment of the colonies, with varying degrees of success. As with all standardizing projects, the colonial project never achieved its goals. Although the settler colonies of the Americas and Australia came to resemble the metropolitan nations in many ways, most of the peoples of French Africa and Indochina never became French citizens. Most Indians never became British. The project was always hampered by finances inadequate to the tasks at hand. Indeed, the very fact that the standardization was imposed from outside rather than emerging out of indigenous desire meant that it had to take on different forms from those of the Western powers. As one observer of colonial schools observed, "Schools for the colonised superficially resembled metropolitan schools; often they bore names associating them with metropolitan institutions. King's College, Lagos, when it was organised, bore little resemblance to British colleges or grammar schools; the *lycées* of French colonies were *lycées* in name only, giving a variant of vocational, rather than academic education" (Kelly 1979, 210).

The same might be said for virtually all the colonial institutions. They were in many respects pale reflections of their metropolitan counterparts. The standardization policy in the colonies, much as it was zealously pursued, met with only mixed success.

Social Movements

The great social movements of the eighteenth, nineteenth, and twentieth centuries were also linked to standardization. They drew their ideals from a long line of utopian imaginings, beginning with Sir Thomas More's ([1535] n.d.) *Utopia*. In that novel, More envisioned a highly standardized, well-ordered, patriarchal world that most of us today would find stultifying. Later utopian socialists, such as Saint-Simon, Fourier, and Comte, would imagine other equally standardized utopian worlds. The American Albert Brisbane even went so far as to imagine a world in which people would be organized into agrarian communal living units called phalanges, each directed by "its Areopagus of experienced and practical men" (Brisbane [1840] 1969, 361). His followers went so far as to construct such a community in the New Jersey pine barrens, the North American Phalanx, where it remained until it burned in the 1930s (Heilbroner 1961).

If the American (1776) and French (1789) revolutions could hardly be called entirely utopian in their aims, the very idea that the old regime could be eliminated and replaced by a new one must have shocked their contemporaries. Moreover, they did establish the notion of the equality of all men as a global goal. Much of the following period up to the present has been an endless series of attempts to expand the notion of human equality.

The labor movement was perhaps the first to take up the call. As Marx (e.g., 1906) and other scholars and activists had observed, capitalist forms of the organization of production had the largely unintended effect of putting workers in essentially the same relationships to capitalists in every capitalist nation. Marx's famous cry, "Workers of the World Unite!," could only be meaningful as the standardized institutions of capitalism spread around the world. Workers would lose control of the means of production, no longer owning the tools of the workplace. They would lose control of the work process, working instead at the speed of the incessant movements of the new machines that were becoming commonplace. They would work not when they wished but at the standard hours demanded by capitalists. They would become alienated from others as the ties to community were severed. And they would become alienated even from their true selves as the creative aspects of work were reduced to mere drudgery. This, Marx believed, would transform the working class from a "class in itself" into a "class for itself." In short, class consciousness emerged for Marx not from the promises of utopian socialists or even from the exhortations by radicals such as Marx himself but from the very standardizing project that was central to capitalism.

In much of the Western world, the labor movement soon became synonymous with the union movement. Beginning with skilled workers, by the end of the nineteenth century it had spread to unskilled workers as well. The leaders of the labor movement insisted that all union members be treated the same, according to a set of rules written into a contract with capital, but more and more written into law as well. Largely as a result of the labor movement, every Western nation and many non-Western nations passed labor laws during the early twentieth century, specifying the length of the workday and workweek, the minimum wages to be paid, the conditions of work, and even the safety of the workplace.

In the former Soviet Union, the triumph of revolutionaries in the name of Marx went even farther in standardizing. While much of the twentieth century was interpreted as a battle between two different views of the future, in fact the futures of the two seemingly different sides were extraordinarily similar. Thus, the institutional apparatus established in the West to standardize labor was employed in the Soviet empire as well. The tests employed to measure achievement of the standards—gross domestic product, output per worker, educational attainment, provision of health care services, and so forth—were essentially the same. As Susantha Goonatilake (1982, 333) wrote a few years before the Soviet Union collapsed, "The [republics of the] Soviet Union in a sense are today the time heirs to the nineteenth century western belief of unending growth. The Soviet Union appears implicitly to seek the technological Americanization of its country." Indeed, what frightened the West most was arguably the very sameness of the goals they shared. If revolution could succeed in the Soviet Union, might it not also succeed in the West by making the same claims? Hence, on the one hand, the Western nations went to extraordinary lengths to prove to their citizens, and later the citizens of what became known as the Third World, the superiority of capitalist to communist approaches to development. On the other hand, they worked hard to stamp out communist movements outside the Soviet sphere.

The women's suffrage movement of the nineteenth and early twentieth centuries, the U.S. civil rights movement, the antiapartheid movement in South Africa, and the more recent feminist movement provide other examples of standardizing social movements. Their leaders took seriously the standardizing claims of equality of the American, French, Russian, or Chinese revolutions and demanded that their members be treated equally under the law.

The latest of these standardizing movements to appear is the gay rights movement, which to date has had variable success in its inclusionary project. The message of gay rights activists is that the standard benefits accorded to other citizens should be extended to them—the right to marry, to raise children, to live and work where they wish without harassment, and so on.

Of particular note for our purposes here, however, is that the anti-gay rights movement is an equally standardizing movement. From the perspective of its members, granting the same rights to homosexuals as are enjoyed by heterosexuals would have the effect of destroying a standardized view of sexual relations, family life, and gender-specific behavior that they believe is God-given. It is the potential for greater sexual and gender differentiation that is to be avoided.

Medicine

The practice of medicine is as much an art as it is a science. Until the early twentieth century, the likelihood of one's being cured by medical practice was rather limited. Even within the Western world, allopaths, homeopaths, osteopaths, naturopaths, herbalists, chiropractors, hypnotherapists, and other less definable schools of medicine proposed different types of cures. Spas and sanitariums were established to which middle- and upper-class people could go for sometimes dubious treatments leading to a "cure." Medical schools were heterogeneous, with different approaches to medicine proffered by different schools. People went to hospitals to die, not in the expectation of getting well. The likelihood of leaving a hospital cured of a disease was low indeed.

But early in the twentieth century, an attempt was made to subject medical practice to a single standard, in part because of increasing state support for medicine and in part because scientific research was beginning to yield some tangible medical advances (e.g., better hygiene, vaccines, tests). In the United States the standardization of medicine was furthered considerably by the Flexner Report (Flexner 1910). In keeping with the general tenor of the times, the report advocated the application of science to medicine and argued that much of what passed for medicine was not only unscientific but resulted in greater harm than doing nothing at all. The report lambasted all forms of medicine except allopathy. Indeed, Flexner's harsh view of homeopathy can be seen in his description of low medical school admissions standards: "The low standard or immature type of medical student must have his medical knowledge carefully administered in homeopathic doses" (Flexner 1910, 177). Ultimately, the report led to the closing of most nonallopathic medical schools and the establishment of strict licensing criteria for physicians. A few schools of osteopathic medicine survived, but largely by adopting most allopathic practices.

Moreover, the last century has been marked by a variety of attempts to make medical data comparable across time and space. However, "[f]or medical data to become comparable, . . . terminologies and communication routes need to be standardized, and technical standards have to be implemented so that the information systems of all these different parties can communicate smoothly" (Timmermans and Berg 2003, 7). At the beginning of the twentieth century physicians generally kept their own casebooks as a means of engaging in clinical research. Hospitals and clinics, for the most part, did not keep records at all. As one might imagine, the information in the casebooks tended to be somewhat idiosyncratic; each physician collected whatever information he—and it was nearly always he—thought was important. Moreover, since the books were bound, it was impossible to reorganize their contents. Hospital record keeping started in the United States and later spread to Europe, such that by 1930 most hospitals had record-keeping systems in place with individual files for each patient. Indeed, the American College of Surgeons saw this as a key feature of their hospital standardization program. Since physicians were reluctant to engage in the low-status task of record keeping, stenographers and record clerks were hired to do the work. Furthermore, standardization was hardly limited to patient records. "Simultaneously, design standards were set for hospitals, performance standards were set for professional training, procedural standards were formulated for professional conduct (no fee-splitting, for example), and standardized accounting procedures were introduced and further refined" (Timmermans and Berg 2003, 50).

More recently, guidelines for clinical practice have been issued by most industrial nations; indeed, the attempt to standardize has produced a proliferation of sometimes conflicting guidelines. Together, these guidelines create a standard notion of good medical practice (GMP)—something that is often used in malpractice cases by plaintiffs' lawyers. Put differently, if a surgeon operates on me using GMP but botches the job, I have a far lower chance of recovering damages than if that same surgeon had developed a unique and nonstandard form of surgery.

Medical equipment has also become standardized over the last century. X-ray machines, scanners, and other medical diagnostic devices have only become routinely useful as the devices and the interpretation of results have been standardized.[11] Moreover, the standardization allows radiographs obtained in one laboratory to be read by practitioners in another laboratory; no familiarity with the specific conditions in a given laboratory is necessary to read the films.

Standardization has also made its way into medical education. A commonplace phenomenon these days is the "standard patient." The standard patient is an actor who plays the role of patient in the training of physicians. The scenarios are usually scripted to some degree, so as to emphasize typical cases for diagnosis. For example, one standard patient, Lynn White, said to be a victim of domestic violence, is a thirty-two-year-old housewife with two years of college and two children who is active in church. The full description consists of four pages of text (Doleshal 2007).

The *International Classification of Diseases* (ICD) is yet another aspect of standardization in medicine. Such a classification was first promulgated in the 1850s, when it focused entirely on causes of death, and the reference work is now in its tenth revision. "The ICD has become the international standard diagnostic classification for all general epidemiological and many health management purposes" (World Health Organization 2007). As Bowker and Star (1999) note, the ICD reflects the tensions between statisticians, who have little interest in rare diseases, and public health officials, who are often concerned about a single patient who might start a pandemic.

Nor does this even begin to tell the entire story of standardization in medicine. Myriad medical devices—thermometers, sphygmomanometers (for blood pressure measurement), bandages, syringes—are now standardized, as are medications, hospital beds, hospital meals, forms, staff uniforms, and bedpans. One could include here attempts to describe the standard human (Epstein 2009). Even organ procurement has been the subject of standardization, although not without considerable difficulty (Hogle 1995).

Agriculture

Agriculture, too, was subject to standardization, largely through an almost obsessive focus on production and productivity. Building on the earlier work of the botanical gardens, breeders worked hard to create grain and oilseed varieties that produced relatively short and even stands, usually by selecting for dwarfing genes. The even rows of corn that one sees driving through much of the Midwestern United States are the result of such breeding programs. Such crops put more energy into the edible parts even as they are easier to machine harvest.

Farmers were also encouraged to plant a single variety over vast expanses to facilitate marketing. Two Stanford economists argued that "[t]he more perfectly a wheat-growing region can standardize its wheat, the more uniform and invariable it is, the greater the sense of security to the buyer" (Alsberg and Griffing 1928, 285). Similarly, a bulletin issued by the Mississippi Extension Department touted the benefits of a "One Variety Cotton Community Organization" (Willis 1938). Such single-variety communities provided greater consistency of cotton staple length to textile manufacturers.

Nor was standardization limited to marketing. To the contrary, everything from seeds and fertilizer to tractors and other farm machinery was subject to standardization. Certified seed programs grew rapidly in the United States, creating uniform cultivars (literally, cultivated varieties). The Common Catalogue in Europe promoted seed standardization while prohibiting sales of less productive varieties. Fertilizers were standardized through the development of standardized labeling, describing the soluble macronutrient value of each sack in equivalent terms. Nitrogen, phosphorus, and potassium (N, P, and K, respectively) content was to be clearly marked on each package. And, though it took a substantial period of time to coalesce, eventually virtually every farm tractor contained a (standard) power takeoff, hydraulic lift, and balloon tires, no matter what the brand (Sahal 1981). Moreover, the implements pulled by tractors could and did become assembly lines on wheels, moving workers through the fields harvesting and even packing fruits and vegetables (Friedland, Barton, and Thomas 1981).

Arguably the most extreme example of the attempt to standardize agricultural production was the planning and development of a single wheat farm of 100,000 acres on the Soviet plains in 1928. But it was not the Soviets who planned this vast expanse; it was none other than M. L. Wilson (1885–1969), a distinguished agricultural economist who later became U.S. undersecretary of agriculture and director of federal extension work. Wilson was already familiar with Soviet agriculture, having spent six months there advising on wheat production, and where he had already helped to create a tractor experiment station. He and several others (who had no other experience in the Soviet Union at all) spent two weeks in a Chicago hotel room that year designing the farm down to the last detail. Not surprisingly, the entire experiment was a failure. Peasants had no idea how to operate or maintain the farm equipment. Spare parts were in short supply. Machinery that was expected to last a decade was no longer functional after two years (Fitzgerald 1996).

What was true for agricultural production was equally true for food. Prior to the mid-nineteenth century, legislative bodies occasionally passed laws defining various food products. Perhaps the best known is the *Reinheitsgebot*, a Bavarian law of 1516 that required that beer contain only barley, water, and hops. Other legislative bodies defined bread, wine, butter, and other staples of the diet. But, as Stanziani (2005) notes, starting in the late nineteenth century, the first food safety laws began to appear. These laws differed in that they were generally enforced by regulatory agencies and involved the application of scientific tests to the food itself. However, often the tests were not quite up to the legal requirements. For example, in France, if 10 percent margarine were added to butter, it was largely undetectable. In the United States, the watering down of fresh milk was generally undetectable (for a reasonable cost) until the creation of the Babcock milk fat test in 1890. Nevertheless, numerous products were removed from sale as they did not meet the new laws. Today, virtually every industrialized nation and many others have food safety laws that set minimum standards of acceptability of food products for human consumption. In addition, the Codex Alimentarius Commission, jointly operated by the World Health Organization and the UN Food and Agriculture Organization, maintains an ongoing set of global food safety standards. The Codex standards are specifically cited in the Sanitary and Phytosanitary agreement of the World Trade Organization as authoritative (see Lee 2009).

At the same time that food safety was being standardized, nutritionists tried (ultimately unsuccessfully) to convince Americans that meat and potatoes were the staples of the diet and that consuming vegetables, especially in dishes that combined a range of ingredients, was unhealthy. Furthermore, by the late nineteenth century,

Improved transportation opened up possibilities of national markets for mass-produced food products, but it also exacerbated a marketing problem that resulted from improved processing technology: the advances which allowed for the production of items such as flour, biscuits, sugar, salt, and canned goods on a massive scale also produced foods that were absolutely uniform in appearance, quality, and taste. Since most manufacturers used essentially the same technology, there was little to choose from among the products. (Levenstein 1988, 34–35)

The solution to the problem lay in the promotion of brand-name products. This was accomplished through a mix of packaging and advertising that distinguished one brand from another.

In short, the movement to standardize food and agriculture met with considerable success. Farm practices became more regular. Crops and livestock acquired similar characteristics. Much of the daily food supply—especially in the Western world—now came packaged, and the contents of those packages varied only slightly. World War I helped the process along as French soldiers were fed standardized Camembert cheeses and American soldiers learned to eat standardized canned products. The modern consumer was portrayed as one who purchased prepared foods and ate a standardized diet.

Public Education

Public education was also the subject of standardization. In the United States, although the formula changed over time—from elementary (grades 1–8) to high school (9–12), to the current elementary, middle, and high school—there was widespread insistence on the use of this formula for school organization. The monitorial or Lancastrian system was introduced by the early nineteenth century. Modeled in part on military hierarchies, it involved using more advanced students to impart knowledge to less advanced ones to keep the cost of instruction to a minimum. Hence, the classroom was organized much like a factory of the day, with the teacher as the head, the various assistants as foremen, and the students as workers. Moving up in the hierarchy occurred as one learned more. As one observer noted, "The method followed the lines of a factory model with knowledge being given to students on an assembly line basis" (Spring 1972, 45). While reasonably effective in their own right, in order to accomplish their aims the monitorial schools had to provide standardized knowledge to (create) standardized pupils.

At the same time that Taylor was urging efficiency in the factory (see below), William Bagley (1874–1946) was particularly influential in standardizing education around the goal of efficiency. Bagley (1907, 4) could hardly have been more direct: "The school resembles a factory in that its duty lies in turning a certain raw material into a certain desired product. It differs from a factory in that it deals with living and active, not with dead and inert, materials." The key for Bagley was to make schools more socially efficient by eliminating waste. This was to be done by standardizing most aspects of the school. Class periods were to be divided into standard lengths for each age group. Standards for school attendance were to be set. Disciplinary measures were to be subject to a set standard: "There should be a 'standard' method of inflicting corporal punishment" (Bagley 1907, 126). Teachers would show their adherence to strict professional standards by resigning immediately were school board members or parents to attempt to intervene in the classroom. Pupils were to be encouraged to "acquire a standard method of attacking the printed page" (209). Even the posture which students were to assume while sitting was to be standardized. By engaging in these activities, the classroom teacher would develop the professionalism characteristic of other professions; in short, the teacher's standing would be enhanced. And somewhat later, with the creation of the junior high school, vocational guidance would be used to more efficiently steer students into courses of study deemed most suitable for their intelligence (Spring 1972).

Similarly, class size and classroom organization were to be standardized as well, with the ideal of thirty students per teacher. Furthermore, the size, shape, positioning, and type of desk were standardized. Desks would be in rows all facing the same way, while the teacher's desk would be at the front of the room, facing the class and with a blackboard on the front wall. The curriculum was standardized through the dominance of what came to be known as the three Rs—reading, 'riting, and 'rithmetic. Later this was expanded to include "social studies," the sciences, and (less frequently) foreign languages, but other subjects were and remain offered only sporadically. U.S. students would all either walk to school or ride on standard yellow school buses. Finally, teachers were standardized through certification. In principle, although frequently not in practice, only certified teachers—who had a prescribed amount and content of college education—would be allowed to teach in public schools.

One notable aspect of educational standardization is the ever-increasing role played by standardized tests. Such tests are so commonplace these days that they would appear barely worth mentioning. They can be found at every level of schooling, from elementary to postgraduate. They appear to be objective, that is, they provide the same opportunity to every person taking the test and sort those who do well from those who do not. Yet early proponents of such standardized testing, such as Edward L. Thorndike (1874–1949) and his pupil Henry E. Garrett (1894–1973), were supporters of eugenics and linked high intelligence test scores to moral superiority. Garrett was still arguing for inherent racial differences as revealed by the tests in the 1960s (Karier 1973).

Of course, many things had to be standardized besides the tests themselves: the time allowed to complete the test, the conditions under which it was taken, the types of materials that might be brought into the test room, the procedures for grading the exams, and so on.

In the United States during the last decade, as a result of the No Child Left Behind Act (07 P.L. 110, 115 Stat. 1425 [2002]), standardized tests have received a considerable boost. The act provides for regular testing of all elementary and secondary pupils in several core areas of competency. Schools not performing well on the tests or not demonstrating sufficient improvement over a fixed period of time are denied federal aid. Though the program has been widely criticized, there is little doubt that it serves to further standardize school curricula.

Civil Religion

Nor should the rise of civil religion be overlooked. Arguably developed most in the United States and the former Soviet Union, civil religion advocates sought to standardize national identity through the invention of rituals, symbols, and other paraphernalia designed to evoke patriotism. In the United States, the notion of Americans as a chosen people extends back to the nation's inception (Bellah 1975). Much later, standards were developed for flying, folding, and disposing of flags. The Pledge of Allegiance was written and is now recited by nearly all schoolchildren each morning while they face the flag and hold their right hand over their heart.

In the late nineteenth century, the American ideal of the melting pot became widespread. In 1782 John Hector St. John de Crèvecoeur ([1782] 1904, 55) wrote, "Here individuals of all nations are melted into a new race of men, whose labours and posterity will one day cause great changes in the world." The metaphor was popularized by Israel Zangwill's play, *The Melting Pot*, first performed in Washington, D.C., in 1908. Zangwill's character David informs us that "America is God's Crucible, the great Melting-Pot where all the races of Europe are melting and re-forming!" The metaphor soon caught on and became popular with both the general public and intellectuals, meshing nicely with other extant standardizing tendencies.

According to the then prevailing wisdom, regardless of their national origin, upon emigrating to the United States, all people would become part of the same homogeneous population of Americans. The National Americanization Committee actively worked to create homogeneity among Americans and to rapidly transform all aspects of the lives of immigrants so that they would "unite in a common citizenship under one flag" (Hill 1919, 630). Immigrants would be encouraged to eat American food, raise their children as Americans, and abandon ethnic and racial proclivities. Americanization was claimed to be necessary in order to eliminate disorder, unrest, and even sedition.

Similar notions prevailed in most European nations, even as they had seen relatively little immigration and more likely had to deal with indigenous minorities. Hence, Scots, Welsh, Irish, and English were all taught that they were British. Bretons, Provençales, Corsicans, Alsatians, and others were all taught that they were French, and urged to speak the lingua franca. Catalans, Basques, and others were taught that they were Spanish. In many nations, indigenous peoples were forbidden to speak in their native tongue, and in some instances to dress in their native clothing.

The former Soviet Union presented a particular problem for standardizers since only about half the population was of Russian ethnicity. Hence, Soviet policy at once constructed nationalities for each of the sixteen republics and pursued a policy of modernization, technical change, and Russification. Party and state officials in each republic were drawn from the major ethnic group there, simultaneously making leadership dependent on Moscow and reducing ethnic strife (Roeder 1991). This was particularly the case in the more urban areas. "Thus, whether one examines the impact of urbanization at a single point in time or over a longer time span, the trend definitely favors the gradual Russification of the non-Russian nationalities" (Silver 1974, 65). As we know now, with the benefit of 20/20 hindsight, the Soviet approach ultimately led to failure as ethnic strife increased. Today, Moscow finds itself in the awkward position of having to use force to maintain its reduced national boundaries.

Families

Somewhat later, in the period immediately following World War II, an attempt was made to standardize families as the (white) middle-class nuclear family: mother, father, son, and daughter living together in a private home in the suburbs, with a car in the garage. In the United States, from 1954 to 1962, the television series *Father Knows Best* portrayed the family in these idyllic terms. The family was decidedly patriarchal. Father worked to earn a living, while mother took care of the home.[12] Teenage pregnancy and sexual abuse were systematically denied. Since all families were supposed to be middle class, poverty was largely ignored. Levittown became the model suburb. Although later the subject of satire (Pete Seeger sung of the "little boxes made of ticky-tacky"), the uniform suburban ranch home with attached garage remains symbolic of the American family.

Furthermore, the status hierarchy among American families was (supposed to be) singular. Keeping up with the Joneses became for a time a national pastime. In practice, this meant a need to continually acquire more and more consumer goods, especially those visible around the home. Thus, particular brands of automobiles, kitchen appliances, and consumer electronics—first radios, then televisions, then stereo sets, and so on—became symbols for which one was to perpetually strive. In the immediate postwar period, rapidly growing incomes among the middle class, combined with larger homes in the suburbs, permitted the purchase of a seemingly endless cornucopia of consumer goods and simultaneously provided a place to store the purchases. This trend toward the accumulation of more and more things continues largely unabated, although the range of goods has increased enormously, and keeping up with the Joneses is no longer a national pastime. In recent years self-storage units have been built throughout the nation, and the space available has been growing at 9.5 percent per year for the last twenty years. Indeed, according to the Self Storage Association (2006), by the end of 2005 an astounding 6.86 square feet of storage space was available for every person in the nation. Indeed, the entire population could stand in the available space!

Fashion

Perhaps nowhere has standardization been more prominent than in fashion—in clothing, hairstyles, housing, and furniture. In the premodern world, clothing, furnishings, housing, and even behavior were prescribed and rigidly regulated for each class. Certain colors and types of cloth were reserved for the nobility, others for the clergy. Chairs were reserved for nobility, while those of lesser status sat on benches. Today, the remnants of those fixed differences remain with us in name only in the notion of the chairman of the board, the fellow whose privilege it was to sit in an ornate chair at the end of the board, while others sat on benches.

Indeed, modern societies destroyed those differences. As Arjun Appadurai (1986a) has noted, in the modern world fashion is the means by which status differences are identified. Today, anyone with sufficient wealth can purchase what he or she desires. However, this has hardly eliminated the quest for status differences. Seemingly paradoxically, the endless novelties of fashion both standardize and organize the process of status differentiation. Indeed, in the ever-changing world of fashion, it is the fashion leaders who set the standard for others to follow. However, as more and more people accept the standard fashion, the fashion leaders change the standard again, in a never-ending race to produce and demonstrate status differences. In short, what we call fashion is a form of standardized change. The examples of the width of men's ties or women's hairstyles are instructive: in both instances, certain styles are in fashion and others are passé. And regardless of how rapidly one changes one's style, other, newer styles will emerge to replace those currently in vogue.

This pursuit of status in a world of standard goods was actually seen as a boon for consumers. Home economist Jesse V. Coles (1894–1976) (1932) saw three advantages to standardization for consumers: (1) the elimination of unnecessary variety, (2) the production of uniform products and manufacturing through the use of standard materials and equipment, and (3) the facilitation of marketing. Thus, standardization would be advantageous in that it would make consumer choice easier and clearer. Perhaps one needed to keep up with the Joneses; however, one would be able to do it with far less difficulty.

Markets and Economies

Perhaps the grandest project of the late nineteenth and early twentieth centuries was the attempt to standardize markets and the economy as a whole. Throughout the industrialized world, a newfound interest emerged in (re)making the economy, in standardizing it in ways that would eventually resemble the textbook cases of neoclassical economics. It is unclear whether neoclassical economics arose as a result of changes in the economy or whether those changes were themselves the result of neoclassical theorizing. Likely, to some degree, they were both. Some of the key standardizing policies and projects that were put into place just about a century ago include the following:

• A single standard currency for each nation, itself backed by a national bank that was usually state-owned.

• Systems for checking the precision and accuracy of standard weights and measures used in commerce.

• Antitrust laws that blocked firms from obtaining too much market power.

• Stock markets with clear trading rules, on which shares of companies could be traded in a controlled setting.

• Marked prices for consumer goods, and the gradual suppression of haggling.

• National standards bodies, beginning in 1900 with the formation of the British Standards Institution.[13]

• Mass production of a vast array of intermediate and consumer products.

Each of these policies and changes gradually improved the ease with which people and goods could be exchanged over vast distances. Put differently, the very standardization of the market—its goods and its institutions—was directly linked to the increase in world trade. The grain elevator is an excellent example of this process in action (box 2.1).

Box 2.1
Standards in the grain trade

Until the mid-nineteenth century, virtually all grain was traded in small lots. In most industrial nations grain was traded by the bushel, although the weight or volume of a given bushel might vary. (Witold Kula [1986] reviews the many European measures for grain and other commodities.) Thus, the price of each bushel was often the subject of negotiation between buyer and seller. Moreover, grain was usually shipped in sacks. William Cronon (1991) recounts how the streets of Chicago were littered with sacks of grain awaiting sale following the harvest. Each would have to be inspected by the potential buyer and the price negotiated. Moreover, quite a few hours would be required for a team of persons to fill a rail car or ship with sacks of grain.

About the mid-nineteenth century, it became evident that this system was no longer workable. Moreover, high labor costs and technical change brought an alternative and Chicago led the way. "By the end of the 1850s, Chicagoans had refined their elevator system beyond that of any other city, leading the way toward a transformation of grain marketing worldwide" (Cronon 1991, 111). Not only did Chicagoans invest in elevators, elevators fundamentally changed the way that grain was handled. What had been treated as a solid with individual lots was now treated as a liquid that could be poured into (or out of) elevators, ships, and rail cars. Instead of individual lots, grain of the

Box 2.1
(continued)

---

same grade was put into a given silo and treated as a homogeneous commodity. In addition,

The practice of bulk handling of grain by grade made possible the use of uniform warehouse receipts to represent the commodity. In conjunction with the elevators, these receipts facilitated trade between markets, expanded credit, caused the Chicago Board of Trade to become an outstanding exchange, and made futures trading possible. This machinery, copied by many of the world's central markets, made one big world market as far as supply, demand, and price were concerned. (Lee 1937, 31)

In short, as with all standards, the invention of grain standards involved a complex and heterogeneous infrastructure in which elevators, farmers, railroads, ships, grain dealers, bankers, telegraph operators, and others took part. Moreover, while this process was hardly free of conflict—as with all standards, there were winners and losers—it created a new form of market, one in which physical presence and inspection were no longer necessary, in which warehouse receipts could be bought and sold, in which futures trading could go on in an orderly manner, in which distant buyers could confidently transact huge purchases, and in which supply and demand could be equilibrated based on comparable information publicly available. In that sense it became emblematic of the market uniformity and standardization that swept the entire world over the following century.

Of course, not all objects of commerce could be as standardized as grain. Many, to one degree or another, resisted standardization efforts. Standardization remained a project, with varied success. Thus, despite considerable efforts on the part of Chicago processors and merchants, lumber and meat never achieved quite the degree of standardization imposed on other raw materials such as oilseeds, salt, sugar, metal ores, coal, petroleum, and milk (Cronon 1991).

By the early twentieth century the move toward ever greater standardization of the market and the economy was fairly evident. No matter where one stood on the political spectrum, standardization was the order of the day. Perhaps no one summed it up better at the time than did radical economist Thorsten Veblen (1904, 10) (1857–1929) when he wrote, "modern industry has little use for, and can make little use of, what does not conform to the standard. What is not competently standardized calls for too much of craftsmanlike skill, reflection, and individual elaboration, and is therefore not available for economical use in the processes." From his perspective, not only did modern industry have little use for any nonstandard goods to sell, but accounting, invoicing, contracting, and other services were rapidly becoming homogenized, standardized. Workers were standardized by the constant "supervision and guidance" provided by the standardized machines. Even the workers' thinking demanded standardized units as employed by the machine process. Furthermore, workers had to be as replaceable as the parts of machines they tended: "The working population is required to be standardized, movable, and interchangeable in much the same impersonal manner as the raw or half-wrought materials of industry" (Veblen 1904, 326). Finally, consumers themselves were becoming standardized: "As regards the mass of civilized mankind, the idiosyncrasies of the individual consumers are required to conform to the uniform gradations imposed upon consumable goods by the comprehensive mechanical processes of industry. 'Local color' it is said, is falling into abeyance in modern life, and where it is still found it tends to assert itself in units of the standard gauge" (Veblen 1904, 11).

In short, for Veblen, standardization was not limited to the creation of interchangeable parts for a given machine; it was the increasing interchangeability of everyone and everything. That appeared to him to be a new but central feature of industrialized societies.

Some years later, Veblen (1921) described the situation right after World War I as one in which three distinct classes were emerging: capitalists, engineers, and workers. The capitalists, absentee owners of the growing, ever more standardized industrial machine, were increasingly preoccupied with finance, while losing their competence in the "industrial arts." At the same time, in large part owing to the enhanced productivity brought about by the use of machines to produce standardized goods, there was a constant danger of overproduction and economic collapse. Capitalists had responded by creating all sorts of waste, including unemployment and idle resources, sales forces to convince people they needed things, the production of superfluous goods, and the systematic dislocation of people and communities.

Moreover, since they had lost the technical competencies necessary to run the factories, capitalists were forced to employ legions of experts, of engineers, whom they distrusted and who were quite capable of developing a class consciousness of their own.[14] A general strike of the engineers, Veblen quipped, would rapidly bring the entire industrial order to a halt. After all, only the engineers understood both the management of existing machinery as well as how to improve it. The engineers, as designers and maintainers of the machines and the abundance they foreshadowed, would replace the capitalists and usher in the new age. Veblen believed that if they were to take over, they would be able to rapidly raise productivity by 300 percent to 1,200 percent, in large part through the gains made possible by standardization.

At the same time, even though the craft-oriented American Federation of Labor had capitulated to the existing order, standardization made *trade* unionization inevitable since it created the necessary discipline among all workers. Workers, Veblen argued, need to align themselves with an "Executive Council of Engineers," a "Soviet of technicians," and push the obsolete absentee owners of capital aside, ushering in the new order. But all this had to happen rather quickly: "With every further advance in the way of specialization and standardization, in point of kind, quantity, quality, and time, the tolerance of the system as a whole under any strategic maladjustment grows continually narrower" (Veblen 1921, 122). In other words, standardization was drawing everything into one grand megamachine that had to be monitored, calibrated, and adjusted with ever greater care if it was to bring with it abundance.[15]

All of this might appear to contemporary eyes to be rather fanciful, wishful thinking on Veblen's part. Although the Fabian socialist William Robson (1926) argued in a pamphlet entitled *Socialism and the Standardised Life* that standardization was antithetical to socialism, Alexei Gastev, founder of the Central Labour Institute in Moscow, proclaimed with enthusiasm that workers in the new Union of Soviet Socialist Republics would become increasingly standardized (Bailes 1977). (Somewhat later, in 1936, Charlie Chaplin would spoof this view in his film *Modern Times*.)

But standardization was also embraced by persons of a far more conservative bent than Veblen. Indeed, it was arguably Herbert Hoover (1874–1964) who most enthusiastically embraced the new age of standardization. An engineer himself, Hoover was equally concerned about class divisions. However, unlike Veblen, Hoover was convinced that standardization would resolve class antagonisms by eliminating waste, reducing prices, and ushering in a new age of abundance. Hoover (1952, 28–29) explained it best himself in his memoirs:

It involved increasing national efficiency through certain fundamental principles. They were (a) that reconstruction and economic progress and therefore most social progress required, as a first step, lowering the costs of production and distribution by scientific research and transformation of its discoveries into labor-saving devices and new articles of use; (b) that we must constantly eliminate industrial waste; (c) that we must increase the skill of our workers and managers; (d) that we must assure that these reductions in cost were passed on to consumers in lower prices; (e) that to do this we must maintain a competitive system; (f) that with lower prices the people could buy more goods, and thereby create more jobs at higher real wages, more new enterprises, and constantly higher standards of living. I insisted that we must push machines and not men and provide every safeguard of health and proper leisure.

Moreover, unlike Veblen, who labored in various academic jobs, Hoover was in a position to advance his views, first as food administrator during World War I, then as secretary of commerce, and finally as president. Hoover's view of standards was both complex and nuanced. Let us briefly examine some of its central features:

• Hoover understood that standards for persons as well as things were required. Increased population density and greater education posed new moral problems, some of which required new laws but others of which might be better addressed through "voluntary forces," that is, through standards developed by industry. Moreover, new standards of human conduct were needed to keep pace with the endless variety of new inventions, forms of production, and consumer goods and services (Hoover [1924] 1937).

• Hoover saw standards in evolutionary terms. The analogy used was that nature produced great variation but then, through natural selection, reduced variety to a few standard types. As he and his co-author, Albert W. Whitney, put it, "Variation is creative, it pioneers the advance; standardization is conservational, it seizes the advance and establishes it as an actual concrete fact" (Whitney and Hoover 1924, 3). Standards freed people from the details of the moment so that they would be able to conceive of new technical advances. As such, Hoover was supportive of Taylor's time and motion studies: "I can scarcely believe that motion that will produce the greatest physical results with the least fatigue will not be correlated with both bodily and emotional satisfaction to the worker" (Whitney and Hoover 1924, 6).

• Hoover was not an ideologue of industry or of the market or of the state, but saw each as necessary to moral, social, and economic progress. From his vantage point, voluntary associations would instill in persons the necessary moral obligations they had to society, thereby solving most problems if sufficiently encouraged; the state would need to intervene only to deal with those persons and organizations that were somehow incorrigible.

• Hoover, unlike most of his more conservative contemporaries, was well aware how standards could be stifling to workers. Hence, he was an advocate of collective bargaining and of fair treatment for workers, farmers, and consumers. His administration witnessed the first attempt to stabilize farm prices, thereby attempting to improve farm incomes (Hamilton 1986). And he placed consumer representatives on the National Standards Board (Office of Technology Assessment 1992).

• Hoover believed that standards had to be produced via democratic means. Standards that benefited one group were to be avoided. The key was to include all of the relevant parties in standards making.

• Finally, Hoover saw standards as eliminating what he perceived as the growing waste of industrial societies and the rechanneling of those savings into rising levels of living. Hoover likened the waste to a disease (Whitney and Hoover 1924) in which things were overproduced; the treatment was greater and more effective standardization.

Hoover had the opportunity to put many of his ideas into practice, and he did so with great zeal. In particular, he went to considerable lengths to convince industrialists of the value of standardization. However, fearing that some would think that the term meant that everything would be the same, he encouraged the use of a less pejorative term: simplification. To that end, a Division of Simplified Practice was established at the National Bureau of Standards in 1922. By 1927 some seventy-nine simplified practice recommendations had been completed. Moreover, the Navy Department became the first to simplify its purchases, leading to what would eventually become a government-sponsored system for establishing specifications for virtually everything purchased by the government. According to one report, efforts at simplification reduced costs to producers by over $293 million (Hudson 1928).

Even sociologists such as Robert E. Park (1864–1944) and Ernest W. Burgess (1886–1966) saw standardization as a central feature of the new market landscape: "Standardization of commodities, of prices, and of wages, the impersonal nature of business relations, the 'cash-nexus' and the credit basis of all human relations has greatly extended the external competitive forms of interaction. Money, with its abstract standards of value, is not only a medium of exchange, but at the same time symbol par excellence of the economic nature of modern competitive society" (Park and Burgess 1921, 556).

For still other observers the need for standardization was to combat the disconnect between private and public morality in the marketplace. As Yale president and economist Arthur Twining Hadley (1856–1930) put it, "The man whom you could trust to help a weaker neighbor will nevertheless go to all lengths to hurt a weaker competitor for money or for office. A man who in private life would despise snobbishness and servility of every kind will in business or politics cringe to the stronger power for the sake of his own personal advantage" (Hadley 1911, 4).

He attributed this to the rapid pace of change, to the difficulty in trans-ferring ethical lessons learned in one situation to other, radically different ones, and to the clarity of "public sentiment" in the former case and its obscurity in the latter. The solution to the problem, according to Hadley, lay not in stricter laws or harsher enforcement but in industrial leaders stepping forward to further the public good over and above their own interests, thereby raising the moral standards of the nation as a whole. Although Hadley said little about industrial standards, it is clear that well-defined, perhaps even codified, standards of morality in the marketplace were essential to his program.

The National Industrial Conference Board (1929, 40) also waxed enthu-siastic about the coming age of standardization: "Standardized stationery, letterheads, pencils, filing equipment, requisition blanks, invoices, checks, instruction sheets are the sine qua non of standard office practice and production control. Standardized machines with interchangeable parts, and standard tools are of fundamental importance in the shop and factory likewise."

On the other hand, it was less enthusiastic about the moral tone set by Hoover and others: "As an engineer, one does not discuss the social sci-ences, draw codes of ethics, morality, or good conduct, discourse on aes-thetic values or introduce legislative measures" (National Industrial Conference Board 1929, 61).

The advantages of standardization were explained in innumerable papers, pamphlets, technical articles, and books. Standardization would help producers, manufacturers, retailers, and consumers (in today's busi-ness language, everyone in the supply chain) by making price and quality comparisons possible. Consider the case of electrical machinery, as explained by an observer at the time:

In the early days there was no agreement as to what was meant by a ten horsepower motor, and fair competition was impossible. The ratings of the several manufactur-ers varied as much as 30 per cent and the customer was at the mercy of the persua-sive talents of the salesman. It took the Standards Committee of the American Institute of Electrical Engineers five years to develop a system of rating satisfac-tory to all concerned and capable of reasonably accurate checking by commer-cial tests. The results of this work have proved to be of world-wide value (Adams 1919, 291).

The author goes on to praise the National Bureau of Standards for its work in helping to produce such standards.

Other observers saw standardization as needed in agricultural markets as well. At the time, the product quality of grain, fruits, vegetables, and meats varied considerably from locale to locale, with the same or similar terms used to denote rather different product qualities. According to one observer, in 1919 twenty-three different grain bushel measures were in use in the United States despite the standardization in Chicago markets. Meats with the same labels often were found to be entirely different cuts as well (Gephart 1919). Standards would provide order to the growing national agricultural markets.

Furthermore, standards making was heralded as a new form of self-government, analogous to that found in small towns in the United States. As P. G. Agnew (1926, 95) put it, "The standardization method has all the directness and vitality of elementary local self-government." Standardization was an improvement over legislation since it permitted greater flexibility. Industrial leaders were urged to step forth and pick up the banner of standardization, thereby relieving legislatures of the "impossible load which they are now facing" (Agnew 1927, 259). Moreover, an agreement was signed between the American Standards Association (now the American National Standards Institute) and the National Bureau of Standards to provide assistance to small businesses and industries that lacked the means to engage in standardization themselves (*Science* 1929).[16]

It should also be noted that what Hoover was enthusiastically pursuing in the United States was being pursued in Europe with equal enthusiasm. As in the United States, the industrial lessons of World War I were driven home: standardization was not merely a matter of eliminating waste; it could be a matter of winning or losing a war. German movements toward "rationalization," British efforts to introduce industrial planning, and the creation of Gosplan in the Soviet Union were all designed to standardize and, by so doing, to eliminate waste and inefficiency. One British observer proudly noted that the British Engineering Standards Association had reduced the number of tramway rails from seventy to five. For him, "standardization as a coordinated endeavor is bound increasingly to benefit humanity at large" (LeMaistre 1919, 252).

In sum, as the National Industrial Conference Board (1929, 34) put it, "The issue before the modern business executive is not standardization, but how much standardization."

Not everyone was equally enthusiastic about the great project of standardization, however. Among others, economist Homer Hoyt (1895–1984) was an avid critic. Hoyt's arguments against standardization were twofold. First, Hoyt argued that standardization would tend to focus the terms of competition solely on prices, since quality would be determined by the standards. The result would be that only those firms selling at the very lowest price would remain in business, as the others would be forced out by price competition. Even the National Industrial Conference Board (1929) had to agree.

Second, Hoyt argued that standardization would lead to a stultifying uniformity: "A society in which everyone rode in Ford cars and lived in uniform cement houses would be monotonous even though it were the most economical" (Hoyt 1919, 273). Finally, since the end result of standardization must ultimately be the global homogenization of taste, fashion, and even culture, the individualism characteristic of American society would be obliterated.

In short, regardless of where one stood on the political spectrum, the grand project of standardization was the order of the day. But it was in the factory that standardization was the most enthusiastically received.

Factories

Although standardized coinage, weights, and measures date from antiquity, the factory—a triumph of standardization—is a modern invention.[17] Moreover, the usual story of the development of the factory tends to get the order of events backward. Factories were not so much made necessary by machines as machines were made possible by the gathering together of workers in factories. This is evident from the older term, manufactory (*manu* = hand, *factory* = place where items are fabricated), a place where people got together to make goods by hand. Indeed, it appears that the first factories were not places where machinery was used so much as places where people could be grouped together and closely supervised. Put differently, they were places where people could be subjected to factory discipline. Indeed, Andrew Ure (1835) credited Richard Arkwright for establishing the factory system by disciplining workers, and not for his introduction of machinery into textile manufacture. That discipline had two critical components. First, workers were expected to work for a standard period of time measured by a clock—thirteen hours each day—rather than be paid by the piece. Those who were paid by the piece, after all, could work at their own pace, and could decide (within the limits of necessity) whether and when they would work. In contrast, hourly workers would be expected to work continuously throughout the workday. Second, the factory allowed the minute division of labor noted with some admiration and some concern by both Adam Smith ([1776] 1994) and Alexis de Tocqueville ([1835–40] 1956). It permitted the standardization of each part of the manufacturing process so as to maximize output per unit of labor. In short, under the factory system workers would work at the prescribed speed, for the prescribed number of hours, using the prescribed methods. Work would be standardized.

Adam Smith's ([1776] 1994) account of the pin factory is a case in point. Although his work is usually referred to for its explanation of the division of labor (each worker in the pin factory takes on a different task), its description of standardization is rarely discussed. Yet standardization is critical to the division of labor he describes. On the one hand, Smith's choice of the pin factory is revealing. Had he chosen most other industries at that time, he would have been unable to make his case. For example, furniture was made on a piece-by-piece basis in response to requests by buyers. As such, the division of labor in furniture shops was rudimentary at best, usually consisting of a master, some journeymen, and perhaps a few apprentices. Furthermore, the drawers on one chest did not fit and did not need to fit other chests produced by the same furniture makers; each chest was crafted for a specific client and not "for the market." On the other hand, the pins produced by the workers in Smith's pin factory were standardized. All had to have a rounded top, a shank that was neither too thick (such that it would not pass through the cloth) nor too thin (such that it would bend on touching the fabric), and be pointed at one end. Their lengths, too, needed to be within certain tolerance limits. In short, Smith provides us with one of the first accounts of widespread standardization of both labor and materials.

Furthermore, all of this could be accomplished without the need for much machinery. Adding machinery, as was done later, reinforced the standardization of the manufactory. While the complex tasks of clothing production or metal fabrication could not in their entirety be subjected to machine production, the very minute division of labor permitted the mechanization of various parts of these tasks by breaking them into their "essential" components. As a result, either workers were made into machine tenders (as was the case in the development of the textile industry) or machines were increasingly used to monitor and standardize their work. By the beginning of the twentieth century, even the factory building was undergoing standardization (box 2.2).

Box 2.2
The standardization of factory buildings

> By the early nineteenth century factories dotted the landscape in most of the Western world. By the end of the century significant standardization was already apparent: "The three most common types of factory in 1900 were the

Box 2.2
(continued)

one-story general utility building; the one-story building with sawtooth roof; and the multistory factory building, of either light or heavy design" (Slaton 2001, 133). But in the early twentieth century, further uniformity emerged as the quintessential standardized factory building came into being. The key to standardizing building was the invention of reinforced concrete and the passing of common ordinances permitting and standardizing their construction (*Cement Age* 1904). Reinforced concrete was a technoscientific invention that not only made possible standardized factories but also transformed the building trades. In particular, it transferred control over the labor process from brickmasons and carpenters, among others, to engineers. Standard molds, and even precast elements bought off the shelf, allowed the creation of post-and-beam concrete structures of several stories in height. Since the posts bore the entire weight of the structure, the walls might be made of any available materials of sufficient thickness and strength to protect the occupants from the weather and to prevent theft. Architects could be dispensed with, since building size was reduced to a multiple of the sizes of the molds.

Not only were the buildings standardized, so also was the labor used to construct them. Historian Amy Slaton (2001, 73) explains:

Most tasks were reduced to simplified series of repeated actions, determined to the most minute detail before the project was begun. This allowed construction companies to pay lower wages to individual employees, even if overall numbers of employees were not reduced, and also to avoid using union labor if they so chose since so many workers on the concrete site could be hired off the street (and be easily replaced if found to be unsatisfactory).

By 1921 standards produced by the American Society for Testing and Materials (ASTM) were incorporated into most textbooks on concrete construction. Preprinted specifications were available for use on most sites. Building codes soon followed suit. As a result, the factory itself became a mass-produced, uniform symbol of the industrial world. "By 1930, buyers of reinforced-concrete factory buildings had subsidized a transformation of large parts of the American industrial landscape into an environment of stark and uniform structures that celebrated the standardized cultural product" (Slaton 2001, 171).

Susan S. Silbey and Patricia Ewick (2003, 81) tell a similar story with respect to scientific laboratories. As they note, "Laboratories have developed into vast, prototypical, universal products with interchangeable parts. . . . Their contents have been so standardized that contemporary laboratories are designed and built Lego style: a pattern module is composed of stock materials, then arranged in various configurations."

Arguably, the standardization of work and workers reached its zenith in the work of Frederick Winslow Taylor (1911), the creator of scientific management. Taylor made several claims of particular importance to us here:

1. There was one best way to accomplish each task, and science—through time and motion studies—would determine what that was. This required the reduction of work to two variables that could be easily measured—time, with a stopwatch, and motion, with a variety of instruments. In short, work could be standardized much more carefully than in Smith's pin factory.

2. Science would replace the idiosyncratic, personalized relations between workers and management with the standardized anonymous relations of the operations manual.

3. Workers and managers would be trained to follow those rules and regulations. Thus, standardization would permit and even require the elimination of what Taylor referred to as "soldiering," or going through motions that made it appear that one was working in order to avoid punishment. Soldiering was to be replaced by training in the best method for performing a particular task.

4. Through standardization of work and workers, unparalleled abundance would be created. Economies of scale were seemingly infinite.

5. Science would resolve disputes between workers and management by transferring them from the world of interpersonal relations to that of scientific facts. The abundance produced in doing so would eliminate any opposition to this system, which from Taylor's view was clearly superior.

Taylor's approach was widely embraced by large segments of the business community soon after World War I. Henry Ford in particular adopted Taylor's theories. Ford would go on to produce standardized automobiles using standardized workers. Moreover, true to Taylor's promises, Ford was able to produce and sell the cars at extremely low prices, even as he was able to pay his workers what at the time was a top wage of five dollars a day.

Indeed, Ford went even further; he established a "Sociological Department" at his firm in 1913. Its job was to determine whether workers' personal lives and habits were such that they *deserved* the full wage. In particular, it investigated every potential and actual Ford employee to determine whether that person displayed sufficient thrift, had good personal habits and home conditions, and was otherwise fit to participate in Ford's profit-sharing plan. Single men and all women did not qualify for the program, and having British ancestry was an unwritten but often necessary requirement. As one historian puts it, "To a large extent then, the mission of the Sociological Department was to reform (i.e., American-ize) the 'ethnics' so that they might become good 'Ford men,' and therefore qualify for participation in the profit-sharing plan" (Hooker 1997, 48). Put differently, it determined whether workers were standardized off the job as well as on it. In sum, for Ford there would be standardized vehicles—any color as long as it was black—as well as standardized work processes, stan-dardized workers, and standardized wages.

But Ford and American manufacturers were not the only ones enamored of Taylor's system. Lenin promoted its widespread use in the Soviet Union. German and Japanese manufacturers were equally impressed. The Hungar-ian state developed a profession of "agrarian work science" (Lampland 2009) in pursuit of Taylorist ideals. Standardization seemed to be the answer to many of the most pressing social, technical, economic, and moral problems of industry.

Ford's Sociological Department and Pullman's company town[18] were examples of attempts to impose standardization on workers even when they were not at work. This approach, however, was probably nowhere more widely used than in the former Soviet Union and China. There, it was not at all uncommon for workers in a given factory to be required to live in the same community, frequent the same community centers, send their children to the same schools, and purchase the means of subsistence in the same stores. While the particulars varied from place to place, there were a number of common elements to these attempts. First, they each defined certain behavior as moral and other behavior as immoral. Second, both positive and negative sanctions were used to enforce these standards. Model workers would be rewarded not only economically but also by the prestige attached to recognition. In contrast, those who violated the stan-dards would be punished, in extreme cases by banishment from the com-munity. This is not to suggest that resistance was never encountered. Indeed, in every location where it was instituted, such attempts at stan-dardization were *projects*, not faits accomplis. As such, they in no instance fully achieved their goals.

Nor was Taylorism confined to the workplace. Farmers were urged to run their farms in this manner. One author argues that certain strands of modern ballet were influenced by Taylorism (Delinder 2005). The class-room was subjected to it as well. Home economists enthusiastically embraced Taylorism as a means to reduce the amount of time spent on housework (Jones [1916] 1917). Ironically, standards for household clean-liness rose to fill the time "saved" (Cowan 1983). At the same time, the standard height of kitchen counters and stoves that we take for granted today is one of the fruits of scientific management in the household.

Laboratories were also altered based on Taylor's theories. Milton J. Greenman, designer of the Wistar rats, was inspired by Taylor. He had read Taylor's work on shop management and realized that its principles could be applied to laboratory practice as well. Indeed, just as Taylor urged the development of standard conditions in the workplace, standard tools for workers, and standard work practices, so Greenman began to refer to his Wistar albinos as "material of standard type" (Clause 1993).

Taylorism was also linked to the development of the first standardized tests. The so-called IQ tests claimed to scientifically measure one's "intelligence quotient." These tests were used first by the U.S. Army during World War I to determine what jobs would be assigned to what soldiers. Later their use became widespread in public schools to determine which students would be placed in what programs. Despite continuing conflict over their validity and reliability, they remain in widespread use, having been joined by other standardized tests as a means of "scientifically" determining who is fit for what career.

Finally, Taylorism was eventually introduced into the fast-food industry (Reiter 1991; Schlosser 2002). Indeed, while street food has been around for thousands of years, fast food is in large part the result of applying Taylorist principles and those of quality control to the production of standardized food by standardized workers and its consumption by standardized consumers. As George Ritzer (1993, 100) puts it, using McDonald's restaurants as a paradigm case, "McDonaldization involves the search for the means to exert increasing control over both employees and customers."

From Quality Control to Continuous Quality Improvement

Parallel to but slightly later than the standardizing efforts of Taylor and his followers was the rise of quality control. Unlike Taylorism, which focused on workers and the work process and only paid attention to things to the extent that they could be modified to speed up work, quality control focused at its inception on standards for things. Those manufacturers wishing to ensure that their products were of consistent quality would check each item as it was completed to see that it met expected specifications. In short, quality control was all about ensuring that standards were met and variation was kept to a minimum.

Initially, quality control could hardly be formalized since each individual object was unique. Clearly, the discerning buyer could determine (some) differences in quality, and even rank objects in terms of quality.

However, no means was available to control quality in a rigorous way until standardized parts, the so-called American system, came into widespread use. By 1870 gauges were available that allowed the measurement of upper and lower tolerances for objects such as axles (Shewhart and Deming 1939). These "go/no go" gauges could be used by inspectors, for example, to determine whether axles were too large in diameter to fit within the bearings, or too small such that they would wobble after they were put in use.

But perhaps the key invention that made quality control a part of the quest for precise standards and a means for standardization was the quality control chart. The chart, invented by Walter A. Shewhart (1891–1967) while he was employed by Western Electric (AT&T's telephone handset manufacturer) in 1924, transformed industrial production and the jobs of some industrial workers. Figure 2.2 provides an example of such a chart. A worker charged with inspection would examine every part or every nth part, measuring one or another dimension. For example the worker might use calipers to measure the diameter of each axle. Each measurement would be marked in the appropriate box on the control chart starting from the left and going to the right seriatim. In so doing the inspector would produce a statistical map of the quality of this particular aspect of the production process.

Perhaps the desired standard diameter is precisely one inch. Deviations of up to four thousandths of an inch are acceptable for this particular purpose. Hence, an axle measuring anywhere between 0.996 inches and 1.004 inches will pass inspection. As can be seen from a visual observation

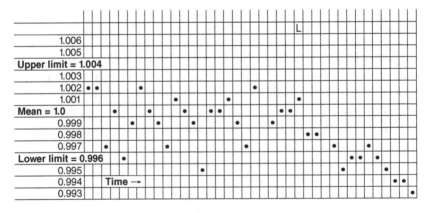

Figure 2.2
A simple quality control chart.

of the chart, all of the observations to the left of line L except one fall within the desired limits. However, those observations plotted to the right of line L deviate farther and farther from the mean value and eventually exceed the accepted limits. Based on the plotted observations, the inspector could determine that some piece of equipment was now misaligned and needed immediate attention. Depending on the organization of the factory, the inspector could either identify the problem and attempt to resolve it or report it to management.

In addition, the quality control chart could be used by an industrial statistician for several related purposes. First, quality could be monitored over time and the variance from the desired mean value could be regularly calculated. Second, by analysis of the deviations from the mean, it could be determined whether errors were random or formed some sort of pattern that might be attributable to a particular piece of manufacturing equipment. Ideally, over time, all nonrandom sources of error could be eliminated and quality could be markedly improved.[19]

It is also worth noting that the quality control chart is a statistically based example of Aristotle's notion of standards. It provides a "golden mean," a norm to be strived for by both individual workers and the factory as a whole—the mean value desired for the item being manufactured. By this definition, a "good" product is one that is as close as possible to the mean value desired.

Shewhart also noted that three different kinds of control charts could be developed. The first is the one illustrated in figure 2.2, where both upper and lower limits are specified and may be varied depending on the expected use of the product. A second type of control chart has its lower limit at zero and its upper limit fixed at a maximally acceptable point. Fuses are a good example; an ideal fuse would break the circuit instantaneously, as soon as the current exceeded the desired maximum. However, it could also take some maximum fixed amount of time, perhaps 5 nanoseconds, before breaking the circuit.

The third type of control chart would have the lower limit fixed but would have no upper limit. The example provided by Shewhart is the tensile strength of steel. For whatever use the steel is to be put, what counts is the lower limit; the steel can always be stronger than the minimum required.

After Shewhart invented the chart in 1924, it took several decades for its use to become widespread. During that time Shewhart himself spent a considerable amount of time and energy promoting his ideas. In 1931 he published a textbook, *Economic Control of Quality of Manufactured Product*, in which he showed how industrial manufacturing, engineering design, statistical methods, and economics could be integrated. The key to this integration was what he called "statistical control." As he explained,

For our present purpose *a phenomenon will be said to be controlled when, through the use of past experience, we can predict, at least within limits, how the phenomenon may be expected to vary in the future. Here it is understood that prediction within limits means that we can state, at least approximately, the probability that the observed phenomenon will fall within the given limits.* (Shewhart 1931, 6, emphasis in original)

Such control, he argued, had five advantages over older methods: (1) a reduction in the cost of inspection, (2) a reduction in the cost of rejection, (3) maximization of the benefits of large-scale production, (4) the achievement of uniform quality, and (5) a reduction in tolerance limits. To realize these advantages a great deal of care needed to be exercised in sampling, that is, in determining how a product should be measured, how many products needed to be measured, and how those to be measured should be selected. In addition, careful attention was to be paid to the analysis of deviations from the ideal measure, so as to sort out "assignable causes" from random error.

Shewhart's work was well received. The British Standards Institution held a conference in 1932 to promote industrial statistics. Shewhart was the guest of honor. The institution later produced a manual, *The Application of Statistical Methods to Industrial Standardisation and Quality Control*, to encourage the use of industrial statistics. As noted in the introduction, "The main object of this publication is to set out as clearly as possible why statistical technique is needed in the solution of many industrial problems, and the kind of assistance it can give" (Pearson 1935, 12). Not content to limit himself to industrial applications, Shewhart gave a series of lectures at the Graduate School of the U.S. Department of Agriculture in the late 1930s. These lectures were published as well (Shewhart and Deming 1939).

Both engineers and managers responded positively to Shewhart's ideas. They fit well with engineers' understanding of the world. As one engineer put it, "Far less man power is required to maintain a system of quality control, allowing, say, 1/2 per cent defectives to slip through the inspections, than to enforce 100 per cent inspection of all processed piece parts, and this saving in labour cost more than outweighs the cost of the 1/2 per cent of defective items rejected at the final assembly stage" (Rissik 1942).

Engineers began to recognize that in addition to understanding how to design the perfect object, attention also had to be paid to the errors of manufacturing (Darwin 1945). Engineering schools began to provide students with training in statistics and the basics of quality control.

Similarly, management journals lauded his approach as a means of increasing managerial control. Said an observer in the newly founded journal, *Industrial Quality Control*, "Particularly, control means a method for determining when the activities of business are deviating more than they should from the course laid down by Management" (Rice 1946, 6). Soon, Shewhart's methods could be found applied not only to the production of telephones and their parts but to items as diverse as aircraft and canned food.

Shewhart's ideas received a big boost during World War II. Quality control was a major concern for both the munitions industry and the military itself. Indeed, one historian notes that during World War II, "Statistical quality control became a way of life for the military and those who supplied the military" (Klein 2000, 36). The U.S. War Department requested that the American Standards Association develop quality control standards (Gaillard 1942). *Fortune* (1943a, 1943b) both praised Shewhart and noted the challenges facing business as a result of the exacting standards of the military. The military, aware of the challenges, responded by training some eight thousand persons in the private sector in quality control by 1945. In 1946 the American Society for Quality Control (now the American Society for Quality) was established by many of the engineers trained by the military, and in 1948 it began to confer an annual Shewhart Medal to honor his achievements. Shewhart's insights were later refined by Abraham Wald, as noted in chapter 1. Wald's work made statistical control considerably less costly, so much so that by the end of the war, at least six thousand firms had sampling plans based on sequential analysis (Klein 2000).

Perhaps Shewhart's star student was W. Edwards Deming (1900–1993). Ten years his junior, Deming also had a doctorate in physics and an interest in the application of statistics to industrial production. The two met while Deming was working for the U.S. Department of Agriculture as a statistician. Deming also served on the committee of the American Standards Association (1941, 1942), which developed the war standards. But whereas Shewhart focused his efforts nearly entirely on things, Deming quickly realized that the same ideas could be applied to management.

Indeed, it is fair to say that Deming constructed an entire philosophy of management around quality issues. Deming understood that "[q]uality to the production worker means that his performance satisfies him, provides to him pride of workmanship" (Deming 1982, 1). Deming first applied his work in Japan in the 1950s—indeed, his work arguably had more impact there than in the United States—thereby transforming the devastated postwar Japanese economy into an economic powerhouse.

Deming (1982, 23–24) proposed fourteen points for management, which together would apply statistical quality control to the entire enterprise:

1. Create constancy of purpose toward improvement of product and service, with the aim to become competitive and to stay in business, and to provide jobs.
2. Adopt the new philosophy. We are in a new economic age. Western management must awaken to the challenge, must learn their responsibilities, and take on leadership for change.
3. Cease dependence on inspection to achieve quality . . . by building quality into the product in the first place.
4. End the practice of awarding business on the basis of price tag. Instead, minimize total cost. Move toward a single supplier for any one item, on a long-term relationship of loyalty and trust.
5. Improve constantly and forever the system of production and service, to improve quality and productivity, and thus constantly decrease costs.
6. Institute training on the job
7. Institute leadership. The game of supervision should be to help people and machines and gadgets to do a better job.
8. Drive out fear, so that everyone may work effectively for the company.
9. Break down barriers between departments. People in research, design, sales and production must work as a team.
10. Eliminate slogans, exhortation, and targets for the work force asking for zero defects and new levels of productivity. Such exhortation will only create adversarial relationships, as the bulk of the causes of low quality and low productivity belong to the system and thus lie beyond the power of the workforce.
11. Eliminate work standards (quotas). . . . Eliminate management by numbers, numerical goals. . . . Substitute leadership.
12. Remove barriers that rob the hourly worker of his right to pride of workmanship. . . . [R]emove barriers that rob people in management and engineering of their right to pride of workmanship.
13. Institute a vigorous program of education and self-improvement.
14. Put everybody in the company to work to accomplish the transformation.

While one may view Deming's fourteen points as a bit hokey, they have been widely—and successfully—adopted by firms around the world. They can and should be viewed as a set of standards for management of a wide range of organizations.

Of particular note is that many of Deming's fourteen points stress standardization of products and processes as a means of controlling quality. Yet Deming was almost painfully aware of a rapid change occurring in the world: the use of statistical quality control as a means not merely of producing the same thing with ever greater precision and accuracy but of harnessing the creative powers of everyone in the organization toward the improvement of its products or services. Thus, Deming's fourteen points, as well as other managerial innovations that followed it, could use standards either to standardize or to differentiate.

At the same time, Deming was more than a little skeptical about standardizing people, whether managers or production workers. Indeed, he saw the use of quotas and merit ratings as undermining the kind of continuing improvement he aimed to bring about. On the one hand, he saw such ratings as instituting a climate of fear, in which conformity to bureaucratic rules was valued above all and creativity was stifled. On the other hand, he emphasized that most problems were systemic. Attempting to trace them to the actions of particular individuals was misguided at best. I take up Deming's statistical critique of individual evaluation later in this book.

Deming's work eventually spawned total quality management, continuous quality improvement, Six Sigma, and other management tools. Both the private sector and government commonly make use of these tools. For example, although health care is generally not seen as a governmental activity in the United States, the U.S. government has three agencies that deal with quality of health care: The National Quality Measures Clearinghouse (2007) collects and maintains a database of evidence-based health care measures. As noted on their webpage, "The NQMC mission is to provide practitioners, health care providers, health plans, integrated delivery systems, purchasers and others an accessible mechanism for obtaining detailed information on quality measures, and to further their dissemination, implementation, and use in order to inform health care decisions." In addition, the National Guideline Clearinghouse (2007) "is a comprehensive database of evidence-based clinical practice guidelines and related documents." Finally, the Agency for Healthcare Research and Quality (2007) "is the lead Federal agency charged with improving the quality, safety, efficiency, and effectiveness of health care for all Americans."

The value of any of these tools and the effectiveness of these agencies is beyond the scope of the discussion here. Moreover, I have no means to gauge their effectiveness, no measure of quality for these measures of quality. However, it is important to note that Deming's lasting contribution was (1) to develop a set of standards by which organizations might be judged and (2) to provide an approach to linking standards for things to standards for humans.

Organizational Communications

While giant organizations were not unknown in antiquity—Lewis Mumford (1967) has described the massive organizations needed to construct the pyramids—organizational size was limited by the difficulties involved in keeping track of seemingly innumerable persons and supplies. The creation of larger organizations was limited by the lack of standardizing technologies for organizations. But centralized large-scale organizations became critical when railroads emerged. Here was the first large-scale *networked* technological system. Engines, tracks, rolling stock, stations, and fueling points had to be distributed across the landscape. Goods and people had to be moved vast distances. Without standardized bureaucracy, railroads would have been virtually impossible to operate.

Critical to the standardization of organizations was the development of standard means for storing and retrieving information. To get an idea of the complexity of this problem, one need only look at the chest carried by Christopher Columbus and used to set up the rudiments of Spanish authority in what is now the Dominican Republic. The chest consisted of a collection of pigeonholes, each of approximately the same size. Into each hole was put correspondence folded and tied with string. Apparently, that was sufficient to contain the entire correspondence with the new Spanish colony.

But as organizations grew in size, building larger pigeonhole cabinets hardly offered a solution to the growing paper glut. A variety of systems were developed in the early nineteenth century, none of which proved particularly satisfactory. Eventually several systems became widespread. One involved the use of a card catalog. Each individual item would be numbered and filed in numerical order, while the catalog contained the index, using geographic or author names to sort. The other system involved the creation of what would eventually be called filing cabinets—large cabinets into which letters and documents could be filed upright in alphabetical order by author name. The latter, which required far less work, soon became the dominant system in the United States. However, even today, a trip to offices around the world reveals a staggering array of different formats, including small, letter-sized drawers or compartments in which papers are stored flat, upright boxes that sit on shelves, and large expandable folders into which papers may be inserted and which are then tied shut with string or elastic bands.[20]

While filing systems aided in retrieval of vast quantities of paper correspondence, they did not solve the problem of the growing need for multiple standard copies. For this, a host of other technologies was developed, including carbon paper, the hectograph, the ditto and mimeograph machines, and most recently the photocopier. Carbon paper allowed one to write on or to insert multiple sheets (usually no more than three to four) of paper at once into a typewriter and to type multiple copies of a letter or memo. Carbon paper was particularly handy in keeping records of outgoing mail, which otherwise required laboriously copying each letter by hand.

The hectograph machine—a tin tray with a layer of gelatin onto which was transferred a carbon imprint—made possible the creation of multiple standard copies of a memo, for circulation to various departments in a large firm. By moistening the gelatin slightly, putting a blank sheet of paper over the inked surface, and gently rubbing, the operator transferred an inked image to the blank sheet. The process could be repeated some twenty to fifty times before the ink became too light to read.

The hectograph was replaced by the ditto and mimeograph machines, each of which involved a drum on which a master stencil was placed. Ditto machines worked until the ink on the master was exhausted; in contrast, mimeographing allowed production of hundreds of copies as long as ink was available and the stencil did not tear.

The photocopy machine replaced the earlier technologies, producing in principle an infinite number of standard copies, each with the same message. Moreover, technical improvements and cost reductions soon made the photocopy cheaper than the older technologies.

More recently, email has partially replaced the memo. It has the advantage of having virtually no marginal cost of production, and it can be easily stored and searched electronically. On the other hand, as all users of email know, the very ease of its use poses problems: far too many of us are now confronted daily with hundreds of unwanted email messages, most of which are deleted unread at the stroke of a key.

But we should not let our fascination with the technology detract from the very necessary standardization of the messages themselves that had to take place. Prior to the use of copying technologies, letters were often

written with a bold flourishing hand and embellished with elaborate openings and closings: "My most gracious and kind sir, please allow me the
honor of writing to you to make my humble request. . . . It has been my
great pleasure to write to you. I wish you the best in all your future endeavors, and remain your loyal servant." Or some such language. Indeed, in
many Romance languages the art of letter writing—now honored mainly
in the breach—is quite elaborate. One book provides advice to French letter
writers, with some five hundred exemplary letters (Sandrieu 1988). It
devotes several rather obtuse pages to the appropriate way to end a letter
based on the status relations between sender and recipient. *Egalité*, perhaps,
but not when it comes to writing letters. According to the manual, the
proper standard for addressing the pope is as follows:

Prostrate at the feet of Your Holiness and imploring
the favor of your apostolic benediction,
I have the honor to be,
Very Holy Father,
with the most profound veneration,
of Your Holiness,
the very humble and very obedient servant and son.
(Sandrieu 1988, 18, my translation)

But interoffice memos could not contain all of these bells and whistles,
which have gradually been eliminated and replaced by much shorter
formalisms:

To:
From:
Subject:
Date:
(Body of message)

This format is commonly used and is widely accepted. That said, it is
worth noting that when email first became widespread, this format was
considered by some to be far too curt and even offensive. Gradually, as
the medium has come into common use, most of that formality has
been abandoned, leaving the bare-bones standardized email message as the
norm.

These technologies helped standardize interoffice communications.
When combined with standardization of the messages themselves, they
have made possible the large-scale organizations we take for granted today.

The Standard (Auditable) Organization

In times past, organizations were largely free to develop their own internal structures as determined by their members. Authority within an organization might be set in numerous ways, but there was little or no oversight in the design and structure of an organization. This is no longer the case. Since 1934, public U.S. firms must register with the Securities and Exchange Commission (SEC) and file quarterly and annual reports. To do so, their structure must conform to the standards established by the SEC. In particular, the SEC requires that all publicly traded companies meet its requirements for disclosure by making public certain information that might be relevant to an investor's decision to buy, sell, or hold shares in that company. For example, most companies must file Form 10-K annually. That form requires, among many other things, the following information:

- Date of the end of the firm's fiscal year
- Types of securities registered by the firm
- Risk factors for securities' holders as defined by various acts and regulations
- A list of properties held by the company
- Information on legal proceedings in which the company is involved
- A list of matters subjected to a vote by securities holders
- Various types of financial data on the company
- Quantitative and qualitative disclosures about market risk
- Disagreements with accountants on accounting and financial disclosure
- A list of directors and executive officers of the company
- Compensation paid to executives
- Costs of auditing of the company's finances

In addition, as Power (1997) has suggested, the pressure for audit has grown over the last century, leading to severe constraints on organizational structure.[21] It has shifted from strictly financial audits to virtually all aspects of organizational governance (Power 2003). In particular, organizations must adopt structures that permit audits to take place. For example, to meet the rules listed above, firms must have fiscal years, as well as types of securities that are recognizable to the SEC. I pursue this issue at length in chapter 6; suffice it to say here that audits are no longer limited to financial issues. More and more they are formative of organizational structure.

Management

Over the last century business schools have been created in most industrial nations. More recently, low- and middle-income nations such as China, India, and Brazil have also developed business schools. Whereas in the past, the leaders of firms came from a wide range of professions, today it is not uncommon for most managers to be graduates of MBA programs. Such programs emphasize certain features of management, marketing, and finance, and graduate professionals with training in these areas. In particular, business schools tend to include the basics of (usually neoclassical) economics but *not* to include the details of any particular industry. The training provided to someone who will become a manager in a food-processing company is no different from that provided to someone who will be a manager in a steel mill. Thus, for better or for worse, as business schools have spread worldwide, the nature of management has lost much of its heterogeneity and has become standardized. To a perhaps surprising degree, business school graduates are likely to think and act alike, having been provided with a particular set of tools for management and a particular (partly implicit) set of values to be maximized.

The widespread use of English in international trade has likely increased the degree to which management skills have been standardized. In the largest multinational corporations, managers now need to deal with persons in a common language. English has become that common language, so much so that many business schools in nations whose mother tongue is something other than English now conduct classes in English. Indeed, INSEAD, the premier French business school, offers its MBA degree only in English. Of course, the faculty also tend to assign readings to students in English, since it is the common language of business. This in turn means that articles written in English are more widely read than those written in other languages. Not surprisingly, this gives considerable advantage to English-language perspectives on management.

Today we find a combination of standardized communications, standardized audits, and standardized management, all of which are combined in multinational corporations. This means that most large organizations are now sufficiently standardized that corporate executives frequently move from one to another and are able to learn the ropes quickly.

Law and Politics

It is said that the law is fundamentally reactive. Put differently, it is only after something has happened that was unintended or unexpected or surprising or appalling that legal redress is sought. That redress may be through the courts, but it may also be through new legislation—legislation designed to remove the loopholes revealed by events, or by changes in the world that only now have come to light. Since the law is reactive in this sense, the move toward standardization has come rather late to this field.

But of course, law cannot be standardized in the sense that one law would cover all circumstances. Nor would a merely formal standardization of law—say, by requiring all laws to be of equal length—be much desired by anyone other than a handful of formalists. No, these approaches would hardly inspire people to act. They lack the substance that marks the other standardization endeavors noted above.

But what if one could develop a single *principle* on which all law would be based? What if justice were to be redefined in such a way as to decontextualize it, to desituate it, such that justice might become rather formulaic? What if a test could be developed that could be applied to all laws—by legislators, administrators, and judges—that would once and for all define and standardize justice?

This is precisely what has been attempted in two related fields of endeavor. The Chicago law and economics (CLE) movement, originating at the University of Chicago Law School, has attempted to develop a singular, standardized conception of law for use by judges. Public choice theory has attempted to develop a singular, standardized conception of law for use by legislators and administrators.[22] These two approaches have transformed law and political science over the last several decades, promising to cut the Gordian knot of legal doctrines and descriptive analyses that previously constituted law and politics, respectively. Of importance, both are based on several assumptions that have their roots in the Enlightenment. From Bacon and Descartes they borrow and update the notion that absolute knowledge is possible—in this instance, absolute knowledge about justice. Of course, this knowledge is not absolute in the sense of complete certainty; rather, it is that the tendency toward certainty is approximated. From Hobbes and Locke they borrow and update the notion of autonomous individuals coming together to create a constitution, as well as using the market and marketlike rationality to achieve their ends. Let us briefly examine each in turn.

As we all know, judges are those persons called on to hear and adjudicate in an endless number of situations where someone has been (or claims to have been) aggrieved. In general, judges in Western societies rely on two sources in making their decisions: precedent, or what decisions were made in similar cases in the past, and statutory law, or the record of legislation made by that body having jurisdiction over a given territory or substantive area of law. (In a smaller number of cases, judges may also rely on administrative codes of one sort or another.) In each instance judges have to interpret the law, and in so doing they make (or remake) it as well.

It would seem at first blush that each case would have to be decided on its merits within the constraints imposed by both precedent and statutory law. Moreover, it would appear that some notion of fairness would need to be applied to produce a just decision. While numerous approaches have claimed to produce the best approximation to fairness, CLE has suggested the use of another approach entirely.

CLE was initially influenced by the work of Chicago economists F. A. Hayek and Frank Knight and three of Knight's students, Milton Friedman, George Stigler, and Aaron Director. Director was arguably the pivotal figure, as he was hired by the University of Chicago Law School and taught economics there for many years. (Economist Ronald Coase joined him there in 1964.) From Director's perspective, "regulation was the proper function of markets, not government" (Mercuro and Medema 2006, 98).

We need not be concerned here with the numerous perspectives on law in the United States and other legal traditions that preceded CLE. Proponents of each of these perspectives tried to provide guidance to jurists in their own way. But what distinguished that guidance was that it was craft knowledge. It provided rules of thumb, but not any precise standard. In contrast, proponents of CLE argue that one relatively straightforward principle is and should be central: Kaldor-Hicks efficiency. This technical term, borrowed from economics, can be summed up as follows: "A change may be judged to improve society['s] well-being if and only if both the gainers from the change could compensate the losers for their losses and remain better off themselves, *and* the losers could not have compensated the gainers to forgo their gains without being themselves worse off than in their original position" (Mercuro and Medema 2006, 93).

This principle, also commonly referred to as "wealth maximization," appears to provide a unique version of what constitutes justice and hence a standardized means of adjudication. Moreover, it enshrines a particular version of economics in legal thought—one that would appear inconsistent with both liberal and neoliberal doctrine (see chapter 3). As Campbell and Lee (2007, xiv) note, "the most striking feature of their work is the extent to which they substitute the imposition of wealth-maximizing patterns for the commitment to free choice, which is at the heart of liberal economics." Yet it is the very same practitioners of neoliberalism who proposed this view of law (Mirowski and Plehwe 2009). In recent years its standard bearer has been Richard Posner, judge, University of Chicago Law School faculty member, and highly prolific writer.[23]

All of this would be academic—in the sense of trivial—were it not for the very profound effects that CLE has had on U.S. law. In particular, the CLE approach has transformed antitrust law. In the early twentieth century bigness itself was suspect, both to the general public and to the courts. Theodore Roosevelt's trust busting set the stage for antitrust activities until well after World War II. But from the perspective of CLE, monopolies are unstable and fleeting; therefore they are not likely to last in a market economy given competitive pressures. Moreover, domination of a market by one or a few firms may produce the most efficient allocation of resources in the economy. Finally, CLE proponents argue that the courts should intervene only in cases where inefficiencies are found; court intervention to redistribute wealth should be avoided precisely because it might tend to reduce efficiency. By the early 1980s, this view of antitrust was accepted wisdom not only by the U.S. courts but also by the government agencies charged with antitrust activities. "In the final analysis, the concepts . . . are now so well accepted that even non-Chicago enforcers of antitrust use them, . . . they do not curtail corporations from engaging in mergers on a scale unthought of thirty years ago" (Mercuro and Medema 2006, 154). This is not to suggest that CLE has succeeded in standardizing justice with respect to antitrust but that it has become hegemonic in this area. The CLE project of standardization remains a project, but it is certainly one that has met with considerable success.

What is true of antitrust is almost as true of tort law. Torts are to be handled using the insights and assumptions provided by Ronald Coase. In particular, they employ his insight into the reciprocal character of torts. As Coase (1988, 112) put it, "If we are to attain an optimum allocation of resources, it is therefore desirable that both parties should take the harmful effect (nuisance) into account in deciding on their course of action." Consider a situation in which a farmer and a rancher share a common boundary on their lands. Occasionally the cattle break through the fence and eat the farmer's crops. Conventionally, one might say that the rancher is liable for the damage and must either repair the existing wooden fence or build a much more secure metal fence. But Coase asks how their joint income might be maximized, arguing that this would be the most efficient solution.[24] This might involve the rancher compensating the farmer rather than building the more secure metal fence. This would be the case if it was more expensive for the rancher to build the metal fence than it was to leave the wooden one and compensate the farmer. In that instance, total income would be maximized.

Of course, as Mercuro and Medema (2006) note, this assumes that the losses can be easily expressed in monetary terms, that the undesirable activity can be reduced by devoting scarce resources to it, and that the individuals involved respond to prices in the marketplace. We might also add that it assumes that the price system gives an accurate measure of value to the parties and that all property rights should be negotiable in order to maximize the value of production. Regardless, it is clear that Coase's insight, as embodied in CLE, offers judges a very different approach to adjudication than does demanding that the rancher build a better fence.

Furthermore, CLE practitioners reinterpret common law so as to claim its efficiency. In so doing they employ an evolutionary approach. Mercuro and Medema (2006, 124) explain: "Simply stated, the hypothesis is that the development of the common law, especially the law of torts, can be explained *as if* the goal was to maximize allocative efficiency—that is, *as if* the judges who created the law through decisions operating as precedents were trying to promote efficient resource allocation."

This argument, which enshrines efficiency as the central value in common law, also has as a consequence the downgrading of statutory law. After all, if common law is already efficient or tends toward efficiency, and if efficiency is the goal of law, then statutory law is far more likely to create inefficiencies than it is to further the goal of efficiency.

It should be noted that to date, while CLE is known in Europe, it has had little direct impact on legal decision making. It remains largely an American phenomenon. Whether its impact will be felt in Europe or elsewhere or whether it will go down in what might be called a "bonfire of the verities" remains to be seen.

Moreover, what I just described for judicial decision making has its obverse in the legislative and administrative arenas. Public choice theory models all of politics as a market (Bromley 1997), applying the same economic logic to these nonmarket phenomena. It also rejects both the notion that the state might have some properties that are not merely the sum of the individuals who are its agents and the view that government officials of any kind might act to further the common good (Mercuro and Medema 2006).

It replaces these notions with that of autonomous individuals who—in either the legislative or political arena—are vying to maximize or optimize their utility or self-interest. It then proceeds to critique the outcomes of various democratic and bureaucratic processes from the perspective of

Kaldor-Hicks efficiency. In other words, its practitioners, like those of CLE, claim to have found the Archimedean point from which to critique current political practice, to have developed a standard that allows judgment to be passed on all political processes. They clearly see themselves as realists who are describing political processes, warts and all. Indeed, Mercuro and Medema (2006, 191) note, "to the extent that public choice theory is an accurate description of the machinations and outcomes of the political process, it is something less than a pretty picture." Yet it is worth noting in passing that their realism is based on a Lockean assumption of individual autonomy, and on voting as the quintessential act of the political process. About political dialog and debate they have far less to say.

Public choice theorists have become quite influential in both economics and political science, with several adherents winning Nobel prizes. However, public choice theorists have been somewhat less influential in practice than their colleagues in CLE. That said, there is little doubt that their perspective on politics has entered into the political lexicon itself, both in terms of increased questioning of (and perhaps cynicism about) various political processes and in terms of the standardized political litmus test public choice theory appears to provide.

In sum, together CLE and public choice theory can be understood as late-appearing attempts at standardization. Both attempt to employ economic notions of efficiency to phenomena that would appear not to be markets. Equally important, each claims to have identified a single criterion by which all laws and all policies can be evaluated.

Knowledge

Knowledge itself became the subject of standardization as well. From the standpoint of the French Encyclopaedists, knowledge was not to be the secret domain of the guilds but was to be publicly available to all. The highly controversial *Encyclopédie* of Denis Diderot (1713–1784) was an attempt to comprehensively organize all knowledge of the eighteenth century, ranging from philosophy to the mechanical arts. Construction of the *Encyclopédie* required the creation of standardized categories of knowledge, such that all knowledge might be easily organized. And Diderot and his colleagues certainly did try for comprehensiveness. The second edition of the *Encyclopédie* contained an astonishing 166 volumes when it was finally completed.

The creation of standard categories for knowledge was also driven by more practical and less ambitious concerns. The development of standard library card catalogs became imperative as the size and scope of large libraries expanded far beyond the memories of librarians. In the United States, two major cataloging systems emerged, the Dewey Decimal system and the Library of Congress system. For many years librarians at every library would consult large volumes that provided standardized guidance to cataloging whenever a new volume arrived. One ironic result of this was that the same book might have a different catalog number in a different library. This problem was drastically reduced when the Library of Congress organized a system whereby catalog cards might be ordered from a central distribution center. Thus, once a book was cataloged, it would have the same number in all subscribing libraries. Furthermore, the advent of International Standard Book Numbers ensured that every book received a unique identifier. More recently, the Library of Congress has also required American publishers to print the catalog information on the obverse of the title page of each new volume.

Quantification and Standards

Much of the standardization of the late nineteenth and early twentieth centuries focused on precise measurement of things. Such measurements were nearly always quantitative.

Soon after, the quantitative measure of people came into practice as both the now largely discredited notion of phrenology and as the somewhat less discredited notion of intelligence testing. Today, the use of myriad quantitative measures is commonplace. Government agencies in particular frequently use such measures as a gauge of effectiveness, and firms use such measures to gauge quality. Indeed, as Theodore Porter (1995, ix) notes,

Since the rules for collecting and manipulating numbers are widely shared, they can easily be transported across oceans and continents and used to coordinate activities or settle disputes. Perhaps most crucially, reliance on numbers and quantitative manipulation minimizes the need for intimate knowledge and personal trust. Quantification is well suited for communication that goes beyond the boundaries of locality and community. A highly disciplined discourse helps to produce knowledge independent of the particular people who make it.

Numbers are undoubtedly particularly handy tools that allow us to do all sorts of things that would be impossible without them. But numbers are also the most abstract "things" that we know. To say that $1 + 1 = 2$ is both very straightforward and extraordinarily complex. On the one hand, it is true by definition. It cannot be disputed (except by completely rewriting the rules of arithmetic). On the other hand, it applies to nothing in particular, and applying it to something particular necessarily involves a judgment.

At least since Galileo, we have been operating under the debatable assumption that there is some underlying validity to numbers that does not exist in other domains of knowledge. Both Galileo and Descartes went so far as to argue for the *mathesis universalis*, the notion that the form of the world was itself mathematical. It is possible that this is the case, but it is entirely unclear how one might go about proving (or disproving) it. Moreover, it is manifestly untrue that the world is *only* mathematical. To the contrary, without knowledge of the phenomenal world—the world of everyday life—the abstractions of mathematics remain just that, abstractions. Even the most arcane research in atomic or nuclear physics cannot go on unless the researchers (at the least) (1) talk to each other using ordinary words and everyday syntax, (2) refer to and describe many of the objects in their world using thick descriptions rather than numbers, and (3) tell a *story* of which numbers and other mathematical relations are only a part. To do otherwise, to talk merely in the abstract symbols of mathematics, would simply be unintelligible.[25]

That said, numbers are a particularly effective means of standardization. For example, numbers in the form of statistics have allowed the formation of nation-states. Indeed, statistics, in its earlier meaning as a particular form of statecraft, had its origin in the need for rulers of nation-states to understand what was going on in their nations. Apparently, the term first came into use in the mid-eighteenth century by virtue of the work of the German philosopher Gottfried Achenwall, who wrote about *Staatswissenschaft* (literally, state knowledge). As James Scott (1998) has argued, statistics of this sort allowed one to "see like a state," permitting among other things the more efficient and orderly collection of taxes. Gradually, the statistics of statecraft became the field of applied mathematics concerned with probabilities.

However, as with all other forms of standardization, merely creating the numbers does not provide much useful knowledge. Instead, both the data that are counted and the persons and things implicated in the process of counting must also be standardized. This is a process that is both expensive and laborious. Consider, for example, the work required to collect data on unemployment. Briefly, this requires defining who is unemployed. A naive view would suggest that any adult not working would be counted. But in fact, one must also be part of the labor force and actively seeking a job in order to be unemployed. So-called discouraged workers who have given up looking for employment are not counted. Moreover, one must have an employment history so that one is noticed, by virtue of having, for example, a Social Security number and a record of paying taxes. These rules and many others must be consistently reported across multiple jurisdictions, and summarized and reported in official statistics. Even with multiple safeguards, there are always errors and disputes in calculating the rates. The same is true for most other governmental and privately collected statistics.

Porter (1995, 37) goes so far as to argue that "the concept of society was itself in part a statistical construct. The regularities of crime and suicide announced in early investigations of 'moral statistics' could evidently not be attributed to the individual. So they became properties instead of 'society,' and from 1830 until the end of the century they were widely considered to be the best evidence for its real existence."

Hence, crime rates, suicide rates, unemployment rates, and hundreds of similar measures are now taken for granted and loop back (Hacking 1999) into our everyday world. One need only pick up today's newspaper to learn that the lead in parts per million in the water supply is above acceptable levels, that unemployment rates are unacceptably high, that the prime interest rates on loans is falling, that x percent of elementary school students perform below grade, that the divorce rate is rising, that teenage pregnancy rates are falling, or that any of thousands of similar quantitative measures are changing to understand how standardized measures simultaneously create and report societal conditions. "The bureaucracy of statistics imposes not just by creating administrative rulings but by determining classifications within which people must think of themselves and the actions that are open to them" (Hacking 1991, 194).

Yet the use of standards to standardize everything and everyone may well lead to the denial of personhood (Boltanski and Thévenot [1991] 2006), to mechanical objectivity (Porter 1995), to a collapse of creativity. Perhaps, as M. Clement Chardin des Lupeaulx, a general secretary of a minister in one of Honoré de Balzac's ([1838] 1993, 242–243) novels, argued:

Well then, although statistics are the childish foible of modern statesmen, who take figures for calculations, we must make use of figures to calculate. Shall we, then? Figures are, moreover, the convincing argument of societies based on self-interest and money, and that is the sort of society the Charter has given us, at least in my opinion. Nothing convinces the "intelligent masses" more than a few figures. Everything certain, so say our statesmen of the Left, can be resolved in figures. So let's figure.

But we need to ask just what it is that is happening when we attempt to quantify everyone and everything. Consider the case of the small child who would prefer to have four pennies than a nickel, since four is far more than one. A fair-minded adult might point out to that child that the number of coins is far less important than their value. This simple morality tale is often forgotten when discussions shift to other quantitative measures.

Moreover, the obsessive demand that all standards be quantified actually has several highly undesirable effects. First, it insists that in all instances one kind of story is better than another. It is certainly clear that standards for certain physical objects are far more helpful if they are described in quantitative terms. Therefore, one would fully expect that tolerances for the diameter of an axle would be quantified. But it is also the case that many phenomena are severely impoverished when they are quantified. For example, endocrinologists can now measure various human feelings by examining hormonal activity. They can even quantify those measures, thereby producing standard quantitative measures of such things as sexual satisfaction, rage, or aggression. But even as such measurements might be useful in understanding endocrine function, these feelings are impoverished if reduced to quantitative form. Can we really say that someone whose testosterone levels are twice as high as another person's is twice as masculine or twice as aggressive? Will persons with the same levels of testosterone express their aggression in the same way? Will the effects on others be the same? Clearly, even as the quantification of testosterone levels is useful and a "normal" range may be established, the phenomenon of aggression is impoverished if considered only in quantitative terms.

Second, quantification often assumes that the things quantified are in fact the (only) relevant factors. Contemporary society is replete with examples of such measures. Do we really mean to argue that scores on standardized educational tests are the most or only relevant factors predicting success in life? Is there only one path to leading a productive, meaningful life? Similarly, there are dozens of studies that purport to show that Americans are scientifically illiterate. But those studies are flawed by the fact that everyone is a novice outside his or her field of competence. Hence, most particle physicists are rather ignorant of even the basics of rain forest ecology and are not likely to be able to tell you much about house carpentry. Most sociologists would not be able to tell you much about the engineering of computer modems or the proper venting of toilets. Most plumbers would not be good authorities on electrical wiring, and so on.

Third, precise quantified standards may suggest precision where little or none exists. For example, a plant breeder who found that two rice varieties grown on standard plots produced 2.815 and 2.816 tons per hectare, respectively, would be hard-pressed to argue that one was more productive than the other.

Fourth, quantitative standards applied to people may encourage working to the measure. For example, if I am rewarded based on the number of journal articles I publish each year, then I may decide to divide my research into many small pieces, each of which can be published separately. Similarly, if elementary school teachers are rewarded based on the proportion of their students who attain a given score on a standardized test, they may devote most of their class time to test preparation and ignore other important skills in the classroom.

In short, the quest for ever more precise measures may undermine the very phenomenon for which one wishes to develop a standard. But please do not misunderstand me. I am not suggesting that quantification is *necessarily* problematic or wrongheaded. I am suggesting that the quantification of a given measure should be undertaken with very considerable care.

Conclusion: Creating Standardized Cognition

Edwin Hutchins (1987), Jean Lave and Etienne Wenger (1991), and others have argued that cognition is not what goes on in one's head but is distributed across people and embedded in things.[26] Thus, my confidence in what I know is bolstered by (1) other persons who confirm (or deny) that knowledge and (2) the things with which I interact. Consider the following example: I wish to go from Victoria Station in London to Blackheath via the railway. I arrive at the station, buy a ticket, slide it into a machine, board the train, and continue in this fashion. My knowledge is confirmed or modified by the action of other persons—the ticket agent, the conductor, and so on—as well as by the physical objects with which I come into contact—the stations, the train, the ticket machine. Were any or all of these to act in unexpected ways, I would have to modify my knowledge. At the least, I would have to say that the ticket machine was malfunctioning, the conductor was confused, or the workers were on strike. At the extreme, I would have to conclude that this was not Victoria Station at all, or that the railway did not exist but was a strange dream. In short, what I know is never what I know but always what *we* know, and it is always situated and always subject to revision.

Figure 2.3
Paris travel agency window, still from *Playtime*.
Source: Jacques Tati, *Playtime*. Paris: ©Les Films de Mon Oncle, 1967, restored 2002,
www.tativille.com.

The great Enlightenment project in its various manifestations was fundamentally a project of standardization of distributed cognition, of creating a world in which people would find themselves in similar if not identical situations. The frame from Jacques Tati's satiric film *Playtime* (figure 2.3) illustrates how this could be so. Finding himself in a Paris travel agency, Tati, as Monsieur Hulot, is confronted with posters for various cities, each of which contain a picture of the very same Bauhaus-style building. Similarly, the economic project first spelled out in detail by Adam Smith involved the use of market structure as a device for channeling the highly unpredictable and nonstandard passions into productive activities (Hirschman 1977). Markets would be designed to encourage price competition; eliminate joint stock companies (what we today call corporations), unions, professional associations and other means of collective action; and put all actors in the marketplace in the same relationship to each other. Through the nineteenth and twentieth centuries this project was frequently tinkered with and became hegemonic in the form of neoclassical economics (Lazear 2000). When applied by practitioners, neoclassical models encourage standardization of people, things, and markets.

Similarly, the class struggle envisioned by Marx required a standardized world with standardized workers (who would recognize their common plight) as much as did the standard world of work later envisioned by Taylor or the "melting pot" envisioned of America by de Crèvecoeur. In short, the "satanic mills" of the nineteenth century distributed cognition across a diminishing number of worlds.

To this one could add the various projects of the "psy" disciplines—psychology, psychiatry, psychoanalysis—as recounted by Nikolas Rose (1996, 197). Each posited the existence of a "normal" individual and participated in a project to create a largely standardized regulatory regime of the self. Paradoxically, each asserted that the very project of regulating the self would emancipate us, "assist us in the project of being free from any authority but our own."

Put differently, in the nineteenth century and well into the twentieth, cognition was being distributed in such a way that it appeared to many observers that it would be fully standardized, such that one or only a few sorts would remain. This, some argued, would lead to revolt and revolution, while others argued that it would lead to harmony and homogenization, even to greater freedom. This was not the result of some mysterious force but a consequence of the collective projects of standardization enacted at that time.

Somewhat later, the Soviet Union and Maoist China attempted to bring into existence a version of the kind of uniform, standardized world that, despite its Marxist rhetoric, resembled more that of the Utopian Socialists such as Henri de Saint-Simon, and required a great deal more force than Marx envisioned using. Hitler would attempt to bring racial uniformity to the Third Reich and the entire world. Senator Joe McCarthy would attempt to stamp out dissent by engaging in a witch hunt in the name of freedom and democracy. Put simply, in a standardized world, cognition would be distributed along standardized lines. The establishment of a monopoly of force by the state (Weber [1922] 1978) was another part of the project of standardization that saw its zenith in the twentieth century. Armies would be raised only by nation-states. Only nation-states would have a body of both criminal and civil law and the means—both material and moral—to enforce that law. Arguably, this project of standardization reached its height in the great dictatorships of the mid-twentieth century—those of Hitler, Stalin, Mao, Franco, and Mussolini—although a wide range of ideologies proclaimed the need to move in that direction.

Not everyone was equally enthusiastic about these standardizing tendencies. The dangers of this standardizing tendency were duly noted by dystopian novelists such as Aldous Huxley ([1932] 1998), Franz Kafka ([1926] 1998), George Orwell ([1949] 1990), and Yevgeny Zamyatin ([1924]

1993). Max Weber ([1905] 1958), summoning the image of an "iron cage," suggested that rationalization (Weber said little or nothing about standards) would lead to an increasingly oppressive society. Similarly, Robert Michels's ([1915] 2001) "iron law of oligarchy" suggested that democratic governance was all too likely to lead to something quite undemocratic. Both suggested there was little one could do to stop this unfortunate tendency.

But this was, after all, a project and not a foregone conclusion. There were many actors on stage, and more than a few forgot their lines, failed to show up, acted unpredictably, or otherwise performed poorly. The attempt to produce a standardized world didn't quite work out as planned. Despite attempts to get everyone to use the same recipe, some cooks insisted on using procedures not to be found in the standardizers' cookbooks. Still others used the authorized cookbook but modified the recipe to fit their circumstances. Yet others were simply incompetent cooks, their attempts at standardization leading to other, often unintended, consequences. On the one hand, the project produced a reaction on both the right and the left. On the other hand, a response to standardization began about a century ago: the growing use of standards to differentiate. This is the subject of the next chapter.

3 From Standardization to Standardized Differentiation

Variety is what constitutes organization; uniformity is mere mechanism. Variety is life; uniformity, death.
—Benjamin Constant ([1815] 1988, 77)

The nice thing about standards is that there are so many to choose from.
—Andrew S. Tannenbaum, (2003, 235)

Arjun Appadurai (1986a) notes that in modern capitalist societies the medieval marks of class—special clothing, types of housing, forms of association—are shattered and that fashion—in clothing, but also in housing, automobiles, books, and other consumer items—replaces the rigid class divisions of the past. But what Appadurai only notes in passing is that it is a succession of *standard* goods that marks class and status in modern society. Indeed, standard goods permit keeping up with the Joneses, the striving among the middle and upper classes for status goods. As those goods are attained, they are rapidly succeeded by other goods. The treadmill of consumption turns ever more rapidly. Thus, already built into modern society is the notion of using standards to differentiate. But until recently, neither the social organization nor the technical means existed to differentiate other than rather slowly over time.

Indeed, as strange as this concept may seem at first, some standards are designed to standardize, while others are designed to differentiate. In a superficial way, it appears that all standards standardize and all differentiate. After all, they attempt to make beings the same in some way, but in doing so they distinguish those beings from others that are differentiated in some other way. For example, all hamburgers are made of beef (a claim to standardization), but they must be distinguished from similar products made from ground pork or lamb or even soybeans (a claim to differentiation). Similarly, all Russians are citizens of Russia and are said to have certain characteristics, obligations, and rights (claims to standardization), but Russians can be distinguished from Kazaks, Canadians, Danes, and so forth (a claim to differentiation).

We can nevertheless distinguish (in most instances, but always with some irreducible degree of ambiguity!) among those standards (or perhaps aspects of standards) that are *designed* to standardize and those *designed* to differentiate. Simply put, the former make claims that are (largely) universal in character, while the latter make claims that are (largely) particular.[1] A universal claim is one that proclaims sameness for the characteristic of relevance across some defined time and space. Although it may be limited for practical reasons, the goal claimed for it is universal. Thus, the standards of equality before the law and of the uniformity of weights and measures are universalist claims. They brook no exceptions, although undoubtedly exceptions may and do exist as the result of limits to a given authority, illegal behavior, or inadequate sanctions. In addition, universalist claims often have sanctions attached to them that are designed to produce uniformity. Thus, the accounts in the previous chapter of attempts by various nation-states to erase subnational movements and of grain merchants to standardize the grain market had clear negative sanctions attached to them.

In contrast, a particularist claim is designed to distinguish among persons or objects using a particular measure. Contemporary concerns about illegal immigration, for example, are designed to distinguish between those having such-and-such a nationality and those who do not. Similarly, the brand Ford is designed to distinguish automobiles bearing that brand name from those that do not. The statements "everyone is French" or "every car is a Ford" are surprising precisely because they take what are normally particularist claims and attempt to turn them into universals. The reverse is true as well: The statements "only people born on Sundays are equal" or "only houses in Govingia may be built with wooden studs" are equally startling because they take universals and give them particular form.

That said, it should also be noted that standards can be seen as universal or particular only within a given situation or network of relations. Therefore, much as standards themselves may never be fully articulated, so standards for standards must remain incompletely defined. With that as a starting point, let us examine the phenomenon of standardized differentiation.

Using Standards to Differentiate

Even as some have attempted to use standards to standardize, others have attempted to use standards to differentiate. Some attempts to differentiate have arisen in response to attempts at standardization. Others have arisen out of a desire to replace the existing standard. Still others have employed differentiation as a strategy in itself. Of note, standardized differentiation is quite different from the nearly infinite differentiation that was commonplace centuries ago and is still found in much of the world today. Thus, I can go to Mali and find vast quantities of handmade similar, but actually significantly different, craft items. Similarly, even in the most industrial nations I can find persons producing (usually very expensive) handmade craft items. These items are best understood as singularities; they are each one of a kind. In contrast, standardized differentiation involves the production of many things to the same standard template. Let us begin by considering some early attempts to use standards to differentiate a century ago.

Americans over the age of thirty-five will remember the Sears Roebuck catalog. For nearly a century, from 1894 to 1993, this icon of rural America was delivered to countless mailboxes. The catalog massively increased consumers' choices beyond the handful of items available in most small towns. As *Wired* magazine editor Chris Anderson (2006, 42, 43) explains, "What Sears and Roebuck's warehouses and efficient processing operations enabled was nothing less than revolutionary. . . . The 1897 Wish Book was—and still is—astonishing. Even today, in the era of Amazon, it seems impossible that so much variety can exist. Crammed into something the size of a phone book are 200,000 items and variations, all described with tiny type and some 6000 lithographic illustrations."

This was using standards to differentiate on a grand scale. Each of the 200,000 items in the catalog had to be standardized, since the catalog description (along with, perhaps, a neighbor's recommendation) was the only information about the product available to the purchaser. Certainly, many customers bought the cheapest version of whatever product they wanted out of the catalog, but thousands of others chose based on other criteria in addition to price, including color, texture, brand name, and style. Similarly, at a time when most companies had one or two types of canned pickles, Heinz divided up the market into fifty-seven distinct, discernible, standardized types and offered money-back guarantees to unsatisfied consumers (Levenstein 1988). Automobile colors and romance novels offer two somewhat later examples (boxes 3.1 and 3.2).

One could fill an entire text with examples of using standards to differentiate (see table 3.1). In recent years, differentiation has become more and more commonplace, while the growth in standardization has waned. Even the number and scope of literary and artistic prizes have expanded

Box 3.1
Standardized differentiation of colors

In the 1920s, when Alfred P. Sloan began to differentiate automobiles, challenging Ford's highly standardized Model T, he needed to find suppliers of a variety of paint colors. Initially there were thousands of colors that varied from one batch to the next, much to the frustration of the auto companies. By 1932 an astonishing 11,500 different colors were on the market, and each of these had some visible variation. However, DuPont developed the Duco Color Advisory Service to provide technical, aesthetic, and marketing advice to buyers of its paints. But DuPont itself had some 7,500 colors. Howard Ketchum, head of the Advisory Service, reduced the number of colors to 290 consistent and highly standardized ones (Blaszczyk 2007). In short, automobile colors moved from black to uncontrolled differentiation to standardized differentiation.

(English 2002). This is not to say that standardization has ceased or is faltering but that projects of differentiation have gained greater significance in organizing social life. To begin to understand why this is the case, we need to consider what is unusual about contemporary markets, for it is there that the story begins.

The Growth of Non-Price Competition

Prior to modern times, most goods and services were produced to order for particular persons. If I wanted a particular item, I would ask the carpenter or blacksmith or mason—likely someone I knew—to produce it for me. If the good or service was subject to economic exchange, we might haggle over the price. The product or service would be similar to that provided to others, but there would be neither the need nor the demand for standards or uniformity. (Of course, there were always things produced that were in excess supply, and those *residual* items would be sold in the market.) Thus, most goods and services were highly varied, unstandardized, and created for particular persons. Competition, such as existed, was more commonly between buyer and seller over price and other features of the product or service, and not among sellers. During the European Middle Ages, prices and qualities were often fixed by guilds or towns, based on the notion of a "just price." In many places, selling things in the market was seen as an unfortunate necessity. In China, prior to 1949, special markets existed in the countryside where one could trade "without human feel-

Box 3.2
Standardized differentiation of romance novels

It appears that even genre fiction—something that is rather difficult to define with great precision (and arguably the better for it)—can be standardized to some degree. Harlequin does just that in identifying manuscripts for its various romance novels. On their web page (Harlequin Enterprises Ltd. 2009) they define more than thirty romance series that they produce. For example, the Harlequin Historical Undone series requires manuscripts of 10,000 to 15,000 words, with "a high level of sensuality that flows naturally out of the plotline. There should be a strong emotional basis to the heightened attraction. . . . These stories should be hot, sexy and subtly explicit without the lovemaking being vulgar or gratuitous."

In contrast, Harlequin's Kimani Romances imprint is targeted toward African American women: "Told primarily from the heroine's point of view, Kimani Romances will keep it real with true-to-life African-American characters that turn up the heat and sizzle with passion."

Finally, the Steeple Hill Love Inspired imprint "features wholesome Christian entertainment." Thus, the aspiring writer is advised, "These are 'sweet' romances. Any physical interactions (i.e., kissing, hugging) should emphasize emotional tenderness rather than sexual desire or sensuality. Please avoid any mention of nudity."

One might object, arguing that such well-defined standards hamper the creativity needed to produce great literature. Perhaps that is the case, but adherence to such standards does sell books, and arguably Harlequin is in the business of selling books rather than producing great literature. Regardless, the Harlequin standards are an excellent example of the use of standards to differentiate a market. They also reduce the risk for both publisher and reader. The reader who has read a Kimani Romance and enjoyed it will likely buy another, knowing approximately what to expect.

ings" (*wuqing*) (Fei [1948] 1992). But again, to emphasize the point, goods sold in these markets were rarely standardized; they were singularities rather than commodities. They were at best partially commensurable.

However, since the days of Adam Smith, the ideal of economics has been price competition. Smith was particularly prescient. He believed that markets could be designed to ensure price competition and that this, combined with the government provision of essential services and an appropriate legal framework, would lead to what he called the Great Society.

Neoclassical economists have added to Adam Smith's prescription the tools of mathematics and marginal analysis. For example, Paul Samuelson

Table 3.1
Some examples of the use of standards to differentiate

People	Things
People with various rare diseases or unusual medical conditions who hold annual conventions	Artificial logs for gas fireplaces (one company alone has 43 models, most of which come in several sizes)
People who share unconventional sexual practices	Threadless.com (viewers design and review t-shirts, which are then produced in runs of 1,000)
Groups with Yahoo or other Internet sites	Models of automobiles
Owners of antique automobiles	Geographic indicators for food and other products
Collectors of teapots	Varieties of athletic shoes
Breeders of varieties of rare domestic animals	Number of stock-keeping units (SKUs) in a typical U.S. supermarket: 40,000
Cockroach racers	Varieties of soft drinks
Hollering contest participants	Magazines for specialized reader groups

(e.g., Samuelson and Nordhaus 1995), whose introductory textbook in economics was a best seller for decades, used the grain market as an example of a nearly perfect market.[2] In that model, price is the sole factor distinguishing among goods of equal quality. For example, agricultural economist Lowell Hill (1990, 1991) spent his entire career working on the development of uniformity in the grain market. Hill understood that uniformity was not something inherent in the natural world but rather something that had to be achieved if markets were to deliver on their promise of improved welfare for all. Working from this premise, he spent a great deal of time tracking down false claims, making recommendations for grades and standards that better reflected downstream uses, and otherwise helping to standardize the production and postharvest handling of grain.

Hill's observations hold for products as diverse as metal ores, oilseeds, pork bellies, wood pulp, coal, petroleum, fertilizers, salt, cement, and crushed stone. All of these products share these traits: They are produced with an eye to uniformity (usually allowing them to be traded on commodities markets). And they can be made sufficiently uniform that they can be traded based virtually entirely on price.

As Hill recognized, such markets tend to benefit downstream consumers through lower prices more than upstream actors such as farmers and elevator operators. The agricultural economist Willard Cochrane (1993) explained some years ago why this is the case. His argument is fairly straightforward and worth repeating here. In all bulk commodity markets, the commodities are (by definition) highly standardized and can only be differentiated based on price. Moreover, if I am a seller, I am anonymous. Those who buy from me care little or not at all about who I am.

If I am a buyer of such a commodity, I will pay little or no attention to brand, quality characteristics, or other differentiating characteristics, as they will be largely irrelevant. I will simply buy the cheapest lot available (of ten different brands of otherwise identical bananas in the supermarket, most customers would buy the ones that had the lowest price). But as a consequence, if I am a seller in such a market, I cannot in any way raise my price without driving away customers.

This situation forces producers onto what Cochrane termed the "technological treadmill." In short, although I cannot raise my price, I can lower my costs of production—usually by technical change. Thus, if I am a grain farmer, I might grow a higher-yielding variety, or use a more fuel-efficient tractor, or engage in some other activity that will cut my costs. After that I can do one of two things: I can continue to sell at the same price as everyone else and have higher profit margins, or I can lower my price somewhat. If I do the latter, I can attract more customers and expand my market share. But if I do that, I will force all other producers of this undifferentiated commodity to do the same, on penalty of being forced out of the market. The treadmill revolves again, and in order to raise my profits or increase my market share again, I must find some other way to cut costs. This is what General Electric CEO Jeffrey R. Immelt meant when he suggested that innovation was the only way to avoid the "abyss called commodity hell" (quoted in Hof 2004, 129).

In contrast, a world of differentiated standards (re)opens the door to non-price competition. This is not the non-price competition of premodern markets, where every item in the marketplace is slightly different. To the contrary: this is a market in which we can choose among hundreds of different—but equally standardized—varieties of ketchup, automobiles, or airline tickets.

But a simple dichotomy between price and non-price competition is inadequate to understand the situation. Price competition is singular and it is *designed* to be so. Indeed, neoclassical economists are right when they tell us that, in a perfectly competitive market, price provides a summary of all other relevant characteristics. In contrast, non-price competition is plural.

Furthermore, price competition is designed to be coercive. An example found in virtually every introductory economics textbook will illustrate what is meant by coercive. Many sellers have the same item to sell. Most producers price it at $1.00 each. Producer A prices it at $1.50 each. Buyers will clearly avoid producer A unless there is a shortage of supply, in which case producer A will become the seller of last resort. More likely, producer A will be forced by the market (by "market forces") to reduce the price of the good to the going price of $1.00 or go out of business (or perhaps both, if producer A's production costs are above $1.00). In short, in homogeneous commodity markets, the market price exerts *coercive* influence over the participants. It is precisely that coercion that provides justification for the welfare claims of neoclassical economics, that such markets will produce the greatest good for the greatest number.

In contrast, non-price competition may or may not be coercive. In most instances it is not coercive. In fact, participation in any particular form of non-price competition is often voluntary. The range of forms of non-price competition in which supermarkets may engage (table 3.2) suggests that some forms of non-price competition are nearly universally avoided. For example, competition over food safety (our products are safer than yours) is rarely undertaken, in no small part because a single food safety problem can undermine years of building a reputation for safety. Some companies emphasize product variety, building huge hypermarkets with 50,000 or more products on the shelves. Others take precisely the opposite strategy, limiting the number of items on the shelves to keep store size down and to move shoppers through the store quickly. One Norwegian company, Rema 1000, goes so far as to specify in the store name how many different

Table 3.2
Some forms of non-price competition in which supermarkets may engage

Cosmetic quality of produce
Cleanliness
Lighting
Convenience
Checkout speed
Food safety
Fair trade
Organic
Animal welfare
Store brands
Category management
Product variety
Opening hours

but equally standardized products are in the store. Other companies attempt to compete by being open more hours, such as 7-Eleven. Yet other companies choose to compete based on a selection of organic, fair trade, animal-welfare-friendly, non-GM, locally produced, or other products. Some companies build cathedrals of commerce with beautiful carpeted floors, warm colors, and merchandise attractively displayed on expensive shelf units. Others choose to build inexpensive metal buildings with concrete floors on which cases of packaged foods are stacked. Some provide highly trained staff to assist customers, while others only provide cashiers at checkouts.

What is important to understand is that, unless it becomes the norm in the industry—that is, unless consumers overwhelmingly insist on that particular quality—there is little or no market coercion to engage in any particular kind of non-price competition. Companies may and do choose the forms of non-price competition in which they will engage. Moreover, since such competition is not based (solely) on price, it is measurable only in terms of the cost of implementation (not always easy even for insiders to discern) and its consequences, that is, its effect on the bottom line. Furthermore, it is extremely difficult if not impossible for supermarkets (or other sellers) to measure the impact of any particular form of non-price competition on their bottom line. They are all lumped together in an ever-changing set of strategies.[3]

Once again, we can return to standards in understanding non-price competition. Indeed, each of the four types of standards noted in chapter 1 permits and produces a different form of non-price competition. Let us examine each in turn.

Divisions

Division is probably the most commonly used strategy for non-price competition. In the case of supermarkets, the business may be distinguished by its distinctive architecture, by the presence (or absence) of certain services therein (e.g., coffee shops), by having a large selection of private-label products that cannot be obtained elsewhere, or by carrying products that are certified as animal-friendly, organic, or fair trade but that are not necessarily superior in other ways to similar products found elsewhere.

Ranks

Non-price competition by ranks is a bit more difficult but still possible. Creating a reputation for friendly and rapid service relative to other retailers through employee training and standards for employee behavior might distinguish a given chain from others. Or one might emphasize superior store cleanliness and beautifully displayed produce. Similarly, one might advertise that one's stores contain the widest range of fresh produce items.

Filters

Filters are rarely used by individual companies as a means of non-price competition since there are few means of imposing those standards on others. However, they can be used by groups of firms to lobby the state in an attempt to squeeze out companies that do not conform. For example, early in the twentieth century, large canners used state-mandated filter standards to bar companies using soldered cans from the market (Levenstein 1988). They argued their case on food safety grounds, and may even have had a reasonable case in so doing. (The solder contained lead, which sometimes leached into the food.) But the consequence of their success was to drive out firms that could not afford the new crimping equipment with which to make and seal cans. More recently, larger producers of fruit juices have lobbied for more stringent food safety standards in part to drive out the many small producers of unpasteurized fruit juices.

Olympics

Only a very few companies are willing and ready to introduce Olympian standards into their non-price competition. One obvious reason is that this form of non-price competition has a significant upward effect on prices. This in turn has the potential to increase profit rates at the cost of reducing the overall size of the market to include only those consumers for whom price is of little or no concern. High-end food retailers—restaurants, caterers, and supermarkets—do exactly that. They use Olympian standards (e.g., competing to get an award as the best restaurant in the city, carrying only the very best-quality and luxury foods) to engage in non-price competition, segmenting the market in such a manner that only the most well-heeled shop there regularly.

The examples above are for supermarkets, but one can see much the same behavior in other domains. For example, Harvard University maintains its reputation as a top-rated educational institution by virtue of its enormous endowment, the hiring of top-ranked faculty, very competitive admission standards, and very high tuition (with scholarships to keep a modicum of diversity among its student body). In contrast, the private for-profit University of Phoenix focuses on that group of people seeking relatively low-priced, career-targeted college education at particularly convenient hours. Such persons are likely to be far less concerned about getting a well-rounded education, spending time on an attractive campus, having access to a huge library, or associating with professors who are leaders in their respective fields.

The Market for Lemons Revisited

But the dynamics of non-price competition do not tell the entire story. In most economic exchanges of simple things, both buyer and seller have roughly equal knowledge of the items to be exchanged. For example, if I wish to buy a school notebook, I can flip through it to examine the quality of the paper and visually inspect the binding, color, and other quality characteristics of interest. But not all economic transactions have this symmetry of information about the thing to be exchanged.

In what is now a classic article in economics, George A. Akerlof (1970) described the market for used cars. In this market the price is typically known to both buyer and seller, but the quality is largely known only to the seller.[4] Akerlof relaxes the constraint of informational symmetry (that the buyer and seller have the same information about the qualities of the thing to be sold) and asks what happens when the quality of a given product is not easily discerned by the buyer. The used-car market is an ideal example because of the variable qualities of the vehicles for sale—variability usually known to the seller but rarely to the buyer.

But it is important to note that there are at least two aspects to Akerlof's thesis. First, there is the likelihood of informational asymmetry between buyer and seller. Second, there is the limited commensurability of the objects for sale. This incommensurability must be distinguished from product differentiation, which is *designed* to limit commensurability. But, as the success of web-based new car sales illustrates, it is possible to compare highly differentiated products as long as they are in some sense "the same." That is, a particular make and model with a particular set of extras can be found at several dealers and the buyer can choose the lowest price available. In the case of the used-car market, despite heroic attempts to reduce information asymmetry (e.g., through the *Kelley Blue Book*, which provides value estimates for used cars), there is an irreducible incommensurability among the vehicles sold as a result of differences in their unique histories (some of which are perhaps not even known to the seller). This makes price comparisons extremely difficult and sometimes impossible. It also provides the seller with an opportunity and perhaps an incentive to cheat the buyer.

Akerlof's article has spawned a small industry. I shall not attempt to review the various nuances, critiques, and empirical tests generated by it. However, it is clear that his work changed forever the world of neoclassical economics. Yet it is not entirely clear that Akerlof explored all the possibilities.

For example, in some markets quality may be known while quantity is not known. That is, the measure (weight, volume, length) may be false or variable. Therefore, there will be market pressure to give less than full measure. In such instances, price per standard (but unmeasured) unit goes up, while price per nominal unit goes down. This was apparently the case in the United States until the late nineteenth century even in major cities. When Patrick Derry was named to head the Bureau of Weights and Measures in New York City in 1906, he publicized the fact that the accuracy of scales used to weigh meat was quite variable, generally favoring the butcher (Perry 1955). The *New York Times* (1906) reported that the practice was widespread especially in poor neighborhoods. (Ironically, Derry himself was forced to resign some four years later when it was discovered he was not actually doing his job [*New York Times* 1910].) A 1912 review in San Francisco produced similar results (Bureau of Standards 1912, 4). And doubtless what was true in New York and San Francisco was widespread elsewhere as well.

Such situations need not involve solely the exchange of goods but may also include services, where the "measure" is more ambiguous. In the United States, despite legal restrictions, every so often there is a scandal involving itinerant home maintenance services that are contracted for and either never delivered or delivered in an extremely shoddy manner (e.g., lawn care, roofing). In addition, attempts are frequently made to eliminate what are known as diploma mills. But in India, where regulation of this sort is far weaker, numerous schools, colleges, and universities of very low quality remain in existence. Caveat emptor applies there with a vengeance.

But this is hardly the end of the problem. An interesting example of further complexities comes from the sorghum market in El Obeid, Sudan. Some years ago I visited that market, located at the western edge of the arable land in that nation, beyond which lie the endless sands of the Sahara. There one could find a hundred or more people selling threshed sorghum. It varied dramatically in quality. Some was red, some yellow, some white, some brown. Some was clean, while some contained considerable foreign matter. Some had been chewed on by insects, some not. Some had one or another disease on the kernels, some not. The sorghum was displayed in large enameled bowls. More important, each merchant had a measure, usually an old tin can of variable size. And, to bewilder the foreigner even further, some merchants sold by level canfuls, others by rounded canfuls, and still others by heaping canfuls. In short, here we see Akerlof's problem multiplied: Not only is quality difficult to discern but quantity is equally indeterminate. Markets of this sort are still commonplace today.

But there is yet a fourth constant in addition to quality, quantity, and price that may be variable: currency. We take for granted that $1.00 equals 100 cents because all of the cents and dollars are backed in the same way by the government. But in premodern times, not all dollars were the same, nor were all cents. Coins might be "clipped" to reduce their weight and hence their value. Furthermore, since the intrinsic value of the coins was of considerable importance to their owners, coins that were worn from use had less worth than those that were new. Furthermore, goods might be paid for through a combination of currencies that were only partially commensurable. In ancient times, coins were sometimes made of electrum, an alloy consisting of gold and silver. Thus, they bundled two metals of different and fluctuating values (and of varying proportions not easily discerned) in the coins.

In colonial America, one could find pieces of eight (Spanish dollars literally cut into eight pieces), gold, silver, and copper coins, banknotes issued by various banks, and a variety of other forms of currency. Indeed, until the late nineteenth century it was common to have several different currencies in circulation in the same locale, each associated with different levels of public trust.

Moreover, even when the value of a currency is stable at a point in time, it is possible that it varies over time or space. There are numerous historical instances of currencies inflating at a rapid rate, such that one day's value was considerably different from the next day's. Similarly, there are instances of currencies maintaining their value only within certain fixed geographic areas and having little or no value outside those boundaries. Such currencies are usually denoted as "soft" or nonconvertible; they cannot be exchanged for other currencies.

Even today, in our era of standard currencies, various forms of nonofficial currency exist, the value of which varies with respect to official ones. For example, some residents and businesses of the city of Ithaca, New York, have created Ithaca Hours (2008), a local currency accepted by some nine hundred participants and designed to keep cash circulating within the city. Similarly, citizens of the town of Lewes, England, have created the Lewes

Pound (Transition Town Lewes 2008). In addition, many manufacturers and some retailers issue coupons to consumers. These coupons are a form of unofficial currency that has various restrictions on it, only some of which are found in official currencies.

What all this suggests is that in every monetary transaction there are at least four criteria that may or may not be standardized: price, quantity, quality, and currency.

Table 3.3 illustrates the expansion of Akerlof's thesis from lemons to each of four domains of standards that *make modern markets possible*. Most modern markets have fixed standards for

1. quality (standard products),
2. quantity (standard weights and measures),
3. price (standard fixed prices), and
4. currency (unique standardized currency of fixed value).

Table 3.3
Flexibility of market standards

	Quality	Quantity	Price	Currency	Examples
None vary					Modern market = full comensurability, binary decision
One varies	√				Akerlof = market for lemons
		√			False weights = honesty penalized
			√		Haggling = price varies depending on bargaining ability of buyer/seller
				√	Inflation/deflation, "soft" currencies
Two vary	√	√			
	√		√		
	√			√	
		√	√		
		√		√	Medieval "just price" market
			√	√	
Three vary	√	√	√		Sudanese sorghum market
	√	√		√	
	√		√	√	
		√	√	√	
All vary	√	√	√	√	Ancient market = incommensurable exchange

Furthermore, modern markets have background conditions that make them possible. These conditions include in particular contracts that are legally binding and enforceable (de Soto 2000).

In short, in describing the market for lemons, Akerlof has shown how standards may be used to construct markets of a wide range of types. Our modern markets are somewhat unusual in that they tend toward fixed standards. But maintaining these standards itself involves a wide range of costs—costs that may be distributed in various ways among market participants, governments, and even bystanders.[5]

The rise of standard differentiated products should not be seen as analogous to Akerlof's used-car market. While such unstandardized products and services doubtless continue to exist, they are to be found at the edges of the modern market, in unethical and illegal activities, in craft goods and personalized services, or in products that cannot be standardized. Standardized, differentiated products conform to the quantity, price, and currency criteria noted above, but they have a different relationship to quality than either commodities, which are standardized but all of the same quality, or craft objects and services, which are unstandardized and of differing quality. Specifically, they are produced simultaneously to be standard (when compared to each other) and differentiated in space or time (when compared to other products or services), to *create* a niche targeted at some (larger or smaller) group of persons.

But we might still ask how it is that this move toward the use of standards to differentiate came about. Why is it that starting in the late nineteenth century, we see a growing trend toward the standardized differentiation of things, services, people, and practices? What might have driven this process? While no single item can be claimed to be the cause of this phenomenon, several changes in the situations experienced by people over the last 150 years opened the floodgates to standardized differentiation. These changes are easily summarized: improved transportation, improved communication, transformed packaging, and neoliberal institutional reform. Ironically, it was the advent of three standardizing technologies—standard shipping containers, standard means of communication via the Internet, and standardizing neoliberal technologies of governance—that, together with the differentiation permitted by packaging, provided the impetus for differentiation. Each of these technologies created a host of new opportunities for change, opportunities that have reshaped society globally by creating a far greater range of situations through which cognition is distributed. Each created new recipes for realities. Let us examine each in turn:

Transport

The history of changes in transportation is relatively well known. Prior to the early nineteenth century, transport involved traveling on poorly maintained and usually unpaved roads or sailing on navigable rivers. In much of the world there were tolls on both the highways and rivers, and there was always a considerable risk of brigands, more than happy to rob and steal. Hence, such transportation was costly and slow. Canals became commonplace, especially in Europe and parts of North America, in the late eighteenth century, allowing somewhat faster inland transport. Only with the invention of the railway in the 1830s did rapid intercity transport of people and things become feasible. In contrast, ocean shipping relied entirely on sailing ships until the late nineteenth century. Sailing ships were slow, taking a month or more to cross the ocean. They depended on the vagaries of the wind, and were also subject to piracy. This situation in turn changed with the development of the steamship in the late nineteenth century. The steamship made transatlantic voyages possible in a week or less. Later, at the beginning of the twentieth century, came the (standard) truck and automobile and (standard) paved roads. Finally, soon after the end of World War II, air transport became sufficiently inexpensive as to allow large numbers of people and large amounts of perishable goods to be transported great distances at previously unheard-of speeds.

But it was the lowly shipping container—itself arguably one of the great triumphs of standardization—that eased the transition to standardized differentiation. Up until the 1960s, cargo on trucks, rail cars, and ships was loaded by hand, requiring a small army of persons. Ports in particular tended to be dangerous and unsavory places where there was a constant if irregular need for temporary labor to load and unload the ships (Russell 1966). Since most freight was in the form of odd lots, and the ports were cluttered with crates, boxes, and loose items of various sizes, theft was a major problem. While bulk commodities required little labor to load, everything else required careful placement, usually by hand. Goods had to be balanced such that the ship remained stable on the seas. They had to be tied down so they would not shift in a storm. But today all this is gone, and shipping containers are now ubiquitous. We see them on ships pulling into harbors, on rail cars traveling cross country, on trucks in the cities, and on the open road. One need only look at the names painted on them— Maersk, Hapag-Lloyd, Hanjin, NYK—to realize that they transport parts of the rest of the world to wherever we happen to be.

Standard shipping containers of various sorts had been tried numerous times in efforts to improve and speed up rail and truck transport. But it was a trucking magnate, Malcolm Purcell McLean, who successfully led the development of the containers we now see:[6]

Malcolm McLean's fundamental insight, commonplace today but quite radical in the 1950s, was that the shipping industry's business was moving cargo, not sailing ships. That insight led him to a concept of containerization quite different from anything that had come before. McLean understood that reducing the cost of shipping goods required not just a metal box but an entire new way of handling freight. Every part of the system—ports, ships, cranes, storage facilities, trucks, trains, and the operations of the shippers themselves—would have to change. (Levinson 2006, 53)

Of note, this was simultaneously a technical and a social challenge. It was easy enough to design a standard box that could be loaded on a tractor-trailer and removed at the end of the trip. But to design a container that could also be carried on ships and trains was far more complex. For ships, the containers had to be able to withstand being stacked and jostled around while crossing the high seas. For trains, the containers had to fit through tunnels and under bridges, and they had to be strong enough that applying the brakes to the train did not crush the ends of the container. Moreover, if they were to be used globally, they had to be able to fit the trucks and trains in use in most nations. That meant they had to be small enough to meet the narrower road conditions in Europe and Japan, as well as the older and sometimes narrower-gauge rail lines found in some nations.

In addition, there had to be a standardized, efficient, and safe means for loading and unloading the containers. Finally, ships, trains, trucks, and ports had to be redesigned so that containers could be easily loaded and offloaded. Putting containers in the holds of existing ships was soon found to be impracticable; they left too many unfilled spaces and required complex and expensive handling. The ports were also ill-suited for containers. They were usually located close to metropolitan centers, with factory and warehouse space surrounding them. After all, the backlog of goods waiting to be loaded had to be stored somewhere. Thus, for containers to work, the entire ocean shipping industry—ships and ports—had to be rebuilt. New fittings known as twist locks were designed to be put at the corners of each container. Huge mobile gantry cranes had to be developed to unload trucks and rail cars and to load ships. Port warehouses became redundant since containers could be stacked on the dock if necessary. Changing all of this was a seemingly unthinkable task.

In addition, myriad standards for each of the various components—containers, fittings, cranes, ships—had to be developed that satisfied each of the major players in each of the affected industries. Some companies, such as McLean's Sea-Land, established their own standards. The American Standards Association (now ANSI) established standards, based in part on those of Sea-Land. The International Organization for Standardization (ISO) was also invited into the picture. After much negotiation, by 1965 the wide range of container shapes and sizes was replaced by a standard width (8 feet) and height (9 feet), and a choice of several standard lengths, from 20 feet up. Today, container freight is reported in TEUs—twenty-foot equivalent units. Critically, the standardization of the container meant that containers could now be exchanged, rather than being wedded to a particular company.

However, this paled in comparison to the regulations that had to be modified to make containerization work. Indeed, the shipping industry—road, rail, and ship—was a crazy quilt of rules and regulations, most of which favored the status quo. U.S. government agencies, such as the Interstate Commerce Commission and the Maritime Administration, as well as their counterparts elsewhere, had each to agree to the new method of handling freight. Freight rates, previously based on the nature of the cargo, since different types of cargo were easier or more difficult to load, had to be revised to take into account the advent of the container. This required decades of negotiation, but it was helped along by the logistical mess caused by massive shipments of military supplies and the personal effects of U.S. soldiers and civilians to Vietnam. That mess was rapidly mitigated by the construction of a large container port at Cam Ranh Bay and the use of a punch card system to keep track of each container and its contents. That was sufficient to make the U.S. Department of Defense an ally of containerization. In contrast, in Japan it was the legacy of a strong industrial policy that made the Ministry of Transport urge containerization as a means of increasing efficiency in export trade. Finally, in 1980 most U.S. interstate trucking and rail rates were deregulated; shipping rate regulation was abandoned in 1984.

Longshore and other labor contracts also had to be renegotiated as well, promising no layoffs in return for permitting the proposed technical changes. Almost overnight, longshore labor shifted from backbreaking heavy manual labor to the skilled operation of heavy equipment. Today, more than three quarters of the longshore jobs on the East Coast of the United States that existed before containerization no longer exist. Most of the factories that once had to be close to the ports have vanished, along with the jobs they provided. Huge ships can traverse the oceans with just a handful of crew members.

Today, container ships able to carry 10,000 or more 20-foot containers are commonplace. They are sufficiently expensive that they must be kept constantly in use, stopping only briefly to simultaneously load and offload freight. By the 1970s it became evident to all that reducing handling costs at ports saved more money than ensuring that containers were completely full. Hence, pallets are widely used to insert goods inside containers, whereas initially that had been seen as a waste of interior space (Egyedi 2001).

One limiting factor in size is the width of the Panama Canal (see vignette 5 in the prologue), a limit that the government of Panama intends to eliminate. Many of the older ports, such as that of New York City, have fallen on hard times, as they have been replaced by much larger, more spacious ports that can handle the huge container ships, such as that at Port Elizabeth (figure 3.1). Those nations that built large container ports early on (often with government subsidies of one sort or another) have

Figure 3.1
Shipping containers at Port Elizabeth, New Jersey. Note the gantry cranes on tracks that allow easy movement of containers as well as the lack of warehouses.
Source: Photograph by Capt. Albert E. Theberge (ret.), National Oceanic and Atmospheric Administration, Department of Commerce, 2004.

prospered (e.g., Singapore), while nations with small and inefficient ports have seen their relative costs for both imports and exports soar. Cargo theft has declined worldwide, cutting insurance claims and costs as well. Factories worldwide now wait until they have sufficient goods to fill a container, or pay a considerable premium for shipping in smaller lots. Transoceanic shipping costs now make up only a small percentage of the retail price of goods; hence persons, firms, and countries previously not in competition with each other are very much in competition today. In short, the standardization of global shipping through containerization has in part made possible the growing standardized differentiation that marks the world today. In addition, what was once a revolutionary new technology has now become thoroughly embedded in myriad standards, practices, institutions, and specifications (Egyedi 2001). In short, even as containerization has transformed the world, it has also created a new form of path dependence, one that will be with us for some time to come.

Although each of these transport innovations took decades to perfect, it is hardly an exaggeration to say that together, they transformed the landscape in much of the world. In particular, together these innovations decreased transportation costs, opening the way to the transport not only of a larger volume of people and things but of a greater *variety* of people and things. On the one hand, they opened the world to relatively inexpensive global commerce by shrinking time and space. In the days of Columbus, only the most expensive items could be traded globally—spices and silks. Today, not only expensive luxury goods and stable bulk commodities can be cheaply moved around much of the world but even relatively inexpensive things such as fresh fruits and vegetables can be profitably traded worldwide. On the other hand, these technologies made ocean transport possible not only for the relatively wealthy but also for those much farther down the social ladder. Steamships and later airplanes now carry not only long-term immigrants, but also short-term visitors traveling both for business and pleasure.

Communications

A second set of changes that favored the use of standards to differentiate consisted of improvements in communications technologies. Prior to the advent of print media, communication was restricted to face-to-face communications, handwritten letters, and hand-copied manuscripts. The advent of print media meant that for the first time, it became relatively easy for a scholar to compare multiple accounts of an event or multiple drawings of an animal in the same room. It also meant that books could be produced in multiple copies, and one could have great confidence that all the copies would be the same—that is, they would be standardized. Libraries of considerable size could be created, even by persons of relatively modest means.

Yet even after print media became well established, both handwritten letters and printed items depended on a steady supply of expensive paper. The belles lettres of times past were not merely a literary genre. They were the result of beautifully formed handwriting and carefully phrased sentences, a practice linked to the high value of paper. In fact, until paper from wood pulp was developed in the 1870s, paper prices rose in response to increasing demand and limited supply (Smith 1964). But paper from wood pulp, combined with the standardized technologies of organizational communication discussed in the previous chapter, made it possible to have multiple copies of newspapers and mundane business correspondence and other records in multiple places.

Of course, other means of communication emerged as well, mainly during the nineteenth century: First, early in the century photography was invented, permitting the documentation of phenomena previously available only through texts and drawings. The telegraph was invented in the 1830s and by mid-century permitted nearly instantaneous communication across the Atlantic. It was followed roughly half a century later by the telephone and the radio. Television did for images what radio had done for sound. Each of these advances in communications extended the potential reach of individuals and organizations. As Bruno Latour (1987) suggested some years ago, they permitted action at a distance. But they also permitted more rapid and more nuanced comparisons of one's own situation with that of other persons in other places.

The next big wave of technical change in communications was the development of the digital computer. Charles Babbage began to build the world's first analog computer—what he called a "difference engine"—in the 1820s, but the precision required to build a functioning model with the funds at his disposal was beyond the machinists of the day. However, analog computers in the form of card sorters were already in use for the 1890 U.S. Census. They made possible fairly accurate tabulations that would otherwise have been too laborious to compile.

Digital computers allowed both faster results and more complex calculations. Initially they were like ENIAC, the huge mainframe computer at the University of Pennsylvania completed in 1946 to calculate artillery firing tables for the army. Forty years later computers had become considerably smaller and began to appear on people's desktops; the modern age of computing was born. As the cost of computing declined and the user interface improved, it became apparent that computers could be used to automate work previously done by hand in manufacturing, banking and finance, education, and research.

A modest example underscores what the advent of high-speed computing did. In the 1930s the U.S. Department of Agriculture, at the time the nation's premier public research institution, had one of the largest statistical laboratories in the world. Not only did many of the nation's top statisticians work there for a time but many practical statistical procedures were developed there. This should come as no surprise. On the one hand, agronomists needed to develop means to calculate differences between treatment and control plots for thousands of experiments. On the other hand, the USDA was entrusted with the collection and analysis of data on the production and yield of all farm commodities, the characteristics of farm households, eligibility data for farm programs, and data on the well-being of rural America. Collecting all the data was at best a daunting task; analyzing it was even more difficult.

In the late 1950s, one member of the USDA statistical staff, Richard J. Foote (1958), wrote a manual to help novices with the complex statistical calculations needed to do such things as multiple regression analyses. Foote described what was necessary to perform statistical analyses. First, the problem had to be broken down into smaller pieces, each of which could be worked on by a different person. Then a small army of statistical clerks, each sitting at a desk with a large mechanical calculating machine, would engage in simple mathematical calculations. Each calculation would be done by two persons to ensure accuracy. Finally, a statistician would assemble the results obtained by each clerk into a single equation. The process was lengthy, tedious, costly, and error-prone, such that only very large organizations could engage in it.

The mainframe computer vastly reduced the cost and time required by these tasks, but it was the personal computer and accompanying statistical software that made it possible for even small organizations to perform complex but highly standardized statistical analyses. Thus, a wide array of analytical tasks that previously could be done only by the very largest companies and government agencies now became feasible for small firms.

Of course, the final group of communications technologies that must be added to this list is the Internet. Initially developed in the U.S. Department of Defense as a means of communicating among computers, the Internet has become a means for global communication on an unprece-

Table 3.4
A perspective on communications technologies

	Broadcast	Narrowcast
Centralizing	Print Radio Television	
Decentralizing	Internet	Letter Telephone Telegraph

dented scale. Today, approximately 19 percent of the world's population has access to the Internet, and the figure is growing rapidly (Miniwatts Marketing Group 2007).

To sum up, over the centuries, successive standardized communications technologies have been developed: printing, the telegraph, the telephone, radio, television, computers, and the Internet (table 3.4). They have not supplanted each other; each is still around in considerable abundance. But it should be noted that printing, radio, and television are each means of broadcasting. They are fundamentally centralizing technologies as well. They permit and even encourage the production of standardized, one-way, authoritative communications from some central source to a wide audience. Of course, transmitting printed information is more difficult and slower than doing it by radio or television. But in its day, Thomas Paine's pamphlet, *Common Sense*, stirred up the American colonists and helped to foment a revolution. Orson Welles's 1938 radio broadcast of H. G. Wells's *War of the Worlds* led millions of people to believe the United States was being invaded by Martians. Today it is unlikely that a pamphlet or radio show could have the same effect. But as every actual and would-be tyrant knows, control over the media is a means of limiting, standardizing, and restricting the kind of information people receive.

The telegraph and telephone, though equally standardized in form, are narrowcasting and decentralized technologies. They allow communication between point *A* and point *B*. However, the Internet (including the standardized computer technologies that make it possible) is profoundly different from the other technologies mentioned in that it is simultaneously a broadcast technology *and* a decentralizing technology.[7] It is a technology that, as the late Jacques Derrida (1978) suggested in another context, decenters the center. In principle, anyone with access to a linked computer can post written or visual materials to the Internet. No particular claim to authority need be made. Hence, anyone (for the minimal investment necessary) can become an instant pundit by establishing a blog. Anyone can use the anonymity of the Internet to challenge public or private authority or both—and from virtually any perspective across the political spectrum. Anyone can advertise only to persons who click on web pages about Chihuahuas, chapattis, chili peppers, or chimpanzees. Anyone can sell a bewildering array of goods and services, including many of dubious value. By establishing a page on MySpace, anyone can rapidly amass three thousand "friends." Anyone can even take on a different (perhaps even nonhuman) persona on sites such as *Second Life*.

If you remain unconvinced of the change taking place, consider the number of otherwise intelligent people who have been fooled by various urban legends found on the web. These include messages from Bill Gates offering free merchandise, claims that Barack Obama is a radical Muslim, warnings that various innocuous messages contain computer viruses, or notifications that the U.S. Congress is about to pass legislation that is not even on the docket.

In short, while all the more traditional forms of communication continue to exist, new forms have emerged that permit several interrelated effects: (1) a growing number of differentiated niches for news, entertainment, sports, music, and sales of a vast number of new and used products to a rapidly growing audience; (2) a concomitant growth in the number of content providers for these niches, some commercial, some nonprofit, and some personal; (3) an explosion of electronic forums that are eclipsing face-to-face forums and that provide opportunities for dialog on virtually every facet of social life, from sex to senility, from profits to prophets; and (4) a temporal shift in watching and listening brought on by the growth of new technologies that allows users to do with visual and audio media what has always been possible with print media—- set things aside until one is ready to use them.

Packaging

A third opportunity for standardized differentiation can be found in the growing interest in and importance of packaging over the last century. As Cochoy (2002) notes, packaging solves the problem posed by the story of Buridan's ass. This story, attributed to a medieval French monk, Jean Buridan (ca. 1300–1358), can be summarized as follows: Buridan's ass is hungry and finds itself between two equidistant bales of hay. Unable to make up its mind which bale to eat from, it dies of starvation. We may consider this story to be the flip side of Cochrane's account of the technology treadmill mentioned in the previous chapter. In highly competitive markets for standardized goods and services, the consumer is faced with a choice between products *A, B, C . . . n*, all of which are essentially identical. Of course, most consumers do not die of starvation while deciding which apple to eat. Instead, since the apples are standardized and hence essentially identical, consumers randomly choose one or another and eat it.[8]

But how does packaging resolve the problem posed by Buridan's ass? By differentiating among standardized products. And this is the entire focus of the modern field of marketing. Marketing arose in the early decades of the twentieth century in large part as the application of Taylorism to sales (Cochoy 1998). It received a considerable boost in the United States during the Franklin D. Roosevelt administration with the development of market surveys. By inserting the discipline of marketing at precisely the point where a consumer might hesitate in choosing among essentially identical goods, marketers resolve both the problem of Buridan's ass by presenting one as clearly the most desirable choice, and the treadmill problem posed by Cochrane. Let us examine this important and growing phenomenon more closely.

Cochoy notes that all modern economic exchanges are about the exchange of "packaged" goods and services, yet the packaging is ignored by most schools of mainstream economics. Moreover, packaging simultaneously conceals and reveals. For example, the label on a can of peas *conceals* the peas in the can but *reveals* things about those peas that I could never learn from looking at them (e.g., weight, nutritional value). Furthermore, the package may contain information put there by the seller (e.g., a supermarket), processor (e.g., the brand), and the state (e.g., net weight, nutritional value).

Packaging also underlines the differentiating character of contemporary standards. The packaged item comes in various predefined sizes, shapes, weights, colors, and so on. While I may chose among them, I cannot chose between them. In other words, if peas come in 12-oz. and 24-oz. cans, I cannot buy just 14 oz. of peas; I must settle for 12 oz. or buy a 24-oz. can.

Similarly, disparate things often come packaged together now. You can purchase a refrigerator with a television set and voice recorder embedded in the door. Or perhaps you would like a combination coffee pot and weather station. Melitta actually offers such a product!

But packaging cannot be limited solely to wrappings. For example, if I wish to buy a new car, I must also rely on packaging. Most car dealers will normally allow me to drive a new car for a few miles to test it out, but this

Table 3.5
Some aspects of modern packaging

Physical packages (boxes, cans, wrappings, etc.)
Money-back guarantees
Free or discounted repair services
Manufacturers' claims about product or process on labels
Legally required information on labels
Claims of private voluntary organizations on labels
Owner's manuals
Aesthetics of the sales setting
Certifications
Branding

---

tells me virtually nothing about the behavior of that car after 20,000 miles, or about fuel efficiency, safety, or operating costs. For that I need to consult other forms of packaging: A warranty informs me that for the first 36,000 miles I need not worry about repairs. A sticker on the window provides the results of government-certified fuel efficiency tests and allows me to compare its efficiency with that of other cars. The results of a crash test provide an indicator of safety. An owner's manual tells me how often to change the oil and what the tire pressure should be. Finally, the brand name of the vehicle will indirectly inform me of its quality. Some of the many aspects of modern packaging can be found in table 3.5.

Of course, I could open the can of peas to look at them, smell them, or taste one. But I would probably pay a penalty worse than that of Charlie Chaplin (see figure 1.3). After all, in most jurisdictions the law prohibits me from opening the package before purchase. In this and other cases, the very act of use destroys the thing that one wishes to purchase. For example, testing a tissue by using it to blow my nose destroys the tissue. Similarly, lighting a fire cracker to hear the sound and watch the light it produces destroys it.

Moreover, in premodern markets one went to the market as much for conversation as for goods, and would therefore spend time chatting with friends and acquaintances. In markets for standardized goods there is no reason to ponder, but there is also no reason to choose one good or service over another. Today, especially for most small purchases, the act of choosing is not something on which one wishes to dwell. Nor do retailers want you to ponder the purchase, since their incomes depend more on the speed of turnover, or throughput, than on the profit per item. Under these conditions, brands, logos, and certifications (discussed in chapter 4) become convenient means for summarizing a plethora of qualities of a product and allowing rapid decision making. In some instances, as in the case of most household chemicals, they even obscure the actual contents of the package (Haug 1986).

At the limit, retail stores and offices themselves become part of the packaging. Automobile showrooms, hypermarkets, and department stores are remodeled every so often so as to maintain an up-to-date, clean, aesthetically pleasing look. Supermarket executives put special lighting in the produce sections so as to make the items there sparkle. Lawyers and accountants often give their offices a 'substantial' serious look. Banks create vast marble lobbies that serve few functions other than to impress the potential client with the gravity and longevity of the bank.

These efforts at packaging also emphasize the networked character of goods and services. As Cochoy (2002, 91, my translation) suggests, "the ideal product only exists effectively across a network that defines it, that constructs and maintains it as such, through instances of standardization, the technical services of ministries, metrological engineers, testing laboratories, certifying organizations, representatives from industry and consumer associations." Hence, packaging is unavoidable in contemporary society. In the act of exchange, there is no way we can get to "the things themselves."

But what holds for things in economic exchange also holds for humans in both economic and other forms of exchange. Today, at least in much of the world, people are packaged. Management guru Tom Peters (1997) provides advice about "the brand called you." People do not merely live their lives; they have "lifestyles."[9] Of course, as Mark Twain once famously quipped, "Clothes make the man. Naked people have little or no influence on society." Indeed, people have always used their bodies, their dress, their food, and the things they surround themselves with to mark gender, class, and status differences. However, until recently they did not self-consciously package or market themselves. This appears to have arisen in a manner coincident with modern marketing.

Clothing is a good starting point. For centuries, people in the Western world wore their Sunday best on that day of the week. Clothing, including jewelry and other ornamentation, has been a mark of class, status, and religion for centuries as well. But today clothing is often used to reveal or conceal many other aspects of oneself. For example, millions of people wear t-shirts that proclaim their (perhaps dubious) allegiance to brands, companies, sports teams, clubs, or universities. Others wear shirts with flippant remarks or political slogans emblazoned on them, or mottoes proclaiming their beliefs, allegiances, or group membership. For example, one might wear hip-hop clothing, a Rolex watch, or Mudd jeans or carry a Gucci bag to claim allegiance to a certain status group. Magazines, television shows, and more recently websites have sprouted up to advise us on the right way to package ourselves to get the job or to meet mister or ms right. (Some youngsters take this very seriously: fights over high-end clothes in high schools are hardly unknown.) Of course, to this we might add hats, hairstyles, shoes, jewelry, and other accoutrements. Nor should we exclude the myriad pins, buttons, stickers, and other items that proclaim the wearers' various allegiances. Perhaps all that is missing is one for the International Society of Professional Button Wearers.

Clothing is also designed to cover (and sometimes to reveal) bodies. Like clothing, bodies have been used to distinguish people by gender, class, and status for as long as human societies have been around. But for the most part, one was born into a particular body and expected to accept that as a given. Indeed, in the press of producing one's daily subsistence, there was little time to spend considering body shape. Bodily markings appropriate to one's age, gender, class, or status, such as painting, tattoos, and scarification, were accepted largely as a matter of course. In the Western world, during much of the last several centuries, even such markings were frowned upon as remnants of a bygone age. As late as the 1950s there were relatively few persons in Western nations who self-consciously altered their bodies, and the few who did were often held in contempt by others.

But today bodily alteration, decoration, manipulation, and modification are once again acceptable to a considerable portion of the population. Once, sex and gender were seen as immutable. Today, sex change operations are routinely performed, and a significant transgendered culture has arisen to support those who undergo such procedures. Persons born hermaphrodites were once considered in need of immediate gender reassignment; today, hermaphrodites have supporters and support groups (Dreger 1998). Whereas body weight was once considered given by nature or a function of wealth, dieting in an attempt to reduce one's weight (usually to a medically defined normal weight) is commonplace in much of the world. And for those not wishing to diet, there are support groups for the overweight as well as for the underweight. Furthermore, people engage in group exercise not only to lose weight but to build bodies of various sorts— the svelte dancer, the muscular body builder, the athlete.

Of course, people must eat to survive. Here too we can see the rise of standardized differentiation. Just a generation or two ago most of us ate what our parents had eaten. Clearly defined dietary patterns could be found that distinguished ethnic groups, classes, and regions. Only the very wealthy got to eat exotic foods, and even their diet was in most instances considerably more limited than a middle-class diet is today. However, over the last half century we have seen extraordinary differentiation in the kinds of foods eaten by a significant segment of the population.

Today, for most of us in the Western world, with the exception of the very poor, and for a growing portion of persons even in poor nations, food choice is a matter of lifestyle. Hence, in the United States about half of all meals are eaten in restaurants. Some people adore the fare at McDonald's, while others would not be caught dead eating there. Not only has there been an explosion in the variety of "ethnic" restaurants but various forms of fusion cooking have emerged in which dishes traditional to several ethnic groups are mixed together to form a meal. Similarly, a vast portion of the world's population today can decide among thousands of different fresh and packaged goods in supermarkets and hypermarkets. (That said, even I was amazed to find prepared haggis samosas in a store in Edinburgh!) Here, for better or for worse, seasonality has evaporated. If I wish to eat peaches or a mixed salad any time of the year, I can do so rather easily.

Furthermore, thanks to the packaging of food products, I can choose what makes up my diet not solely by the names of the foods I eat but by determining their nutritional value. I can eat a low-fat diet or I can just avoid transfats. I can eat only organic foods, fair-trade foods, or foods grown locally. I can limit myself to whole grains. Or I can subsist on packaged, frozen entrées, either of the high-fat or of the diet sort. I can buy coffee at Starbucks and choose among 19,000 variations (Anderson 2006).

What is true of food is equally true of our cars, our homes, and all the myriad things we purchase to put in them. Each of these now forms part of our lifestyles. Each proclaims various things about us. They tell not only about our income and wealth but about what we care for. Each of these products becomes part of the packaging that identifies who we each are.

However, packaging the self doesn't end there. In more and more parts of the world, each of us carries around a small but growing collection of identifying chits: driver's licenses, credit and debit cards, identity cards, passports; cards of various sorts that mark us as people who frequent various restaurants, bookstores, cafes, airlines, bus and train services, who have accounts at certain banks, who frequent certain libraries, who go to gyms. Those cards package our identity, proclaiming to the world those aspects of us that we deem important in certain situations. Indeed, it is now possible to have one's identity stolen. In the past this required some sort of mystical incantation; in today's world it can be accomplished by hacking into a computer.

Moreover, we are also packaged by documents that reveal the results of educational, medical, and other tests we take. Hence, when I visit my U.S. physician, he pulls out a sheaf of paper that documents who I am. When I apply for a job, I must provide evidence of my citizenship, my education, and my previous employers. All of this forms yet another layer of packaging.

In addition, a growing portion of the population now defines itself through formal résumés or curricula vitae that identify the holder as an academic (we tend to have very long ones), a business person (who tend to have short ones), a professional of one sort or another, or merely a job applicant. Various services provide coaching to help one produce the ideal résumé, one that will show the person in the best light. There is even a Professional Association of Résumé Writers!

In short, this rather vast and growing repertoire of packages that reveals and conceals who we humans are mirrors the packaging of nonhumans. Put differently, just as the packaging of nonhumans is about increasing standardized differentiation, so too is the packaging of humans. To be effective, human packaging must simultaneously be about standardization *and* differentiation. It must show at the same time that I am different from all those others out there and that I am a member of this or that group. If it dissolves into mere idiosyncrasies, then it defines us as unique. That destroys the link between us and others that is central to the packaging. It truly leaves us naked to the world—a throwback to a bygone age, an eccentric, an oddball, a crank, or worse: crazy, insane. And here Mark Twain's dictum clearly applies.

The Neoliberal Reforms

But if improved transportation and communication, as well as new forms of packaging, were necessary conditions for the rise of a new society based in part on standardized differentiation, there was at least one more essential ingredient. This was the rise of what has come to be known as neoliberalism. We all know who the old liberals were—John Locke, Adam Smith, David Hume, Charles de Montesquieu, Jean-Jacques Rousseau. The key tenet that held all of these somewhat different views together was a strong belief in the importance of individual liberty. Hence, they were also united in a belief in the need to delimit the powers of the state and clearly separate it from the market. Both the American and the French revolutionaries were strongly influenced by their views.

In contrast, neoliberalism is a phenomenon that can be dated rather precisely to the 1930s.[10] For our purposes, three events are of particular importance: (1) the publication of an essay by Henry C. Simons in 1934, (2) the publication of Walter Lippmann's *The Good Society* in 1937, and (3) the Colloque Walter Lippmann held in Paris in August 1938. Let's begin with Simons's essay.

Henry C. Simons (1889–1946), an economist at the University of Chicago, was a strong critic of New Deal policies. In 1934, reflecting on the miseries of the Great Depression, by then ongoing for several years, Simons wrote a curious pamphlet with the seemingly oxymoronic title, *A Positive Program for Laissez Faire: Some Proposals for a Liberal Economic Policy*. Critiquing the national planning then very much in vogue, Simons ([1934] 1948, 42) argued for an activist state that would implement a five-point program:

1. The elimination of monopolies.
2. More definite and adequate rules of the game with respect to money.
3. An overhaul of the tax system to make it more progressive.
4. The withdrawal of all subsidies and protective tariffs.
5. Limitations on advertising.

Moreover, Simons insisted that those services not readily susceptible to competitive markets be nationalized, and that corporate power be limited (by federal chartering, prohibiting companies to own stock in each other, and otherwise limiting concentration of corporate power). Finally, he suggested that nonprofit organizations and the Bureau of Standards provide certifications for the quality of all sorts of goods. In sum, unlike the classical liberals, who demanded that the state be barred from the marketplace, Simons argued for a "positive program" whereby the state would intervene to provide a legal framework for the market and defend it from its enemies.

The second significant event was the publication of *The Good Society*, by Walter Lippmann (1938), then already a well-known columnist for the *New York Herald Tribune*. In that book Lippmann voiced his growing concern over both the Nazi and Soviet states, as well as over the excesses of the New Deal. In particular, Lippmann argued that the complexities of modern societies demanded that one reject central planning, which would always fail to capture that complexity; after all, complete knowledge was beyond mere mortals. In contrast, greater adherence to a revised classical liberal philosophy that allowed myriad market transactions would permit and guide changes in the ever more complex division of labor.

But Lippmann also worried that since about 1870, liberalism had been in decline. He traced that decline largely to blind adherence to the doctrine of laissez-faire. As he put it, "Liberal thinking was inhibited in the metaphysics of laissez-faire, and the effect was to make the political philosophy of liberalism a grand negation, a general non possumus [we cannot], and a complacent defense of the dominant classes" (Lippmann 1938, 203). This placed legal rights on a "superhuman foundation" barring inquiries into what laws would best promote justice. In contrast, Lippmann argued that the entire edifice of the market was built on law. Limited liability corporations—by 1900 the dominant form of industrial organization—owed their existence to laws allowing them to come into being and to act (only) in certain ways. Without those laws of property and contract, among others, and their adequate enforcement, such organizations, free markets, and the complex division of labor that they enable would simply not exist. Indeed, for Lippmann (1938, 306),

the only essential difference between a large business corporation and a public school is that the business corporation is operated for profit to meet the demands of the market, whereas the public school is supported by taxes to meet a recognized social need which does not have a market price. But even this distinction is tenuous. For there are many business enterprises which would have to be perpetuated even if they were run at a loss. We can see that when the government is impelled to grant subsidies. And there are many public enterprises which are wholly or partially self-sustaining. So both the business corporation and the social service are in effect agencies created in order to conduct certain collective affairs of a community.

Moreover, liberalism needed to be disentangled from the privileges written into eighteenth- and nineteenth-century common law by new statutes that would limit coercive authority, promote fair treatment, support collective goods and services, and protect those who were weak. In short, for liberalism to regain the high ground over "tyrants and exploiters," it needed to be reformed. Liberalism had to become an active philosophy specifically focused on societal transformation through legal reform (cf. Dewey [1937] 1987). Lippmann's volume was one among many written at the time urging a rethinking of liberal philosophy. But apparently it resonated round the world.

The other major event was an invitational colloquium organized in Paris by Louis Rougier (1889–1982). Rougier, professor of philosophy at the University of Besançon, wrote widely on geometry, physics, and logic, as well as on political philosophy. Inspired by Lippmann's work, which he had had translated into French, he organized the symposium on the eve of World War II and had the proceedings and deliberations published the following year (*Travaux du Centre International d'Etudes Pour la Rénovation du Libéralisme* 1939).

Rougier invited a number of leading advocates of liberalism to the colloquium, including Michael Polanyi, Ludwig von Mises, F. A. Hayek, Jacques Rueff, Wilhelm Rüpke, Alexander Rüstow, and Lippmann himself; several bankers, lawyers, and industrialists; and two sociologists, Raymond Aron and Alfred Schütz. Rougier (1939) opened the colloquium by praising Lippmann for (1) showing that liberal regimes do not emerge spontaneously, but require juridical intervention by the state, (2) showing that Nazism and socialism were two varieties of the same phenomenon, and (3) so clearly posing the key questions of the day. He challenged his guests to provide a "Road Code" for the market. The burning issue was what forms of intervention were compatible with a price system and the laws of the market, what forms would bring the masses security while maintaining free thought. In giving his charge to the group, Rougier proposed that the new perspective be called "neoliberalism." The congregants' self-appointed task was none other than to save Western civilization from war and oppression.

After an introductory talk by Lippmann in which he summarized his recent work, the next several days were spent in debate over what needed to be done. Liberalism was faulted for becoming merely a matter of conformism to outmoded doctrine. Economic concentration, it was agreed, was not due to any tendency of capitalism, as proposed by Marxists, but to protectionism by the state. In any case, von Mises (1939) asserted, monopolies were not problematic unless they engaged in monopoly pricing. Rüstow (1939) insisted that a strong and independent state was needed to ensure the "exact observation" of market conditions. Lippmann (1939) observed that central to all of this would be the maintenance of a competitive price system that would ensure the best use of the means of production. At the same time, state funds would be used to support national defense, education, social services, and scientific research; the proportion of state funds to be used for these ends would be decided by democratic means. At the end of the colloquium it was agreed that the International Center for Studies for the Renewal of Liberalism would be formed.

The war intervened. Those who had assembled in Paris in 1939 and who survived the war were scattered. Yet in the last years of the war, while on the faculty at the London School of Economics, F. A. Hayek ([1944] 2007) published an extraordinarily influential book, *The Road to Serfdom*. He also outlined the case for neoliberalism as an alternative to the socialist aspirations of many in Britain. From his perspective, which was doubtless influenced by the horrors of the war, any form of central planning by the state, including the pursuit of Keynesian policies, would inevitably lead down the slippery slope to something tantamount to Nazism, fascism, or Soviet-style communism.[11] At the same time, Hayek was clear: planning was problematic only if it acted to repress competition. In contrast, planning to promote competition was not merely to be tolerated; it was to be actively promoted, since for him, competition was a noncoercive means of coordination. The key for Hayek was to use the rule of law as a means to provide stable bounds to individual decisions. Unlike Lippmann and some of the other participants in the prewar colloquium, Hayek was nearly as suspicious of any form of concerted action not aimed at competition as he was of state power. Indeed, Hayek concluded his book by arguing for an international authority charged with limiting the powers of individual states.

After the war, at Hayek's instigation, Rougier's project was reassembled as what became known as the Mont Pelerin Society (2006). Since that time neoliberalism has been associated especially with Hayek (1973, 1976, 1979), and with Milton Friedman (1962).

This is not the place to enter into a detailed discussion of the various views held by those within neoliberal circles. Suffice it to say that, as with all schools of thought, they were not of one mind, though they were (rightly) appalled by the excesses of Nazism and Soviet-style communism. They were greatly concerned that, if left unchecked, even democratic states would gradually sink into some form of totalitarianism. Moreover, although they admired the classical liberals, they felt that the excesses of the state committed in the first half of the twentieth century demanded a remedy that went far beyond that form of liberalism. In particular, they were concerned that even in democratic societies, the state was taking on greater and greater power and authority.

Although, like any complex school of thought, neoliberalism means different things to different people, the prescriptions of contemporary neoliberalism of concern to us here can be summed up as follows:[12]

1. *Markets have the virtue of permitting goods and services to be distributed without recourse to a central authority; therefore, they should be actively promoted.* The pretense of eventually arriving at complete, absolute knowledge should be abandoned, as it is now apparent that our knowledge is, and will always remain, finite. Moreover, as society becomes more and more

complex, the difficulties involved in having even a glimpse of how it is organized, or what needs and wants people might have, multiply. Hence, no state, no matter how well governed, can distribute goods as well as the market can.

2. *Since markets will only permit the creation of prices that balance supply with demand when they are* made *fully competitive, when they conform to the rules of a formal logical or mathematical model, the social order must be reorganized such that, in so far as possible, it meets the truth conditions established by the formal model.* Put differently, while governments would need to be omnipotent in order to engage in central planning (or effective planning of any sort), the "self-generating or spontaneous order"[13] provided by the market—a concept that Hayek (1973) appears to have made up out of whole cloth—provides an alternative. However, this spontaneous order, far from emerging by virtue of laissez-faire polices, that is, by leaving the market to its own devices, instead requires a helping hand in the form of an adequately enforced legal structure. Yet that legal structure can only be deduced by virtue of its conformity to a formal, logical, even mathematical model of the market.[14]

Moreover, claim the neoliberals, conformity to the rules of the model maximizes the freedom of the individual and ensures peace. Failure to do so leads to oppression and eventually to totalitarianism.[15]

3. *Governmental power to regulate economic, social, and political affairs other than through support of the market mechanism should be sharply curtailed.* For example, Hayek pays due respect to the constitutional division of powers among the legislative, executive, and judicial functions of the state as proposed by Montesquieu in his *Spirit of the Laws.* But he argues that Montesquieu did not go far enough and that, as a result, "Governments everywhere have obtained by constitutional means powers which those men had meant to deny them. The first attempt to secure individual liberty by constitutions has evidently failed" (Hayek 1973, 1). As such, he strongly believed that Montesquieu's project had to be modified by redesigning national governments in such as way as to force democratically elected governments to restrict themselves to a limited range of actions: raising funds for policing, foreign policy, defense, education, protection from epidemics and natural disasters, provision of standards of measure, and other services that cannot be provided for adequately by the market. As Foucault (2008, 137) explained, for the neoliberals, "there is no need to intervene directly in the economic process, since the economic process, as the bearer in itself of a regulatory structure in the form of competition, will never go wrong if it is allowed to function fully."

4. *Previously public services and enterprises should be privatized.* While certain services have to be paid for by the state, since the market is not able to create an adequate demand for them, there is no reason for these services to be provided by government. They might easily be provided instead by the private sector. Friedman's (1998) proposal to substitute school vouchers for public schools, enthusiastically endorsed by Hayek (1979), was an example of the appropriate role for government to play. Similarly, the construction and maintenance of all sorts of public works and the management of prisons, postal services, railroads, and other services should be delegated to private firms through a system of competitive bidding.

5. *Each individual should be encouraged to become an entrepreneur of him- or herself.* Whenever possible, collective solutions to public problems should be avoided. In contrast to the public provision of public goods and services, the private provision of private goods and services should be substituted— emphasizing the centrality of choice. Thus, rather than a common pension fund, there should be a wide variety of individual accounts among which to choose. Instead of publicly provided health insurance, there should be competing individual plans. Instead of public education, there should be a vast array of private alternatives among which to choose. Instead of any and every public service, its individual equivalent. Hence, each individual is to take on the burden of his or her fate. Each individual is to be atomized. Each individual must learn to manage risks, to calculate, to accumulate and invest his or her own human capital, to be an entrepreneur in the literal sense of the term: one who undertakes. Marriage, divorce, employment, migration, even crime are to be the purview of atomized individuals strategically weighing risks and benefits before acting. Thus, for the neoliberals, the rational economic actor is, for all intents and purposes, also the (and perhaps the only) responsible moral actor. "It is a matter of making the market, competition, and so the enterprise, into what could be called the formative power of society" (Foucault 2008, 148).

6. *Free trade should be promoted, with concomitant competitive exchange rates and market determination of prices.* If the role of the state is to be reduced, then the role of the market is to be enhanced. Or, perhaps better, the roles of state and market should be reversed: Instead of the state letting the invisible hand of the market do its work inside the confines of the market, the state should serve solely to extend the market to as many forms of human endeavor as possible. Indeed, from the neoliberal perspective the market is hamstrung by myriad outdated laws and regulations, tariffs and quotas, subsidies, taxes, and government-mandated prices. If the market is to work as Smith and others envisioned, then not only must these be

eliminated but the state must actively promote competition both within and outside its borders.

7. *Global institutions to limit state sovereignty and protect the market should be established.* Only by creating global institutions with clear and limited mandates, built on market principles, will the tendency toward the growth of the state be curtailed. These institutions would form a kind of global constitution, permanently limiting the powers of legislatures to intervene in the marketplace.

Were these prescriptions confined merely to developing an interesting theoretical position, neoliberalism would long since have been consigned to the stacks in academic libraries. But, quite to the contrary, the neoliberals proved effective at putting their theory into action. The late Pierre Bourdieu (1998) put it correctly when he noted, "In the name of this scientific programme, converted into a plan of political action, an immense political project is underway, although its status as such is denied because it appears to be purely negative. This project aims to create the conditions under which the 'theory' can be realised and can function: a programme of the methodical destruction of collectives."

This is perhaps nowhere more apparent than in the transformed role of the state. Until the recent financial collapse, issues of social justice and equality were pushed to the side. Nation-states have subcontracted various aspects of their operations to private, non-state entities; hence, prisons, postal services, military logistics chains, and myriad other functions previously reserved to the state are now often delegated to private companies. Large private security forces are now commonplace; in the United States, for example, the number of security guards exceeds the number of police officers (see Department of Justice 2005; Parfomak 2004). Moreover, the number of troops in Iraq is considerably smaller than the number of private security personnel. While states have hardly deregulated, they have *re*regulated, and they have done so in ways that promote standardized differentiation.

In short, the neoliberal prescriptions, when combined with changes in the material conditions of transportation and communications, as well as the packaging of people, practices, processes, and things, provide the impetus for standards formation, maintenance, and enforcement. But it would be a grave error to attribute the recent interest in and aggressive pursuit of standardized differentiation to the neoliberal prescriptions themselves. As I explain later in this book, this extraordinary proliferation is better understood as a reaction to the lacunae, to the *horror vacui* created by the increasingly circumscribed role of the state.

Standards and the Long Tail

Together, these transformations in transportation, communication, marketing, and liberalism have combined to produce what Chris Anderson (2006) has labeled "the long tail." Anderson's concept is borrowed from statistical analysis. It can be clarified by considering the distribution of the human population by age. Clearly, there is no one younger than zero years of age, but there are many people over age seventy. The numbers diminish for age cohorts over eighty, ninety, one hundred, or one hundred and ten years (and a very small number of people are even older than that). When graphed, the tail on the right-hand side of the distribution for the advancing age cohorts is much longer than the rather short one on the left-hand side. The same kind of distributional curve could be illustrated for sales of various DVDs, or virtually any other differentiated product.

Anderson argues that prior to the changes noted above, it was difficult for producers of goods or services to market items far out on the tail of the distribution. He offers the case of books. The selection of available books in bookstores was and remains considerably bigger in large cities than in small towns. By the late nineteenth century, with improved transportation and communications, it was possible to order a book by mail, but of course one first needed to know that it existed. It also became possible to advertise books for sale—to package them in enticing blurbs or pictures suggesting the pleasures wrapped within. More recently, the Internet has made it possible for companies such as Amazon or Barnes and Noble to offer millions of books for sale to persons scattered around the world. Furthermore, nearly a third of Amazon's book sales come from books not found in the largest of bookstores (Anderson 2006).

Anderson is of two minds when it comes to how people fit into all of this. On the one hand, he asserts that this explosion of market niches, of what I call standardized differentiation, unleashes the "natural" demand that was out there all along. From this perspective, it is merely that bottlenecks in transportation, communication, and knowledge have until recently blocked the essentially infinite variety of human preferences. On the other hand, he shows how viral marketing, Internet search engines, user ratings, MySpace friends, and other social networks function to shift people's preferences away from the masses of the past to "thousands of cultural tribes of interest" (Anderson 2006, 184).

But a third view might be more helpful here: It is far more likely that the availability of a vast variety of goods and services *and* of various social networks and media outlets by which we can find out about them together *create* the demand. If this were not the case, the massive amount of money, time, creativity, and technology devoted to marketing would be easily replaced by simply providing people with raw, undigested, unpackaged knowledge about things. Nor would it be necessary or even desirable for anyone to find out what someone else was doing, buying, playing, eating, or consuming. This is clearly not the case. For better or for worse, we are collectively creating a future that consists of standardized differentiation. In the next section I consider just a few of the changes afoot, including new social movements, alternative medicine, and shifts in the provision of food and agriculture, education, and technoscience, as well as changes in the landscape of standards organizations.

Mass Customization, Slivercasting, Massclusivity, and Other Oxymoronic Truths

New Social Movements

The older social movements required substantial time commitments of members. In contrast, the new social movements that began to take shape in the late twentieth century—those concerned with the environment, consumerism, fair trade, animal welfare, and so forth—have a very different form. Most rely largely on the receipt of funds from members, which are used to hire and maintain a paid staff that promotes the aims of the movement. Hence, movement members need not engage in any kind of solidarity or collective behavior beyond contributing to the organization's coffers. This permits and encourages a very different kind of relationship between organization and members than the older movements. Indeed, unlike older movements, participants in new social movements need not meet other participants. One need not think much about the movement except during fundraising drives. The result is that members—donors—see themselves as expressing commitment through a monetary contribution. Their commitment is entirely individual; they need share nothing else in common with other members. Furthermore, a given individual may belong to a wide variety of social movements, even ones with apparently contradictory goals, without experiencing a problem with those contradictions. One certainly need not act in a manner consonant with the movement, as one had to in the older movements. For example, it is possible to own a gas-guzzling Hummer and belong to the Sierra Club, to shop for goods at the very lowest price and support fair trade, to support human rights and buy goods made in sweatshops.

Alternative and Personalized Medicine

While the early twentieth century saw the standardization of medical practice as allopathic medicine, which nearly squeezed out all other forms of treatment, the last several decades have seen the rise of a host of alternative treatments and therapies. Such treatments range from traditional non-Western forms of medicine such as acupuncture and Ayurvedic medicine to revived and transformed Western approaches such as holistic medicine, faith healing, and herbal medicine. Proponents of alternative medicine have been sufficiently vocal in their demands as to compel the U.S. National Institutes of Health in 1999 to create a National Center for Complementary and Alternative Medicine (NCCAM). The NCCAM does not, however, take an "anything goes" perspective. Instead, the mission of the NCCAM is to:

• Explore complementary and alternative healing practices in the context of rigorous science.
• Train complementary and alternative medicine researchers.
• Disseminate authoritative information to the public and professionals .

In short, the NCCAM is all about standardizing, purifying, codifying, and testing the widely varied forms of and claims for alternative medicine. Moreover, according to a large national survey conducted a decade ago, nearly half of Americans visited an alternative medicine practitioner in 1998, up significantly from earlier in the decade (Eisenberg et al. 1998). Likely that percentage has since increased further since then. And many of these alternative healers are themselves certified as such. Hence, one can become a certified "energy healer," "homeopath," or massage therapist, among other professions.

Furthermore, new forms of social movements have emerged around health care. These include groups that have formed around specific diseases as well as advocates of particular kinds of treatments. They often challenge the medical status quo, as their membership includes activists who work directly with medical practitioners and researchers, and they make the biological body (i.e., a link between the disease experienced and social inequalities) central to their agenda. Each of these groups attempts to create new standards of treatment and care even as it further divides and differentiates the pursuit of health (Brown et al. 2004). This "niche standardization," as Steven Epstein (2007) suggests, has had mixed results. On the one hand, greater attention is paid to various previously neglected subpopulations (e.g., women, children, African Americans). On the other hand, there is a tendency to employ the same categories rubber-stamp-like to all medical research, regardless of whether they appear to apply.

The story hardly ends there. In addition to the various forms of alternative medicine, recent years have seen at least the promise of "personalized medicine," calibrated to an individual's genetic makeup[16] A recent paper in *Science* explains that "[u]nder a personalized medicine scheme, drug prescribing and dosing no longer would be 'one size fits all' but would be carefully tailored to a patient's individual genetic variants" (Katsanis, Javitt, and Hudson 2008, 53).

For such an approach to be effective, patients would need to submit to genetic testing designed to determine whether a particular drug or therapy would be effective for someone whose genetic makeup classifies them as part of a particular subgroup. At the moment, as the article notes, the entire field is in considerable disarray owing to weak laboratory regulation,[17] as well as a lack of drugs of this sort. Moreover, personalized medicine based on a patient's exact genetic makeup raises questions about cost as well as privacy. Should the test results be revealed to insurance companies or employers? But we can leave these questions to the side here. What must be emphasized is that both alternative and personalized medicine promise ever finer divisions of the population, ever more narrowly defined categories approaching but never reaching individuation. And they contribute to the ever-increasing range of choices we must make.

Food and Agriculture

After more than a century of obsession with increased production and productivity so as to provide cheap food as the sole goal for agriculture, in recent years we have seen considerable interest in a range of alternatives. These alternatives include a variety of production methods such as organic farming, biodynamic agriculture, "beyond organic," and the like. They also include more choices with respect to consumption, such as (1) convenience foods, (2) an increase in the range of fresh fruits and vegetables (both species and varieties) available to consumers (in both industrial nations, but more and more in poor nations as well), (3) greater concern for both process (e.g., fair trade, free-range, environmentally friendly) and product (e.g., pesticide-free) standards, (4) a rapid rise in consumption of health foods and "natural" supplements, and (5) the promise of nutrigenomics (e.g., Afman and Müller 2006).

As with medicine, there is both differentiation of various sorts and the promise to tailor foods (or the nutrients in them) to the genomic profile of each particular eater. The new field of nutrigenomics promises to improve our diets by examining the interaction between particular nutrients and various aspects of human health. Thus, for example, diets might be personalized either to avoid certain health problems that are of dietary origin and are specific to persons with particular molecular traits or to enhance health in an analogous manner. As with personalized medicine, numerous economic, legal, and ethical issues need to be resolved (Chadwick 2004). But again, there is also the division of the population into ever smaller groups, as well as an expansion of the range of choices we must make.

Education

The lock-step system of public education, in which students move through a series of public schools from kindergarten through elementary, middle, and high school, then on to college, and in which the curriculum is dictated by the state, is being differentiated as well. The use of school vouchers, charter schools, and similar programs has encouraged the promotion of standardized differentiation in schooling by allowing the use of public funds to create alternative schools of various sorts. Indeed, many proponents of vouchers see themselves as promoting a kind of Darwinian competition in which parental choice will ultimately improve all surviving schools (Friedman 1998). Regardless of the validity of this claim, differentiation is the order of the day.

At the elementary and to a lesser extent the secondary level, homeschooling is growing. Proponents of homeschooling see it as a means of protecting children from disturbing events at schools, as well as providing an individually tailored curriculum. Opponents see it as second rate and inadequate. Although some persons have been homeschooled over the last several centuries, in large part this was a privilege reserved for the wealthy. In contrast, today an entire industry of (standardized) texts and materials has grown up around homeschooling. And in the United States, the number of homeschooled children is growing fast. In 2007, the latest year for which data are available, 1.5 million children were being schooled at home; in 1999 that number was estimated at 850,000 (National Center for Education Statistics 2010).

In higher education we have also seen the rise of private, for-profit universities, such as the University of Phoenix, which make no claim of producing well-rounded citizens but rather "make higher education highly accessible for working students" (University of Phoenix 2007). To do this, they offer degree and certificate programs at many locations and over the Web. Most programs are vocational or professional in character. They note that they "develop . . . curriculum based on the current needs of industry." To do that they employ a largely part-time faculty.

Finally, online education, especially at the university level, is transforming and differentiating education. Educational television never fulfilled its promise because of extremely high production costs, but extant software and growing Internet access are multiplying the diversity of providers, programs, courses, and curricula, even as online education depends increasingly on standardized computer protocols, software, and hardware. Again, how effective such courses are remains highly contested, but the proliferation of programs and the growing diversity in content and quality of the courses are not in doubt.

Technoscience

Technoscience has not been immune to the tendency toward differentiation. Once nearly unchallenged in its authority –virtually the Church of the modern age—science has recently seen a multitude of challenges. As Ulrich Beck (1992) has suggested, much of this has been due to the failures of success. After all, it was the success of technoscience in developing a nuclear bomb that created the anxieties of the nuclear age. It was the success in eradicating insect pest populations that raised environmental and health concerns. It was the success in improving personal transportation that created massive traffic jams and increased environmental pollution. It was the success in developing new pharmaceutical technologies that generated problems of drug interactions and spawned the entire field of iatrogenic medicine.

Engineers and especially scientists have been largely unprepared for the onslaught. On the one hand, most of them have been ardent believers in the alleged autonomy of science, as well as in its claims to be the sole source of truth. On the other hand, they have been all too quick to accept laudatory praise for popularly accepted technologies emerging from technoscientific research even while claiming a sharp distinction between science and technology whenever social, health, environmental, or other problems were encountered. Perhaps the very process of scientific training, the enormous attention paid to detail and method, has blinded many scientists to the negative consequences of some of their work. Or perhaps, like the scientists in Jonathan Swift's *Gulliver's Travels* ([1726], 1947), they truly believe that the disarray around them is only temporary and will soon give way to a golden age.

But regardless of the source of the problem, in recent years the authority of science has been challenged. The challenges have come from the creationism and intelligent design movement, from those opposed to the use of stem cells in biological and medical research, from those opposed to the use of amniocentesis to determine sex for the purpose of aborting female fetuses. They have also come from philosophers and social scientists in the emergent field of science studies, who have—with more or less success—challenged the special status previously granted to science and scientific knowledge. Science studies scholars (e.g., Knorr Cetina 2003; Latour 2005), arguably more than others, have also shown how scientific knowledge emerges as a social process not radically different from other social processes, how science is fundamentally political even when it is apolitical (the politics of no politics), how different sciences employ quite different methods to obtain knowledge (rather than one true and right method), and how the truths of science are always the best we have at the moment, always subject to future revision. Some have misinterpreted this critique as a misguided form of relativism, but the black box of scientific knowledge production has been opened and will not easily be closed again.

Diversity

The contemporary celebration of human and ethnic diversity, especially in the United States, is yet another example of the trend toward standardized differentiation. If the "melting pot" imagery was gospel at the beginning of the twentieth century, a century later diversity has the same positive unquestioned assumptions attached to it. Indeed, the insistence on diversity—companies and government agencies commonly provide employees diversity training—may serve to reify racial and ethnic differences among persons and groups even as it papers over differences in wealth and status (Michaels 2006). It may even serve to invent such differences, thereby simultaneously creating new identities and separating them from countless others.

The widespread use of the term "diversity" is itself linked to the advent of neoliberal approaches to governance. Specifically, U.S. civil rights legislation (the Civil Rights Act of 1964) enshrined equal opportunity, health and safety rules, and fringe benefits regulation in law. However, the legislation was (deliberately?) ambiguous, leaving decisions about implementation to the companies themselves. In response, corporations created offices to deal with each of these mandates—offices which then hired new kinds of experts: "Human resources specialists, benefits accountants, tax lawyers, safety engineers, and equal employment managers saw in employment legislation new possibilities for professional growth" (Dobbin and Sutton 1998, 445). Once established, or perhaps in order to become established, they developed their own corporate standards for compliance based on their interpretations of the law. Equally important, the term "affirmative action" was replaced by the term "diversity" to put a positive spin on the new programs as a means for enhancing the creativity of various parts of the organization. In addition, "Justifications for these offices drifted away from compliance and toward pure efficiency" (Dobbin and Sutton 1998, 471). In short, a legal project initially the result of concerns about social justice and inequality was gradually transformed into one in justified by market share enhancement.[18]

Standards Organizations

Not surprisingly, the use of standards to differentiate has created an entirely new growth area in contemporary societies. Organizations of all sorts that create standards, from formal standards development organizations (SDOs) to consortia to ad hoc alliances to de facto standards, are popping up everywhere, although precise data are extremely difficult to find and likely not too accurate. Every conceivable aspect of life now has standards associated with it. There are standards for plastic bottles, electrical engineers, fuel cells, police officers, busboys, commuter buses, and serial buses.

Moreover, since using standards to differentiate is the name of the new game, and the older SDOs are often slow, other organizational forms are being developed to speed up the process. As Cargill (1997, 29) explains, "The difference between consortia and formal committees is a matter of openness; SDOs must admit anyone who appears at their door while consortia have the right to request money as a prerequisite for participation." Put differently, SDOs maintain something akin to democratic processes: They are open to whoever comes to the meetings (although it can be costly to attend). They claim to maintain a balance of interests among the affected parties. They adhere to rules of due process. They have an appeals process. Finally, they arrive at decisions by consensus among the participants.[19] As one might imagine, as a result of these requirements, SDOs rarely work quickly. Standards can take four to five years to develop.

In contrast, as the name implies, consortia are groups of organizations whose engineers and marketers get together by invitation. "What the consortium provides is a formalized structure (mimicking the SDOs) to which organizations can send people to accomplish a *set task* within a *defined timeframe*" (Cargill 1997, 126, emphasis in original). Consortia also are expensive, with membership often costing $50,000 or more (Cargill and Bolin 2007). While SDOs do not know in advance what form the output will take, consortia generally have clear commercial goals in mind. Given their speed and more focused character, they are particularly attractive means of leapfrogging over the SDOs. According to the Comité Européen de Normalization (2007), there are now more than 250 *international* consortia in the IT industry alone; no data are available on national or regional consortia.

Finally, there are ad hoc standards alliances. These are more fleeting than consortia and are generally organized for the purpose of developing one particular set of standards. They sometimes revolve around what have been called patent thickets (Shapiro 2000) or anticommons (Heller and Eisenberg 1998), situations in which a small group of companies each own critical patents for parts of a given technology. They organize together both to standardize the technology and to cross-license the patents necessary to legally manufacture the product. In principle, others may join, but doing so necessarily involves paying rather hefty royalties for licenses to the patent holders. Moreover, none of the licenses are of any use unless one has successfully licensed the entire set. Not surprisingly, given their fleeting character, even crude estimates of the number of these alliances are impossible to find.

Together, the burgeoning SDO, consortium, and alliance industry is itself an example of the growing use of standards to differentiate. Moreover, the growth of standards organizations has been accompanied by a growth in certifiers and accreditors, a topic I take up in the next chapter. What should be noted here, however, is that while standards may be considered an impure public good (Cargill and Bolin 2007), their value as such is diminished as their numbers increase. After all, a single (consensus) standard makes some product, organization, practice, or process habitual, routine. A few competing standards may cause a person or organization to pause before deciding which one to follow. But a plethora of standards leaves one bewildered, especially as the various available standards may have been developed for entirely different reasons and may be designed to accomplish entirely different goals. Paradoxically, having too many standards returns us to the starting point in our quest. It has the potential of spawning a vicious circle in which vast sums are wasted creating standards that are never or hardly used, that meet no one's needs, that undermine the very social goods that standards were intended to deliver.

Conclusions

The rise of standardized differentiation has profoundly shifted the nature of social relations by redistributing cognition. The global decline in working-class parties, the abandonment of the project of a singular nuclear family and its replacement by a multitude of forms, and the weakening of

Table 3.6
Some ways in which people consciously differentiate themselves

Occupation	Gender
Dress	Sexual practices
Lifestyle	Body adornment
Plastic surgery	Body piercing
Geographic mobility	Gait
Music	Group memberships
Physical appearance (e.g., hair color)	

trade unions are all part of the new standardized differentiation. Even the rise of identity politics is linked to this decline. Of course, this is not a return to the world as it existed prior to the development of factories, complex machines, or large corporations.

Today standards are differentiated over time and (social and geographic) space. Ironically, individuation can proceed only insofar as there is a proliferation of standards that *permit* the differentiation of products, services, working hours, lifestyles. Put differently, individuation requires multiple means of distributing cognition across space and time. Individuation of this sort can only occur through standards, but without standardization! Thus, the world comes to resemble the children's game of Careers, but rather than the three career choices available in the game—fame, fortune, love—the choices and combinations of choices have now multiplied in countless ways, some of which are listed in table 3.6. It should be emphasized that each of the means of differentiation given in the table implies myriad standards choices. For example, the use of hair coloring depends on the range of colorings, each manufactured to a different standard. Similarly, sexual practices tend to be grouped around standardized categories (straight, gay, bisexual, dominant, submissive, etc.). Musical genres are each reproduced in thousands of standard forms with different media and arrangements. In a seemingly contradictory way, we differentiate ourselves by employing a multitude of differentiated things (each of which is standardized) that mark us as conforming to the standards of some group. And, those who set the rules for each of these multiple ways of being, lifestyles, career choices, now have considerable power over others—power that may be benign or malignant (cf. Frank and Meyer 2002).

In ways only vaguely foreseen by Émile Durkheim, and even less so by Adam Smith, the division of labor (as well as leisure, politics, advocacy, etc.) can lead to fragmentation and anomie. Paradoxically, the more we are linked to each other, the more choices are offered to us and the more individuated we become. The rise of the Internet is a case in point. First, it requires a multitude of standards for hardware and software interoperability. Second, it requires conformity to a set of arcane protocols. Third, it allows communication with millions of persons. One can join hundreds of sites such as MySpace that allow instant "friendship" with thousands of people.

But at the very same time, fourth, the Internet permits the formation of highly differentiated communities. Many such communities could not even exist in the standardized world of the last century. Today we can find web-based communities devoted to finding cures for rare diseases, uniting people with unusual sexual preferences, or permitting the exchange of scientific data on an endless variety of subjects; fetishists, terrorists, reactionaries, and revolutionaries of various stripes; advocates of a plethora of political views; enthusiasts for all sorts of products (automobiles, teapots, trains, medicines); traders in illegally obtained credit card numbers; and collectors of virtually everything, even manufactured memorabilia. Thus, the human and nonhuman standards of the Internet simultaneously connect and divide, standardize and differentiate.

The result of all these differentiated standards is that old class lines are now far more blurred than ever before. No longer is there only (a project of) a single hierarchy of status from top to bottom. No longer is there only (a project of) a melting pot that attempts to homogenize new immigrants. Rather, there are innumerable branches, partial orderings, and social groupings. As David Brooks (2000) suggests, we now have a significant number of persons who can be defined as Bobos—bourgeois bohemians—but the phenomenon transcends the old class lines. Whereas in a standardized world one's class is initially determined at birth and class mobility is at best restricted, many of the statuses now proliferating in contemporary society are active choices. One may choose to be an advocate for AIDS cures, to collect figurines, or to join a religious cult. A new immigrant may choose to become British or American or French, but may just as easily choose to become part of a new immigrant community—to become Turkish German, Pakistani British, or Chinese American.[20]

Each of these choices, regardless of whether made consciously for that reason or not, says, "I am distinctive. I am a member of this particular group." But the choices themselves are provided by the ever-proliferating standardized differentiation. Indeed, there are no other sources for these choices. I can decide to become a teapot collector, an advocate for cures for multiple sclerosis, or any other category provided by the contemporary social order. But I cannot decide to become a xlurd, unless I can convince others that xlurding—whatever that might be—is a legitimate category of social life. To do so is to convince some persons that the standards invoked by xlurding—that distinguish it from other categories—are indeed valid for some purpose, whether frivolous or profoundly serious. Moreover, as each of us chooses among a bewildering array of categories and the standards that allow others to define us as one of "those people," older class attachments become weakened, less apparent, even obliterated. The standardized differentiation, in short, brings with it a differentiated and (over) individuated society to some even as it may be liberating to others.

As the late Gloria Anzaldúa (1999, 85) noted, writing of her formerly fragmented identity, "Something momentous happened to the Chicano soul—we became aware of our reality and acquired a name and a language (Chicano Spanish) that reflected that reality. Now that we had a name, some of the fragmented pieces began to fall together—who we were, what we were, how we have evolved. We began to get glimpses of what we might eventually become."

Similarly, AIDS activists were able to construct a common identity. In so doing those who were able to learn the "language" began to participate in medical research—not merely as patients and subjects but as codesigners of medical protocols and treatments (Epstein 1996). Yet another example is provided by the slow food movement, which celebrates the variety of culinary and agricultural practices. It has provided space for those concerned about endangered crops and domestic animals, cultures, and ways of life to voice their common, yet differentiated, concerns (Miele 2008). Finally, La Vía Campesina is an attempt to give voice to the diverse concerns of the world's peasants and small farmers (Borras 2008). Hence, the project of standardized differentiation—like the project of standardization itself—simultaneously brings with it belonging *and* alienation, attachment *and* estrangement.

In sum, even as the project of standardization of the nineteenth and early twentieth centuries continues unabated in some domains, we are witnessing a vast response to it. Whereas in the past, standards were used largely to standardize, they are now also used to differentiate—among people, things, and processes. This is not to say that the older standardization has disappeared or is even in danger of disappearing but rather that standards are now seen more and more as strategic devices for advancing a wide range of goals. Nations use standards to advance geopolitical goals (see box 3.3). Firms use standards to block competition and enhance market share. Individuals use standards to emphasize their distinctness or

Box 3.3
It's hard to give up the GOST

After the 1917 Revolution, the new Soviet regime was faced with the task of industrialization. Production units were provided with a given set of inputs, an annual target developed by Gosplan, and a set of standards known as GOST (state standards). Unlike the standards provided by standards development organizations in Western Europe and North America, GOST standards were and remain considerably more detailed, including items that would be found in business-to-business contractual agreements in the capitalist world, as well as more general terms.

With the collapse of the Soviet Union in 1989, thousands of quasi-firms in the former Soviet republics floundered or were dissolved. But the former Soviet states found themselves with Russia as their major trading partner, and Russia continued to maintain and require a version of the GOST standards (Dragneva and de Kort 2007, 253).

The problem faced by the former Soviet republics is the cost and expertise required to shift to internationally recognized standards. As a World Bank report explains:

Significant intraregional trade perpetuates the existence of the GOST-based system and complicates the transition to a market-based system. A CIS country exporting to another CIS country requiring GOST-based certification must be able to comply, even if the exporting country is changing to a system that complies with international standards. Also, vested interests, combined with bureaucratic inertia, help maintain the GOST system to justify the existence of bureaucracies and, possibly, to protect some domestic industries. (International Bank for Reconstruction and Development 2007, 21)

In addition, the perpetuation of the GOST standards serves Russian interests, keeping the members of the CIS within its orbit and punishing states that resist, such as Moldova. Russian accession to the World Trade Organization would bring this issue to the fore.

(often at the same time) to draw attention to their sameness. New social movements use standards to frame and support their causes.

But in order for this new standards society to function, the means must exist for monitoring and enforcing this seemingly endless abundance of standards. This in itself has become a significant political and business activity. Thousands of organizations have been created to certify, audit, check on, verify, and attest to the veracity of various claims of adherence to various standards. And hundreds of other organizations have been created to accredit the certifiers. This phenomenon of certification and accreditation is the subject of the next chapter.

4 Certified, Accredited, Licensed, Approved

As I write this chapter, I am seated before a standardized laptop computer. It would take several pages merely to list the myriad standards incorporated into that machine. Such a list would include standards for reading programs, executing commands, communicating with other machines, linking into electrical power when needed, ensuring the safety of the user, the shape and organization of the keyboard, and so on. But how do we, the final consumers, know that standards have been met? How does the manufacturer of the computer know that parts bought off the shelf to build the computer also met the required standards? And how does the retailer know that the computer for sale will function more or less as expected?

Today the answer to these questions is often to be found nearby, in the form of logos, seals, and statements that accompany both persons and things. For example, a logo on the laptop I am using indicates that it is certified to meet fire safety standards by Underwriters Laboratories. I have at my side a cup of coffee that is certified as both organic and fair trade. The cell phone in my pocket is of the type certified to work on CDMA (Code Division Multiple Access) systems. In my workspace are cards certifying my membership in several professional societies. I carry in my pocket a Michigan driver's license, certifying that—providing I wear my eyeglasses—I am competent to drive a personal passenger vehicle. These are but a few of the many certifications that can be found in almost any everyday situation in every industrialized nation and in an increasing number of poor and middle-income nations. For example, although data are difficult to come by, one study estimates that food product certifications are growing somewhere between two and ten times as fast as the food market itself (Byers, Giovannucci, and Liu 2008). In short, certification has become a nearly universal phenomenon.

Little more than a century ago, however, certifications were few and far between. The traditional marketplace was one in which persons vacillated between the extremes of caveat emptor and mutual trust. Typically a well-known buyer would receive what had been contracted for at a fair price, in part because a well-known buyer would return again and in part because the seller felt obligated to deal fairly with those she or he knew and saw every day. Even when a contract was necessary to seal the relationship, it was often quite lacking in details. An example is the contract recommended in 1905 (Corbett 1905, 31) between tomato growers and canners for the purchase of tomatoes not yet planted:

This is to certify that we ____ have bought of ____ the product of ____ acres of tomatoes for the season of ____ at $____ per ton, delivered at our cannery at ____. Stock to be in first-class mercantile condition, To be planted about ____.

Such a contract was extremely general, and subject to the vagaries of the weather, but it attested to the trust between buyer and seller.[1] In contrast, a stranger was fair game. One could take advantage of a stranger since that person was not likely to be seen again and because that person was— by virtue of being a stranger—a naïf, a sucker, an easy target, who might not even realize that he or she had been cheated.

The same was true of those who claimed to be professionals, such as physicians, clergy, and lawyers. The hallmark of a professional was autonomy. As late as 1963, sociologist Everett C. Hughes (1963, 656–657) could write, "Since the professional does profess, he asks that he be trusted. The client is not a true judge of the value of the service he receives; furthermore, the problems and affairs of men are such that the best of professional advice and action will not always solve them." In short, interpersonal trust and the special objectivity claimed as part of "professionalism" were commonplace.

Similarly, in forms of noneconomic exchange, in all sorts of interpersonal relations, trust and reputation were, and sometimes still are, everything. The neighbor, the schoolteacher, the clergyman, the friend were and often still are usually to be taken at their word. Yet in a little more than a century we have transformed the world into something that would be nearly unrecognizable to our great-grandparents. The combination of urbanization, much greater geographic mobility (especially in the United States), growth in global trade, and easier, faster communication has simultaneously increased the number of unfamiliar persons and things we come into contact with daily and decreased the depth of our interaction with more familiar persons and things. Certification can be understood as an attempt to compensate for some of these profound changes.

But certification is also intricately related to the embracing of standardized differentiation described in chapter 3. That is, it is not merely that we now engage in far more mediated or fleeting relationships with *people*; we also engage in far more relationships with *things*. In a standardized world, such as Henry Ford was trying to build in the early decades of the twentieth century, there was relatively little need for certification of things. What need existed in the United States was taken care of through the Good Housekeeping Seal of Approval (focused on consumers and established in 1911) and Underwriters Laboratories (founded in 1894 and aimed to a greater degree at industry). Certification of people was virtually unheard of. Anyone could become a doctor or a lawyer, a teacher or a plumber, by hanging out a shingle. Of course, the quality of the services provided often varied markedly as well.

In this chapter I first examine what brought on the desire or need to certify seemingly everything and everybody. Next I describe in somewhat idealized form how certification works. I then ask how certification simultaneously augments and replaces trust, and allows for the management of risk.

The Rush to Certify

The need to certify persons and things can be understood as the consequence of several interrelated phenomena. First, the standardized differentiation described in the previous chapter could not be as easily governed as before. Second, changes in transportation and communication systems made it profitable to manage complex production and exchange relations over great distances. Third, more recently, neoliberal doctrine encouraged— indeed, necessitated—enhanced "freedom to operate" by all organizations, but especially by large companies, leading to the outsourcing of various activities by governments, firms, and other organizations; introduced market relations where they previously did not exist; and made possible the rise of supply chain management as a business strategy and economic theory for handling and standardizing the complexities of acting at a distance. Let us examine each in turn.

In a world in which a single standard commonly prevailed for identifying persons and things, there was little need for certification. Consider the state of *things*: many things were still produced by craft means, but most people were convinced that such forms of production would soon be obsolete. In contrast, mass-produced, standardized things were either produced largely in-house in vertically integrated companies—at Ford's River Rouge plant, with 100,000 workers, even the steel was produced on-site— or were bought on spot markets, as were nearly all agricultural products. Everything was produced just-in-case, since the vagaries of transport and communications might otherwise require shutdowns. Now consider the state of *people* at that time: not only were the skills required by most jobs easily learned in a few days or weeks, considerable effort went into redesigning work to reduce the skills required for most jobs (e.g., Taylorism). Karl Marx (1977, 266) was convinced there was a fairly clear and inevitable path from the deskilling of virtually all work to replacement by machinery. As he put it in an 1849 newspaper article, "If the whole class of wage-workers were to be abolished owing to machinery, how dreadful that would be for capital which, without wage labour, ceases to be capital!" Standardization applied to consumption for all but the very rich as well. Even as the cost of standardized products declined, differentiation in the activities of consumption was limited by the standardized character of more and more goods and services.

But as more and more standards were differentiated and as transportation and communications improved, that situation changed. In particular, standardized differentiation initially posed enormous logistical challenges. The highly centralized companies that manufactured vast number of identical products had considerable difficulty making vast numbers of *different* products, especially as doing so often required coordinating networks of diverse persons and things distributed over great distances. Initially, this problem went unnoticed. Vertically integrated companies continued to operate much as before. However, some companies, such as General Motors under Alfred P. Sloan Jr., one of the leaders in standardized differentiation, began to treat each division as a separate entity, even encouraging intrafirm competition. Ford almost went bankrupt in the face of Sloan's product differentiation. By the 1960s many companies were becoming conglomerates, with large groups of companies supplying an ever-growing collection of often unrelated products and services.

Later, companies began to seek supplies and services farther and farther from home. With more rapid and reliable transportation and communication, even relatively perishable goods could be shipped long distances and be expected to arrive just where they were wanted at the right time. Visionary company executives began to realize it was no longer necessary to have vast sums of money tied up in parts for yet to be assembled products or to pay the huge costs incurred for storage. Just-in-time began to replace just-in-case production.

Still later, with further improvements in transport and communications, larger companies began to shed many parts of their enterprises. The provision of raw materials and internal company services (cafeterias, cleaning and maintenance services, customer billing, even accounting and payroll) could more profitably be farmed out to other companies. In short, outsourcing became commonplace.

In addition, supply chain management (SCM) was invented as a means to make just-in-time production more effective and a way to put a moat around a segment of the market under the control of a given dominant firm. As I have noted elsewhere (Busch 2007), SCM is a new means for performing the economy.[2] Two somewhat different stories of its origin can be identified (Croom, Romano, and Giannakis 2000). On the one hand, SCM appears to have originated in the 1950s as part of the rise of the systems perspective. On the other hand, it has been argued that SCM emerged from Toyota's "lean manufacturing" strategy of the 1970s, itself intimately linked to W. Edwards Deming's work on improving quality. However, the term SCM does not appear in the literature until the mid-1980s. Likely to a significant degree, managers in large firms began to develop more systemic approaches, and that empirical experience helped build the theoretical models. Those theoretical models in turn reshaped SCM into what it is today.

Unlike older (and still very much alive) economic models, SCM is specifically designed to grapple with the problems posed by multifirm, outsourced, and networked people, practices, products, and processes in a neoliberal economy. It does this largely by borrowing strategies freely from both financial audits and continuous quality improvement (Power 2003) and using them to certify the activities and products of actors in the chain. And, like those older models, it can be put into practice well or poorly.

That said, unlike neoclassical economics, which divides the world into the macroeconomics of national accounts and the microeconomics of firms, SCM makes the supply chain its unit of analysis (although with the goal of optimizing returns for one particular dominant firm in that chain). Of importance here is that SCM proponents see laws, standards, certifications, taxes, finance rates, technoscience, and even NGOs—nongovernmental organizations—as modifiable parts of a strategy for maximizing supply chain (and hence dominant-firm) effectiveness. Put differently, SCM is a means of capitalizing on the "freedom to operate" while avoiding many of the uncertainties and risks promised by the neoliberal market. This is both its potential strength and its greatest weakness. Indeed, since SCM must be performed by (1) purchasing agents, (2) business school professors, and (3) consultants of various stripes and persuasions (Gibbon and Ponte 2008), and since it must be performed over the entire chain, effective performance requires a great deal of choreography.

Moreover, proponents of neoliberal doctrine encouraged transforming as many organizations as possible into markets or quasi-markets so as to reap the claimed benefits of competition. Oliver Williamson (1975, 1994), building on the earlier work of Ronald Coase (1988), initially went so far as to argue that all productive processes could be organized as either markets or "hierarchies" (i.e., organizations), and that hierarchies existed only when markets failed, or, put more carefully, when the costs of market transactions were greater than the cost of organization. One observer otherwise sympathetic to both authors observed that the fact that markets cannot produce anything seemed to pass largely unnoticed (Dietrich 1994).[3]

Influential management gurus such as Tom Peters spread a version of this gospel to corporate executives beginning in the early 1980s (Lonsdale and Cox 2000). The central thesis was that firms have core competencies and should stick to them by outsourcing things that might be done better elsewhere. This advice would have fallen on deaf ears had it been offered thirty years earlier, as the decline in the cost of long-distance shipping and improvements in rapid communication (as well as their deregulation) had not yet occurred.

But all of this posed a serious problem, which can be summarized as follows. In a just-in-case economy, goods and services are either produced in-house or bought on spot markets and then stored until needed. A just-in-time market requires far less capital but works poorly when subjected to the vagaries of spot markets. After all, the input (a person or thing) needed might not be available at the appropriate time or price, bringing production to a halt. Moreover, quality—once an issue largely within an organization—would now be distributed across the entire supply chain. A solution to this problem lay in the creation of long-term relationships with suppliers. In a great irony largely unnoticed by the proponents of neoliberalism, producers would bypass the spot market and permit regular supplies of the needed inputs to flow through what later became known as supply chains. However, large firms found themselves coordinating the work of hundreds, perhaps thousands, of suppliers—a Herculean task at best. How would they know if the inputs they wished to purchase met their specifications? How would they know that deliveries would be made on time? How would they keep the costs of inputs under control, especially given that the discipline of the spot market would be (at least partially) removed? How would they know that the workers producing the inputs were using the preferred practices? How would they protect their brand names from being tarnished by the shoddy practices of a supplier? Moreover, how would they monitor their suppliers without being police officers? After all, they had contracts with their suppliers. Wouldn't their suppliers see any attempt to regulate their activities as an attempt to distort or renegotiate the terms of the contract?

For somewhat different reasons, NGOs have also turned to certifications and supply chains (Guthman 2007, 2008). States largely reorganized according to neoliberal doctrine make difficult targets for NGO concerns. As Robert Reich (2007) notes, they are outspent by corporations. Hence, NGOs have turned from lobbying governments to lobbying highly visible large corporations, as well as to establishing their own standards, certifications, and accreditations, in an attempt to advance goals of concern to them. In short, NGOs have become enmeshed in neoliberal notions of governance.

As citizens and consumers, we also find ourselves confronted with a bewildering level of choices. Ironically, in a society in which individual choice is growing at an extraordinarily fast pace, both for goods and services and for lifestyles, the social character of decision making is also expanding: we are more likely to look to friends, acquaintances, search engines, brand names, certifications, or other indicators than ever before. Indeed, such social supports are needed to help to reduce the transaction costs associated with searching.

However, the same bewilderment can arise during searches aimed at narrowing the range of choices. "Getting in with the wrong crowd" is not hard to do—perhaps easier as a result of differentiation—but in addition, one can wind up getting misleading or inadequate advice.

Asking a librarian how to find information about a given event is likely to lead one to the most trusted sources, and most librarians are trained to do just that. But as we know from the placement of web links based on ad revenue, sometimes the links that come up first are not the most relevant but those of companies that paid the highest fees. The same is the case when we seek advice on the use of fertilizers from sales agents for fertilizer companies or advice on nutrition from food manufacturers.

In short, certification, like all forms of advice on choices and decisions to be made, can be performed well or poorly, in a useful or useless manner, with many biases or largely unbiased. It can restrict the range of choices arbitrarily or provide choices only within a certain framework. Moreover, any form of advice remains within the set of standards provided by society. That aspect of the power of standards remains unchanged by certification. But not all certifications are the same. Let us examine the different ways in which certification is organized.

Table 4.1
Possible forms of certification

	Party certified		
Party that certifies	First	Second	Third
First	Self	Franchises, licenses	Unlikely
Second	Second certifies first (often used)	Self	Only approved certifiers acceptable to buyer
Third	Third certifies first (very common)	Possible. Any cases?	Self

The Party

To understand certification, it is useful to distinguish between first parties (sellers), second parties (buyers), and third parties (certifiers) in an economic exchange. Table 4.1 shows the various possible forms of certification. Each is briefly considered next.

First-Party Certification

In perhaps the most common form of certification, the first party certifies the quality of the goods or services being offered. For this reason, many companies offer a no-questions-asked money-back guarantee to buyers if for any reason they are unsatisfied with the product they purchased. Some companies that produce small, inexpensive items may even return twice the purchase price. Still other companies provide warranties that cover repairs needed to the product over a given period of time.

Second parties, too, may engage in self-certification, although this is less common than first-party self-certification. It was once fairly common in small towns around the world, where selected buyers of goods would be given credit by merchants based on their character. Their character was a form of self-certification. It meant they were very likely to repay the sum owed.

Some third-party certifiers also engage in self-certification when they certify that their certification services are reliable and accurate. That said, this is a relatively rare form of self-certification. More likely, third parties will be accredited by independent accrediting agencies.

In some instances, first parties certify second parties. This occurs in at least two somewhat distinct cases, franchising[4] and licensing. Franchisees are usually certified by those who provide the franchise. Though doubtless it has been around much longer in some form or another, franchising is usually said to have begun in the United States in the 1850s with the Singer Sewing Machine Company. The franchise is an attractive form of business contract precisely because it permits the franchisor to use standards to maintain control over many aspects of the franchisees' behavior.

Thus, McDonald's, Holiday Inn, Century 21, and Ace Hardware, to name a few of the 750,000 (Franchising World 2006) companies that operate on a franchise model, each certify their franchisees, assuring that they conform to a set of specific standards required by the franchisor. For example, hotel franchisees may be required to maintain their buildings in a prescribed manner, greet hotel guests with certain phrases, remodel hotel rooms after a given number of years, and employ standard contracts when hiring workers. They are also subject to regular audit of their performance on all these standards by the franchisor. In return, hotel chains often agree to provide access to technical training, telephone and Internet reservation systems, and a variety of branded products ranging from stationery to towels. Virtually every detail of hotel management is outlined in manuals for employees as well as franchisees. Similar arrangements are found in fast-food and automobile sales.

Over time, franchises have become more or less standardized; one observer reports a hotel executive as saying that if one were to remove the cover page from a franchise agreement, one would no longer be able to discern which chain it belonged to (Clancy 1998). One effect of this standardization is that customers who frequent such hotel or restaurant or automobile dealership franchises know what to expect. They rapidly become familiar with the amenities, the food, and the behavior of employees at such establishments—and they adjust their own behavior accordingly.

An important aspect of the franchising arrangement is that it permits companies to engage in sanctions previously limited to nation-states. Franchisees who fail to meet the standards may be fined, or in extreme cases the franchise may be revoked.

Licensing provides a weaker form of certification of second parties by first parties since it is often difficult and costly for licensors to audit the activities of licensees. Under licensing agreements, licensees agree that they will abide by certain conditions of use of a given product or provision of a certain service. In recent years, licensing has become quite commonplace. Thus, all that software in the home was not (according to the licenses on the package) actually purchased by the user but licensed for use by the manufacturer. Those licenses often go on for several pages of dense prose and demand that the purchaser use the software only in particular ways. This sort of licensing is not followed by regular audits, as in the case of franchises. However, software companies will often defend their licenses in court when they hear of significant violations of the license conditions.

Licensing also provides a means of extending the standards society to areas previously beyond its reach. Licensing is commonly used in various franchising schemes, as well as in the production of copyrighted and patented products. It is used to certify that persons in certain occupations are competent to perform in those roles (e.g., physicians, plumbers, teachers), although this form of licensing is usually done by the state. Even the use of certain brands and trademarks may be the subject of licensing.

As in times past, most patents today are used to create products that are then sold to buyers. But in recent years licensing has become widespread, especially with respect to IT, but also in a wide range of other industries. Licensing allows the patent holder to use the limited-term monopoly granted by the patent to require that licensees adhere to standards defined by the patent holder. Most obviously, it blocks the transfer of property rights in the object or service. But it may also require the buyer to follow (or not) a variety of other product and process standards.

Similarly, for the last decade one finds standards imposed through licensing in the seed industry. Genetically modified and other patented seeds are not sold to farmers but are licensed to them for a single season. Monsanto pioneered this approach with a so-called technology agreement required of users of the seed. The agreement requires, among other conditions, that farmers not save seed to replant in the following year and that they use only certain herbicides. Monsanto polices the licensing agreement vigorously, bringing legal action against those who violate it. Recently, entomologists have complained to the USDA that licensing rules are so tightly enforced that it is impossible to test the environmental impacts of the seeds (Pollack 2009). Whether this is legally permissible or not is debatable, but the high cost of engaging in such litigation has allowed the practice to go unchecked to date.

Second-Party Certification

Also common is second-party certification of the first party. This occurs when, for example, a retailer certifies that all of its suppliers meet certain requirements. Such certifications are usually recognized only by the buyer and are a partial substitute for product inspection by the buyer. However, in some sectors this approach is fairly common. For example, as supermarkets have moved to the use of more store-brand or private-label products, many of them have developed means for certifying their suppliers. Similarly, many baby food companies employ their own in-house certifiers to ensure raw product quality (e.g., Heinz 2009). In such instances an employee of the supermarket or processor will visit the supplier and engage in the testing necessary for certification. This role may be contracted out, but the contractor would still be employed by the buyer.

Third-Party Certification

The solution that has been most widely embraced is certification by a third party, one not directly involved in the exchange. Third-party certification claims to ensure that the products produced or the services performed by first parties conform to the requirements of the second party. Third-party certification asserts that the processes used were the "best" ones (read: most efficient and effective available at a given time and place and for a given price from the perspective of the purchasing firm) for the production of a given service or product. Third-party certification claims to ensure that the supplying firm's employees were properly trained. In short, third-party certification claims to ensure that the practices used were in conformity with the best practices used in the industry.

Certifications are now commonly used throughout society. Table 4.2 provides a sampling of the many certifications commonly found. Indeed, one advertisement for JC Penney stores notes, with all due solemnity, that women can "get your complimentary fitting with one of our 14,000 certified bra specialists" (*Ann Arbor News* 2009). Virtually all aspects of public life are now certified by someone for conformity to one or more sets of standards.

Proponents of certification argue that it eliminates surprises by ensuring that people and things act in the manner desired by the buyer. This in turn increases trust and reduces risk. As such, this approach, known to the industrial world as "conformity assessment," appears to resolve a vast number of problems. Let us take a closer look at the issues of trust and risk.

Trust

Certification is a form of advice to others. Taking that advice depends on trust. Part of the normal taken-for-granted socialization by which nearly everyone attains adulthood is a process in which we learn whom and what

Table 4.2
Selected certifications for people and things

People
Certified Journalism Educator
Microsoft Certified IT Professional
Certified Professional Public Buyer
Certified Hospitality Accountant Executive
USA Track and Field Certified Official
Regulatory Affairs Certification
Certified Planner of Professional Meetings
Certified Professional Coder—Hospital
Certified Forensic Claims Consultant Engineer
Certified Child Passenger Safety Technician

Things
Certified Environmental Management System
Certified Leadership in Energy and Environmental Design
Seal of Approval of the Evangelical Council for Financial Accountability
European Union Product Safety Certification
Cradle-to-Cradle Certification (environmentally friendly design)
Biosafety Cabinetry Certification
Cellular Telecommunications and Internet Association Battery Certification
Programme
Greenhouse Gas (GHG) Emission Reduction Product Certification
Wireless Application Protocol (WAP) certification
Vinyl Siding Product Certification
Firewall Product Certification

---

we can trust. A John Dewey or a George Herbert Mead might argue that we develop *habits* of trust in others and in our own trustworthiness. (Persons who grow up trusting no one exhibit pathological behavior.) Infants spontaneously trust their mothers, unless given good reason to do otherwise. Indeed, children are often so trusting as to wander off by themselves, innocently chat with potentially dangerous strangers, walk to the edge of precipices, put their hands on hot stoves, and engage in other acts adults regard as risky. But as we grow up, we begin to differentiate, to be more discerning in whom and what we trust. We are more wary of strangers, dark alleys, wild animals, hot surfaces, and new situations. We also learn to distinguish between two somewhat different forms of trust—trustworthiness and predictability.

Trust as trustworthiness is always dialogical. It usually involves an exchange of words as well as gestures (Mead [1934] 1962), and things

(Latour 1987). Moreover, as Erving Goffman (1993) has suggested, even face is dialogical. Hence, wide-awake normal adults trust people whom they come to know well—through conversations, through gestures, through the exchange of and interaction with things, and by creating (and sometimes losing) face.

Thus, when I interact with my wife of forty-plus years—via conversation, gestures, exchange or manipulation of things, or face—I hardly need any advice from anyone else. I trust her implicitly to never act deliberately with malice toward me and to do things that will illustrate her concern for my well-being. In other words, she cares about (some) others, including me. She is trustworthy—literally, worthy of my trust in her.[5] But at the same time, I know her well enough often to predict her reactions in common situations. This involves not trustworthiness but predictability.

The same cannot be said of all types of relationships with people or with things. Some people and some things are not necessarily trustworthy but are quite predictable, and can be trusted in that sense. Thus, I trust that the staff of the postal services will deliver the letter I mailed. I trust that the members of the fire department will come if I need to call them. I also trust that my car will get me to work and that my computer will store a manuscript I am working on until I need it again. I trust that rocks are hard and that the waters of Niagara Falls flow downward. These things exhibit predictability, but they do not (usually) exhibit trustworthiness.

However, that said, since many things are made by persons, it is possible for some of the trustworthiness of persons to be transferred to things. Thus, in times past and in some places today, one might find a particular baker to be trustworthy; that would mean the bread she baked would also be worthy of my trust. Put differently, the care exhibited by the baker is manifested in the bread. Similarly, I might trust a given auto mechanic or plumber to repair my car or plumbing in a careful manner and to charge a fair price. The trustworthiness of the mechanic or plumber is then to a large extent transferred to the thing repaired. This is what we mean when we say that someone has done a good job: the person has exercised care, and that care is reflected in the tasks accomplished.

Certain animals can also be trusted in both ways. Dogs, for example, may display care for their owners *and* act in very predictable ways (e.g., barking when someone approaches outside our homes). They are therefore not merely predictable but trustworthy. Dogs trained to help the handicapped are an excellent example of animals we can trust in both senses.

Engaging in trust relations—whether of trustworthiness or predictability—implies a willingness to accept a certain degree of risk. For example, if I trust my wife, I must reject as extremely unlikely the risk that she might betray me in some way, that my trust in her is somehow misplaced. Similarly, I must reject the risk that a person or object will be unpredictable if I am to trust that object. Thus, I trust that when I press my foot down on the brake pedal, my car will behave in predictable ways. But also I trust that my old jalopy will have yet another mechanical failure. I trusted that, alas, Osama bin Laden would continue to act and to encourage others to act in violent ways until his recent demise. And in most normal everyday life situations we find that trust is not misplaced.

But it is important to emphasize here the significance of the difference between these two types of trust. Trust as predictability tells us about the consistency of the qualities of persons and objects and thereby allows us to determine (with more or less precision) the risks involved in a particular interaction. Trust as trustworthiness goes beyond that, implying care and character. The trustworthy person will go the extra mile, while the predictable person will simply behave as predicted. Certification, then, can be thought of—when it works as advertised—as a *displacement* of trust as predictability. Certification signals that a person or thing can be trusted in the sense of predictability. Certifications may be used in the marketplace to inform parties to various exchanges, or they may be used in a wide variety of other settings where personal inquiry or intimate knowledge is lacking.

The certifications listed in table 4.2 provide good examples. Each of these certifications allows action at a distance. They tell us that, barring experience to the contrary, the persons or things that are certified will act in a predictable manner. The certified physician will predictably examine me and determine the cause of my illness. The certified cell phone in my pocket will predictably connect me to the phone at my daughter's home. They also tell suppliers to other firms that the products or services delivered will perform as expected. But—and this is important—certifications cannot tell me whether someone or something is trustworthy. In fact, the entry of certifications into realms normally reserved for trust as care and character could well serve to undermine those relations. For example, if my neighbor were to tell me that he was certified by the Good Neighbor Society as a good neighbor, or if my wife were to tell me that she had received the Good Wife Seal of Approval from the National Council for Happy Marriages, I might well become suspicious that neighbor or wife was concealing something.

Likely in part for this reason, many companies that use certifications for their suppliers do not advertise those certifications, relying instead on their brand reputation to maintain customer loyalty. Thus, certification is at best a partial substitute for that form of trust. Put differently, it is a form of trust (though likely not the only one) well suited for a world consisting largely of strangers acting in the marketplace. This is reflected in the poster for the first International Accreditation Day, reproduced in figure 4.1.

Risk

Since certifications are a displaced form of trust, they can also be seen as a means of reducing unwanted risks. Like trust, risk may be thought of in

Accreditation:
Delivering Trust in the Global Market

International Accreditation Day
9 June 2008

Figure 4.1
Poster for International Accreditation Day, 2008.
Source: © International Accreditation Forum. Used with permission.

two, somewhat different senses. Thompson and Dean (1996), reviewing the risk literature, define a continuum of risk, which they see as running from the contextual to the probabilistic. Contextualized risk is usually to be found in everyday settings. In most instances we tend to see this kind of risk as the result of volition. Hence, we can say that Jane took a risk by driving at 110 miles per hour on Main Street. Or we can note that Alice risked her life by jumping into the frigid waters to save Richard. These examples emphasize the situated character of a risk, usually referring to a situation well known to the participants in it as well as easily recognized by those to whom it is described. This form of risk, which I will call situated risk, is analogous to trustworthiness.

However, in modern parlance, risk is also commonly defined as the statistical probability of harm. This approach to risk is used in a wide range of scientific studies and in their application to a host of problems, from diseases to auto accidents to environmental pollution. Hence, if I take the time to gather the information, I can determine my risk of dying from listeriosis, the likelihood that I will be in an auto accident, or my life expectancy (the risk of dying at a given age). The same approach applies to things: I can find out the risk of my computer's hard drive crashing this week or of the mercury in a certain lake rising to dangerous levels. This form of risk is very similar to the second form of trust I identified. Both involve predictability, both are amenable to precise statistical measurement, and both can (in principle, and perhaps in practice) be attenuated through the use of certifications.

But it is worth emphasizing that one need not take a risk to contract pneumonia or be killed by a car or have genes that code for multiple sclerosis. While both engaging in risky behavior and contracting a disease can be construed as a statistical probability of incurring harm, taking a risk often has a positive side. For example, investing in the stock market, betting on horse races, skiing, and skydiving all have risks attached to them. The risks can even be quantified. But risks of this sort are often accepted voluntarily. Indeed, gambling, skydiving, and skiing would be rather dull activities if they did not involve any risk.

In most Western nations and a growing number of others, legal frameworks require companies to internalize some risks. For example, knowingly discharging certain compounds into the water supply or failure to protect workers from hazards in the workplace can result in both criminal and civil penalties. In addition, even where the laws are weak or nonexistent, the publicity created by failure to protect against these hazards can destroy the reputation of both organizations and individuals. This in turn can reduce sales or cause particular individuals to lose their jobs. Thus, persons and organizations are perhaps more averse to these kinds of risks than they were in the past.

Third-party certification is one way to protect oneself or one's organization from such risks. By engaging an independent certifier who certifies to recognized standards, one reduces such risks. Moreover, when problems occur—and there are always "normal accidents" (Perrow 1984)—one can claim due diligence as a defense in court. That said, certifications are nearly always about risk as a statistical probability of harm. That makes them fertile ground for the use of science. Let us look at how certifications employ science.

Science, Standards, and Certification

These days it is not uncommon to hear claims that certain standards are or are not scientific. Indeed, in some quarters there is a mistaken belief that all standards are (or should be) scientific. This is not to say that science is absent from standards making and enforcement but that the role of science is both more complex and more ambiguous than would appear to be the case.

Of course, all sorts of standards existed long before modern science. Weights and measures, coinage, and other standards for things clearly predate anything resembling modern science. The same may be said of moral standards and other related standards for human beings. But over the past several centuries science has been inserted into a wide range of standards for both people and things. Science is used in standards setting and enforcement in at least three ways: (1) as a theoretical frame within which a given standard is implemented, (2) as a scientific method or test employed to enhance precision or accuracy, and (3) to enhance the perceived validity and importance of a given standard (table 4.3). Let us examine each of these uses in turn.

Science as Theoretical Frame
Science can be particularly useful in standards setting and enforcement. Consider, for example, the case of standards for blood cholesterol. According to current medical standards, blood cholesterol for normal adults should be <200 milligrams per deciliter (mg/dL). This number was obtained through (among other things) statistical analyses in which blood samples were taken from thousands of individuals and correlated with the presence or absence of heart disease, as well as through autopsies of patients and

Table 4.3
The uses of science in standards setting and enforcement

Use	Persons	Things
Theoretical frame	Heart stress standards	Blood cholesterol levels, thermometers
Testing	Time and motion studies	Presence of harmless compounds in parts per trillion
Prestige	Intelligence tests	White-coated scientists in ads for various products

the observation of clogged arteries in their hearts. It was found that the statistical probability of having a heart attack rose as cholesterol levels increased beyond 200 mg/dL. In addition, science was used in the design of the tests used to determine blood cholesterol levels. Specifically, science was implicated in the modification and use of the gas liquid chromatograph to analyze and quantify the volume of cholesterol in very small quantities of blood.

Such uses of science can be found in a wide range of fields beyond medicine. For example, the thermometer was originally a scientific instrument. It was designed based on understanding the role of heat in making a column of mercury (later alcohol) expand. While thermometers can be produced to varying degrees of accuracy, they are all scientific in that they are based on a theory of heat, tested based on known values (e.g., the boiling point of water at sea level), and verified by thousands of observations.

Science as Test

Science may also be used for the tests that it provides, regardless of the validity or meaningfulness of those tests. In some cases an extremely sensitive test is used to specify extremely minute levels of a known toxic compound; in other cases the level specified is not particularly low but the importance of the measure is at best debatable.

For example, with current testing technologies, many compounds can be identified at levels as low as one part per billion or trillion. To visualize one part per billion consider that it is roughly the same as one drop in a large tanker truck full of liquid. One part per trillion would involve one drop in one thousand trucks—indeed, a very small amount. Yet equipment designed to measure compounds at such low concentrations is available. For a very few compounds that are quite toxic, this extremely low level may be of some relevance, but for most compounds such low concentrations are virtually meaningless. That said, since they can be measured, some standards have been set that low or lower. For example, for many years starting in 1958 the Delany clause in U.S. food law required that compounds known to cause cancer in animals could not be present in food at *any* level. As the precision of testing improved, the absurdity of this clause became more and more apparent. It was finally abandoned in 1996.

A second example involves the calculation of bacterial counts on fruits and vegetables. Quite obviously, none of us would want significant quantities of *Listeria* or *Salmonella* on our food. But many bacterial species can be found in low concentrations almost everywhere. Others are ubiquitous but harmless. Some food companies, apparently after prompting by their lawyers, who fear damage suits, have begun to require suppliers to meet bacterial count standards (maxima) for these naturally occurring bacteria. Yet in many cases no evidence whatever exists to suggest that such bacteria, especially in very small amounts, are in any way problematic. Indeed, removing them might well create a sterile medium that is ripe for invasion by some far nastier bacteria. And some of the commonly occurring bacteria might well aid in digestion.

Science as a Means of Enhancing Prestige

Finally, science may be used to lend prestige to a given standard. We have all seen the television ads featuring white-coated scientists, intended to lend an air of scientificity to a given product. The same can be done for standards—they can be cloaked in the mantle of science to make them *appear* scientific.

For example, IQ tests were administered to thousands, perhaps millions, of people during much of the last century. They were claimed to measure some clearly defined phenomenon called the intelligence quotient. However, critics pointed out several major problems with the tests: For example, IQ could not be measured independently of the tests, which were themselves curiously correlated with achievement. In addition, the tests appeared to be culturally biased; on average, white middle-class children always fared better on the tests since questions were often worded in ways consistent with their experience (including their experience of test taking) but not with the experience of, say, poor African American children. Finally, many talents were not measured by the tests at all. For example, creative arts ability was not measured by the tests; they would not likely identify the brilliant pianist or painter.

Of course, science may be used in more ways than one in the construction and enforcement of standards. It is conceivable that some standards employ all three uses of science to differing degrees. But this brief examination of the uses of science in standards setting and enforcement should make it clear that there is far more to standards than science. Certifications that make claims of being scientific may do so in each or all of the three ways described above.

Accreditation

To a large extent, the use of certifications merely shifts the problem of trust from one party to another. After all, how do I know that I should trust the certifier? Perhaps the certifier is in cahoots with the other party to the exchange. Perhaps the certifier is incompetent. Perhaps the certifier finds the job of certification to be tedious and boring, and therefore exerts less care than necessary. These and other potential pitfalls to certification must be addressed. In general, accreditation has attempted to come to the rescue. Put simply, accreditation is a process whereby certifiers are themselves certified.

In recent years certification has become so widespread that national general accreditation bodies have emerged to support it (Donaldson 2005). These general accreditation bodies may accredit certifiers of everything from meatpacking plants to nursing homes, using the same general criteria for each. At the same time, several international associations have emerged to support systems of certification and accreditation. The International Accreditation Forum (2007) describes itself as "the world association of Conformity Assessment Accreditation Bodies in the fields of management systems, products, services, personnel and other similar programmes of conformity assessment." Similarly, the International Laboratory Accreditation Cooperation (2007) assists "the developing global network of accredited testing and calibration laboratories that are assessed and recognised as being competent by ILAC Arrangement signatory accreditation bodies." Furthermore, ISO has produced a guide for the consistent, standardized application of conformity assessment to accreditation bodies (International Organization for Standardization 2004).

Accreditation bodies push the problems of trust and risk yet one step further. Why should I trust a given accreditation agency? What certifies that they can be trusted? That they will reduce the risk to me, my family, my organization? The short answer to this question is "It's the network, stupid." In short, the more dense the network of standards, certifications, and accreditations, the more likely it will produce the desired results—greater trust and reduced risks. But let's not end the discussion quite yet.

The Tripartite Standards Regime and Its Problems

For the combination of standards, certifications, and accreditations described above, there is a convenient shorthand: the tripartite standards regime (TSR).[6] A TSR may be used to form alternative modes of governance for virtually any and all aspects of social life. TSRs differ from state-based modes of governance in that they are often a cobbled-together network of persons, organizations, and things, rather than being constructed on a formal hierarchy of status relations. TSRs may be supported and granted special status by nation-states, or they may be an entirely private form of governance, subject to state laws about contracts, fraud, and so forth, but not the subject of any special legislation. Moreover, TSRs are surprisingly unstandardized; TSRs for education do not necessarily have the same form as those for auto parts. Nor is there any particular reason for them to be so. Let us consider several divergent examples.

Some TSRs are dominated by the government. For example, organic agriculture standards in both the United States and the EU are written into law. However, certifiers are not necessarily state agencies. In the United States a number of organizations, such as Oregon Tilth, are accredited as certifiers of farmers. Similarly, in Britain, the Soil Association plays this role.

Other TSRs are organized around the market, with little or no direct government intervention. Cashore, Auld, and Newsom (2004) have written at length on the rise and development of the Forest Stewardship Council (FSC), an international NGO. As they note, FSC is an example of a generally rather effective non-state market-driven (NSMD, pronounced nismoid) TSR. The incentive for participation in this particular TSR is the market advantages conferred by participation.

Yet other TSRs exist with little or no direct government intervention, but are also not market-driven. (Of course, markets are implicated in these as in all TSRs.) For example, the TSRs for higher education in the United States have little state or national government involvement (although the federal government is trying to increase its role) but are largely focused on public and nonprofit universities.

To summarize, TSRs can now be found in virtually all fields of human endeavor. They claim to be characterized not by the vicious circle of the Tweed Ring (figure 4.2), in which each actor claims that the next one is

WHO STOLE THE PEOPLE'S MONEY ? — DO TELL . N.Y.TIMES.   'T WAS HIM.

Figure 4.2
Thomas Nast's political cartoon depicting the Tweed Ring (1871). The notoriously corrupt Tweed Ring, led by William Marcy (Boss) Tweed (left foreground), dominated New York City politics during the 1860s and 1870s.
Source: Nast (1871).

guilty, but by a virtuous circle (or network) in which all the actors in the TSR are involved in checking on each other to ensure that the standards are maintained. Such networks can and do work, producing a growing portion of the qualified persons, products, processes, and practices found in the world today. But they are not without their problems. It is to those problems that I now turn.

The Violence of Audits

The very notion that audits do violence may seem a bit unusual to the reader.[7] After all, audits are supposed to be about protecting us from shoddy goods, incompetent practitioners, unscrupulous salespersons, and dangerous substances. The literature is filled with discussions of these clearly desirable aspects of audits. Moreover, the formality of audits may serve to protect some persons from otherwise capricious attack, and constrain our behavior precisely so that we may better live together. This is well known and is argued—often quite justly—in the preambles to thousands of procedural manuals and legal documents.

That said, even the most careful, disinterested audit may do violence to some degree. To explain how this is the case, I will look at the way "error" has been treated in four seemingly disparate fields: physics, molecular biology, insurance, and marketing. All of these are related to Walter Shewhart's brilliant invention, the industrial control chart. And each highlights how the control of error may do violence.

Shewhart's Physics

Physics has long been the most mathematical of all sciences, so much so that for many years, philosophers of science believed that as sciences advanced, they would all become more and more like physics. As someone put it, with apologies to William S. Gilbert and Sir Arthur Sullivan, physics was "the model of a modern major discipline." For many years, sociologists of science accepted this received wisdom with great enthusiasm. However, today nearly all philosophers and social scientists recognize that there are many approaches to science, of which physics is merely one.

By the time Walter Shewhart received his doctoral degree, physics was already a highly mathematical subject. Its practitioners were able to pose a variety of questions and test a variety of hypotheses using the elegant tools of mathematics, as well as an ever-increasing array of instruments. Physical experiments and physical theory became increasingly mathematical, such that today, even scientists from other fields find it difficult to follow the mathematics in many physics papers.

Moreover, unlike practitioners of less mathematized sciences such as biology and the social sciences, even a century ago physicists had already begun to carefully analyze not only the results of their experiments but the nature of the errors they made. They wanted to know not only whether experiments supported their hypotheses but whether the statistically defined errors—the deviations from the expected results—revealed anything about fundamental physical processes. At the time a great debate raged within physics. On one side was the older generation of physicists, who argued that probabilistic estimates were the result of incomplete information that would later be acquired. Here one found even stalwarts such as Albert Einstein, who famously said, "God does not play dice." On the other side were equally respected supporters of the new field of quantum mechanics such as Niels Bohr, who argued that that some phenomena are inherently probabilistic. But it mattered little which side of the debate one took. In either case, it behooved experimentalists to reduce the size of the error terms, either to eliminate them entirely or to reduce them to a probabilistic remainder. It was this widespread interest within physics in error that Shewhart used to develop the statistical control chart (see figure 2.2).

It is important to emphasize here the critical relation between precision and error. Physicists found it useful to analyze errors only because they could predict results with a high level of precision. This is generally not the case in other fields of science or engineering. For example, if a molecular biologist desires to insert a novel gene in a plant cell, she will employ one of a number of available techniques commonly used in the field. The insertion process is sufficiently difficult that most attempts will be failures. The techniques used often compensate for this by involving thousands of trials, with the expectation that one or a few will be successful. If the process is successful, the results will be written up for publication. But nothing will be said about the attempts that failed. Molecular biology is simply insufficiently precise to learn much by analyzing the errors.[8] A multitude of variables might have intervened to make the process of gene insertion fail, and all are inextricably confounded in the error term.

Shewhart was able to apply what he learned in physics to the creation of statistical control charts, for a number of reasons. First, he was employing a very limited number of measurements of physical objects. For example, bolts might be tested for length, width, thread count, and tensile strength. Other characteristics of the bolts—their color, the geographic origin of the ore from which they were made—were of no consequence whatever to the managers at AT&T. Second, the objects of interest were made standardized. The entire production run of bolts were supposed to be the same, where "the same" meant being the same in certain critical ways of value to the company. Third, since Shewhart knew how the products were made, he realized that the nonrandom errors he identified had to be the product of a limited and relatively well-defined set of causes. The stamping machine was miscalibrated or worn, the die-cutting machine was misaligned, and so on. While finding the precise cause of a group of defects might take some time and detective work, in most instances an engineer or foreman familiar with the production process would be able to identify and correct them. In principle, only random error within accepted tolerance limits would remain.

In contrast, Frederick Taylor, despite his myriad time and motion studies, could not and did not analyze error terms. Taylor certainly argued that there was one best way to do a job, insisting thereafter that the job be performed in precisely that way, and he could "correct" workers who did not follow his procedures, but he had no way of analyzing the causes of the errors they made. Was it because they were tired? Because they were not all as strong? Because they were miserably paid? Because they were, to use his term, "soldiering"? Because the job was extremely boring? Or was it for some other, unimagined reason?

Biology and Error

At about the same time that Shewhart was demonstrating the benefits of statistical control, other physicists were beginning to revolutionize biology. A group of about a dozen persons formed in 1932 in Britain, including, among others, the eminent physicist J. D. Bernal. Together they created what became known as the "biotheoretical gathering," attempting to bring the insights, predictability, and precision of physics to the biological sciences. Warren Weaver of the Rockefeller Foundation heard about the meeting and saw it as fitting well within the goals of the foundation. For Weaver, the aim was clear: "Can man acquire enough knowledge of his own vital processes so that we can hope to rationalize human behavior?" (quoted in Kohler 1980, 263). Between 1932 and 1957 the foundation poured more than $90 million into (what became) molecular biology (Abir-Am 1982), helping to transform biology in some surprising and unexpected ways.

Among other things, the rise of molecular biology attempted to do for the biological sciences what statistical control did for industrial production. It claimed to allow the (partial) replacement of the art and science of plant and animal breeding with the far more precise and more rapid transfer of individual genes. The breeding done by plant and animal breeders had to focus on optimizing the average or threshold levels of desired characteristics while minimizing the undesired characteristics. In contrast, the new molecular genetics and later genomics focused on introducing particular desired genes carefully selected to code for particular traits, as well as on identifying undesired genes and removing them or blocking their expression. In short, much as statistical control allowed the removal of nonrandom sources of physical error, molecular biology allowed the removal of nonrandom sources of genetic "error," or, conversely, the empirical identification of "errors" of particular value.

As an example of the first result, the removal of nonrandom sources of genetic error, in recent years various genomic, proteomic, and metabolomic projects have led to the development of an entire new subdiscipline, computational biology. Computational biologists apply statistical techniques to mine the vast data sets that are generated by modern molecular biology to compare DNA sequences, build mathematical models of biological processes, and identify proteins that appear linked to particular biochemical processes. As the author of one text on the subject put it, "Finding relevant facts and hypotheses in huge databases is becoming essential to biology" (Waterman 2000, xiii). And essential to that effort is developing common standards for identifying, characterizing, and reporting molecular data (e.g., GenomeWeb 2008; Taylor et al. 2008).

As an example of the second result, the identification of specific gene clusters of interest, massive projects such as the Molecular Library Initiative (MLI) of the National Institutes of Health and similar projects mounted by pharmaceutical companies permit the testing of vast numbers of molecules. As observed in a *Science* article by Jocelyn Kaiser (2008, 764),

three state-of-the-art yellow robots are hard at work processing biological assays. They fetch plates that are each dotted with 1536 tiny wells of different small organic molecules, mix in a protein or cell solution, then run the plate through a detector that spots whether any of the chemicals on the plates has triggered some change in the protein or cells. In another room, medicinal chemists tweak these "hits" to improve the strength and specificity of the interaction.

The proponents of the MLI, funded at over $100 million per year, are confident that they can identify the proverbial needle in the haystack—a compound that will eventually be transformed into a highly specific drug for treating a particular disease.[9]

Insurance

The third transformation became commonplace during the last decades of the nineteenth century, although it began much earlier. This was the calculation of risk by insurance companies, banks, and other financial institutions. Insurance was initially made possible by the realization that risk could be determined with a high level of precision if one computed the statistical average and insured based on it (see, e.g., Lengwiler 2009). From death records one could calculate with reasonable accuracy the average age at death of a given population. From records of successful voyages of ships, as well as those lost at sea, one could calculate the likelihood of a ship making it to the next port. And to reduce risk still further, ad hoc consortia of insurers could (and still do) insure a single ship. Similarly, from records of theft and pilfering, one could calculate the likelihood of a given overland shipment making it safely to its destination. In each case, if one insured a sufficiently large number persons, ships, or goods, one could nearly guarantee a tidy profit.

Of course, no insurer was foolish enough to insure a man on his deathbed, a ship whose hull leaked, or a shipment lacking the proper protections. But it soon became apparent that profits could be greatly increased if the specific instances presented to them were more carefully analyzed. As one group notes, "One of the ironies of insurance is that, while it is supposed to pool risks, in practice it tends to unpool them, breaking down the larger pool of potential insured in search of smaller, less risky pools, which are more advantageous for some, while excluding others" (Ericson, Barry, and Doyle 2000, 534). What specifics about a person might give hints as to the likelihood of their dying while young? What characteristics of a ship might indicate its seaworthiness? Answering these questions also involved a careful examination of the error terms, of the seemingly healthy person who died, the seemingly seaworthy ship that sank.

Marketing Error
Finally, the rise of data mining in marketing reflects a concern with error. For most of the last century, marketers relied on surveys, using, for example, average values in particular geographic areas to target marketing campaigns. Marketers might determine that the residents of a given township were particularly fond of a particular detergent, whereas those of another township might prefer a different brand.

But in recent years, in part as a result of high-speed computing, marketers have begun to manipulate vast amounts of information. A quotidian example is the data generated by loyalty cards offered by many supermarket chains. These cards provide precise information on the buying patterns of individual consumers, and since one must sign up for the card, the buying patterns can be linked to one's gender, as well as to census tract data on income, race, and education.

Today, data mining is big business and used in a wide range of applications. For example, it can be used to divide markets into extremely fine segments, such as people who buy milk, diapers, breakfast cereals, and party goods on Saturday morning. Once this is done, analysts can determine what else these shoppers buy. Then coupons can be generated at checkouts that encourage consumers to try related products or other brands. Marketers can also reconsider the placement of various products in a store in an effort to increase sales: Should the milk be placed by the breakfast cereals? Or should it be placed near the cheese and other dairy products? Moreover, analysts can identify highly localized trends in consumption, ensuring that store stocks are neither too great nor too small to meet demand. They may predict (stimulate) new trends or discover new patterns of consumer behavior. They may even run a successful air taxi business by mining data on the travel patterns of potential customers (Fallows 2008, 66).

The key point to keep in mind here is that data mining is all about statistical analysis of the error terms. It is about going beyond means, medians, and standard deviations to make sense out of myriad relationships that distinguish one customer from another, and using that information to increase sales.

To sum up, in four apparently unrelated fields of human endeavor—industrial production, computational biology, insurance, and marketing—a great transformation is taking place. This transformation involves the (re)establishment of the links between and among statistical control, experimental control, and social control. Not only can averages be computed. Each of the individual dots in a scatterplot can be examined to determine why it deviates from some predefined norm. This is true regardless of whether the dots represent the subject of scientific experiments, the products of industrial enterprises, the life chances of individual human beings, or the market basket of goods that people purchase. Indeed, this approach has been applied to monitoring epidemics as they occur(Weinberger 2008), predicting riots (Rogers 2008), tracking airfares (Kuang 2008), predicting crop yields (Paynter 2008), sorting documents in complex legal proceedings (Bringardner 2008), predicting voting patterns (Graff 2008), and identifying the processes involved in bone loss (Goetz 2008).

What does all this have to do with certifications and accreditations? The tests employed to certify or accredit require of both persons and things that they conform to some standard. In the case of physical objects we are not particularly concerned about the "discipline" we might impose on them by enforcing a particular standard, and we may often employ statistical control and sequential analysis to enhance that discipline, to make those objects fit our needs better. But when we do the same to people, when we perform the TSR according to the received standards, we must do violence to them to some degree.

For example, under the name GlobalGAP (previously EurepGAP), a number of European supermarket chains have created a set of common standards, or good agricultural practices (GAP), for fresh produce and meats sold in their stores. They aim to audit their suppliers to determine that they meet certain food safety, worker health and safety, food quality, and environmental standards. Each producer of fresh produce for one of these chains must be certified by an accredited certifier. GlobalGAP provides a lengthy list of "major musts," "minor musts," and recommendations.

Accredited certifiers then send teams to the producers' fields to check whether these conditions are being met.

From the perspective of participating, mainly European, retailers this scheme is highly effective. It generally ensures that products sourced in Africa, Latin America, and Asia are safe and of acceptable quality, and are not produced in a manner that is environmentally destructive or that exploits farmers or workers. This helps supermarket chains position themselves as protectors of consumers and as caring members of the global community. However, for farmer-suppliers who wish to participate in this or another certification scheme—a necessity if one is to gain access to this market[10]—it imposes a rather significant set of costs and limitations on their "freedom to operate." Indeed, while apparently voluntary, these standards often substitute for public regulations (Fouilleux 2008). Often, smaller producers have neither the financial nor the managerial ability to participate (Dankers 2003). The impact on farmworkers is somewhat more mixed, with regular, full-time workers likely receiving greater protections (e.g., from harmful pesticides) and part-time temporary workers, often hired at peak labor periods, largely unprotected (Bain 2010).

Furthermore, since (1) the requirements are set uniformly across all suppliers, (2) the audits are largely paper audits, that is, certifiers audit the paper trail produced and do not (or only minimally) include either field testing or interviews with participants, and (3) they assume that legal restrictions (e.g., on wages to be paid, workers' rights to organize, care of the environment) are uniformly enforced, they cannot easily compensate for local differences. Therefore, the results are mixed. Likely some persons are significantly better off, while others are significantly harmed. At least one observer argues that schemes of this sort can be viewed as having a "neocolonial civilizing mission" (Freidberg 2003). Others argue this is far too pessimistic a view (Jaffee and Henson 2004).

Similarly, in a recent best-seller, Malcolm Gladwell describes seemingly innocuous standards the enforcement of which can and does alter one's life chances. Among numerous examples:

• "At four year colleges in the United States—the highest stream of post-secondary education—students belonging to the relatively youngest group in their class are underrepresented by about 11.6 percent" (Gladwell 2008, 29). In other words, all other things being equal, entering school at a younger age reduces one's likelihood of being admitted to college. Of course, when one enters school is a standard established by school authorities; it is not up to the student or her parents.

• In nations where hockey is a popular sport, admissions to hockey teams are age-based, with some date used as a (standard) arbitrary cutoff. Within any given annual cohort of children, those who are younger tend to perform more poorly. As a result, they are not selected for class teams in the following years.

These examples, as well as the obesity audits described in box 4.1 and the growing audits of teachers, police officers, social workers, health providers, and other professionals (e.g., Strathern 2006; Travers 2007), illustrate the violence that may be done through the enforcement of standards and by certifications and accreditations.[11] While doubtless there are persons who wish to do violence to others, there is little or no evidence of that in the

Box 4.1
Obesity audits and violence

Evidence suggests that in many industrialized nations and in more than a few poorer nations, a considerable and growing portion of the population suffers from obesity and the health problems associated with it. Using a measure known as the body mass index (BMI), it is possible to gauge the degree of obesity in a given population. The BMI is calculated by dividing weight by height squared. It has the enormous advantage that it is easily calculated by someone with minimal training. However, the BMI does not directly measure obesity, and obesity is not invariably a correlate of ill health. In short, the BMI is doubly removed from the phenomenon it claims to measure: BMI → obesity → ill health. That said, although crude, the BMI is arguably a reasonably good measure of certain forms of ill health in *populations*. However, as is typical of medical standards, having a normal BMI is seen as good, as achieving the golden mean, whereas having a BMI either above or below the range deemed healthy (i.e., being either too fat or too thin) is seen as abnormal and bad.

Were this the end of the story, there would be little to say here. But, since the BMI is calculated for particular individuals, and since it produces a statistical distribution, it is possible to analyze the error term, that is, to ask whether the BMI for Anne or Edward is more than a certain distance from the mean. Furthermore, based on the score, it is possible to take remedial action, much as one might were someone to be reading far below the level expected for age.

Evans et al. (2008) have examined this phenomenon in some depth, looking specifically at the clinical measures taken in schools in the UK and elsewhere in an effort to put a stop to the "obesity epidemic." In brief,

Box 4.1
(continued)

through detailed studies over five years of a cohort of forty young women suffering from anorexia or bulimia, they show how treating obesity in this manner actually does violence to many young women.

To summarize, they note the following:

• Recent UK government reports have shifted from concern about the health of youngsters to concerns about the obesity crisis, using BMI measurements to support the claim. This has occurred despite considerable debate within the medical community as to the import of the BMI, both as to what level should be considered obese and the value of the BMI as an indicator of individual health and well-being. Indeed, it is being overweight *around the abdomen* that is correlated with disease; the BMI does not capture this essential aspect of obesity. Moreover, simply shifting the cutoff point from 27 to 25 in 1998 instantly made an additional fifty million Americans overweight (Boero 2007).

• The emphasis in schools adds a clear moral value to being overweight. In some instances it has been suggested that having an overweight child be considered a form of child abuse, to be remedied by removing the child from the home.

• Being overweight has been defined as largely an individual problem: "A blame-the-victim culture . . . interprets fat as an outward sign of neglect of one's corporeal self, a condition considered shameful, dirty or irresponsibly ill" (Evans et al. 2008, 55). Of course, this ignores the incessant advertising of junk food, the cult of the (overly) thin model, the car culture, the increasingly hurried and harried character of family life, and other transformations beyond the control of any individual, especially those who have not yet reached adulthood and are still attempting to figure out who they are.

• Teachers, willingly or not, find themselves caught up in the obesity frenzy as it has become part and parcel of the larger "audit culture" to which teachers, schools, and students are subjected each day. Hence, even as individual teachers resist, they are themselves graded on, among many other things, how their students fare on BMI testing.

In sum, through the use of obesity standards, as measured by the BMI and enforced as part of the audit culture of the schools, violence is done to the young women interviewed by Evans and colleagues, and doubtless to countless others. Even if they are aware of it and take actions to resist, the violence is nevertheless done to them. Moreover, this violence is no less painful for the fact that those doing it doubtless believe they are doing it for the good of the youngsters affected.

examples described above. The problem is far more subtle. Standards appear to do violence to persons when

• The audits used to measure conformity to the standard are weakly related to the phenomenon of interest,
• The audits employ overly narrow measures that direct attention away from other important but difficult to measure aspects of the phenomenon of interest,
• The audits employ what Porter (1995) calls "mechanical objectivity," relieving the certifiers of the need to make informed and considered judgments, and
• The demands made on persons by the audits significantly intrude into their pursuance of whatever goals they had.

The problem noted here with respect to certifications is exacerbated when such standards are written into law. The No Child Left Behind Act of 2001 (P.L.107–110) is a particularly egregious example.[12] Ostensibly, NCLB was designed to improve the quality of American elementary and secondary education by instituting mandatory standards and testing. Schools in which more than a certain percentage of students consistently perform unsatisfactorily on the examinations are required to make major organizational reforms. If these changes do not result in school improvement, federal funds are withdrawn and, in the vast majority of cases, the school is shut down.

The violence done by the act, regardless of whether it was intentional, is considerable. Consider the situation in which the tests are given. Administrators and teachers at suburban schools that enroll largely middle- and upper-middle-class students have little fear that the test results will be problematic. Hence, they see them mainly as an annoyance that takes time from regular classroom instruction. In contrast, administrators and teachers at inner-city schools are painfully aware that all eyes are on them. They know they work in poorly funded, often run-down schools[13] with a disproportionate number of students from low-income and immigrant families. They also know that continuing failure to maintain average grades above a certain level will likely mean loss of their livelihoods. It comes as no surprise that occasionally administrators engage in questionable activities designed to increase the school's scores. For example, schools might market themselves to parents. As one observer puts it, "More time and energy is spent on maintaining or enhancing a public image of a good school and less time and energy is spent on pedagogic and curricular substance" (Apple 2004, 23). Schools may also discourage poorly performing students from enrolling. They may devote more attention to those students performing just below acceptable levels and less attention to those at the top or the bottom. School officials may encourage poorly performing students to be absent on the day of the test. In at least one state, extraordinary improvement in math test results was achieved by reducing the difficulty of the test (Bhattacharjee 2007).

In addition, the tests themselves emphasize (English) language and mathematics. Other subjects such as the sciences, foreign languages, art, drama, music, social studies, and civics are unmeasured. This has little impact on those schools that will do well on the tests, since they continue to offer these subjects. But it is devastating for the weaker schools. At best, it encourages a narrowing of the curriculum to focus on the subjects included on the tests; at worst, schooling is reduced to endless drill on items likely to be on the test. Hence, an unintended impact of NCLB is that even those poorly performing schools that succeed in meeting the minimum standards may actually do a poorer job of educating students as a result (Apple 2006).

Finally, the NCLB does violence to students in several ways. First, there is only weak evidence that test scores correlate with any future life chances, which are far more influenced by grades, study habits, and parental income than by test results. Second, the standards demand that all students learn the same subject matter regardless of abilities or interests. Nor are students consulted in any way as to their views on the matter, a rather strange approach for an avowedly democratic society (Noddings 1997). Third, the standards erroneously assume that all students should learn everything the teacher teaches, and they make no allowance for the likelihood that students will learn things that were not taught at all. Put differently, the students are taken to be passive recipients of information provided by the teachers.

In short, under NCLB, students who take the tests have violence done to them in that the students who are having the most difficulty—usually through no fault of their own—are the ones most likely to be presented with an impoverished curriculum and a dreary classroom experience preoccupied with testing. NCLB does violence to teachers in that it wrongly assumes that poor grades received by students are always the result of incompetent teaching, conveniently forgetting the considerable class and race divide that is found in American schools. Finally, NCLB does violence to schools themselves as communities of parents and teachers, taking control of the schools away from those persons who care and making it the province of a faceless bureaucracy located far away.

Ironically, perhaps no one understood the limits to this kind of testing and statistical control better than W. Edwards Deming (1982, 105). What he said of manufacturing plants applies to all organizations: "One of the main effects of evaluation of performance is nourishment of short-term thinking in short-time performance. A man must have something to show. His superior is forced into numerics. It is easy to count. Counts relieve management of the necessity to contrive a measure with meaning." Unfortunately, people who are measured by counting are deprived of the pride of workmanship, even as managers avoid the necessity of making judgments.

Indeed, Deming (1982) amusingly shows how the job of taking marbles of one color out of an urn holding marbles of multiple colors will produce differences in worker performance. As he notes, these differences will say nothing whatever about the quality of the work performed.

However, Deming failed to see the limits to the entire quality movement (total quality management, continuous quality improvement). As I see it, there are at least three fundamental limits: First, proponents assume that quality is always the central goal for everyone, that, as the Ford ad once said, "Quality is job one." This is simply not the case. Many other aspects of organizational life are also important. These aspects include but are not limited to the self-respect of those who work there, and the loyalty they have to the organization.

Second, despite the change of name, quality is still largely about control; it is about the assessment of conformity. Few would argue much about the control of things. But there is a certain ill-defined point at which control of things becomes control of people. And although the quality literature talks of empowering people, it tends to frame that empowerment in the context of a given and unchangeable set of organizational goals. If people are to be truly empowered by quality, audit, and certification, they must be genuinely able to change the goals of the organization.

Finally, the single-minded focus on quality has the unintended consequence of ignoring all those facets of human life that cannot be fitted under the quality heading. As Ulrich Beck (1992, 134, emphasis in original) argues, "Institutions act in legally determined *categories of standardized biographies, to which reality conforms more or less.*" Thus, quality in the workplace or classroom tends to ignore the social lives of those who inhabit these organizations—or, worse still, intrudes into it such that everyone is on call twenty-four hours a day. As one observer put it, "It is not enough to acknowledge that there is a human side to TQM, let alone merely to give lip service to it. People have to be put first—ahead even of quality" (Connor 1997, 507; see also Walzer 1994).

The Limits to the Tripartite Standards Regime

While TSRs are clearly desirable and even necessary in many instances, several major weaknesses may arise from their use. For example, TSRs may certify conformity to the wrong standards, to standards that are outdated, inadequate, irrelevant, or insufficiently (or overly) stringent. To prevent this, standards for certifying physicians are constantly being modified; the combination of course work and practical experience that would have been acceptable fifty years ago is no longer acceptable today. The same may be said of things: Manufacturing telephones to the standards of fifty years ago would be both costly and unacceptable today. This was a particular problem for the Soviet bloc nations, where little incentive outside certain military and aerospace industries existed to update standards to meet changing situations. It remains a problem today in many parts of the world.

Moreover, even if the standards are adequate and up-to-date, the means for certification may be perfunctory. The literature is replete with examples of certification procedures that consist of little more than checking boxes on forms indicating the presence or absence of certain qualities of persons or things. It is all too easy for such certifications to become merely formal "rituals of verification" (Power 1997) in which mindless devotion to the rules of the audit itself is the subject of the audit. Furthermore, auditors might ask easy questions: Did the organization have the required number of toilets? Did the student know some particular group of facts? Did the fruit-processing plant have a safety officer? Were the rules posted? Such questions beg more important questions: Were the toilets accessible, clean, and functional? Was the student becoming educated? Was the processing plant safe? Did everyone know and follow the rules, and were the rules the right rules?

In addition, the procedures of certification and conformity assessment have a potentially insidious aspect to them: They may and often do require that organizations be reorganized to be more easily auditable. But that very reorganization may undermine the values promulgated by the organization. Their widespread use may actually undermine trust and accountability by substituting predictability for trustworthiness, care, and character, as well as by invoking an illusory sense of well-being.

Furthermore, certification may turn interesting and meaningful work into drudgery. This is especially the case if the certification procedures appear unintelligible or absurd to those persons who (or whose work) is to be certified. Then the certification of adherence to a given set of standards becomes a reified, unintelligible, arbitrary set of requirements that must be fulfilled for the benefit of some faraway client, buyer, or agency.

Moreover, even if the problems noted above are adequately addressed, certifications may ignore deliberation over the big questions while emphasizing deliberation over the trivial ones, all in the name of expediency and efficiency. Ironically, this may well lead to precisely the sort of rigidity noted by critics of central planning.

Finally, the violence that can be and often is done by certification, accreditation, standards promulgation, and audits needs to be recognized as neither an unfortunate but inexorable result of social betterment nor as a desirable outcome in an increasingly rationalized and scientific world. Instead, it should be understood as a form of conquest analogous to that noted several centuries ago: "It pursues the vanquished into the most intimate aspects of their existence. It mutilates them in order to reduce them to uniform proportions. In the past conquerors expected the deputies of conquered nations to appear on their knees before them. Today it is man's morale that they wish to prostrate" (Constant [1815] 1988, 77).

Conclusion: Governing the Market Society

The rapid proliferation of standards, certifications, and accreditations of everything and everybody, in the context of improved communications and transportation, combined with the widespread use of SCM, returns us to the very problems of governance that the neoliberals sought to resolve—though not in the form of central planning and central control. Instead, while the enactment of neoliberal reforms transforms the market, regulating it (more or less) according to the rules of the neoliberal model, it also promotes new and sometimes highly problematic forms of governance.

Even as managers and owners of firms relish the freedom to operate offered by the neoliberal marketplace, they are dismayed by the *horror vacui* that is thus created. No small firm can afford to shelter itself from the instabilities and risks associated with the free market. No large firm can afford to subject itself to those very instabilities and risks. SCM and TSRs are means by which segments of the market can be bounded, protected, restructured, and stabilized so as to be made subject to governance by dominant firms in supply chains.

Of course, this is not the kind of governance administered by either democratic or totalitarian states. It is largely a new form (or forms) of governance, although in some ways it is reminiscent of mining and plantation economies in colonial empires, and in other ways it is reminiscent of absolute monarchies. It is the contractual governance of myriad quasi- or pseudostates that complement or supersede much of the governance of the state, that produce a wide range of complex, often difficult to perceive recipes for realities. Furthermore, unlike the governance of democratic states, the new pseudostates—whether firms or voluntary organizations— usually lack judicial functions. Disputes are often handled administratively, by decision of an inspector or some other functionary in the dominant organization in the supply chain. Those who are the subjects of disputes are usually absent at the negotiations where the rules are legislated.

But one may still rightly ask what alternatives there are to audits. To answer that question it is necessary to examine the links between and among standards, ethics, justice, and democracy. This is the subject of the next two chapters.

5 Standards, Ethics, and Justice

Moral science is not something with a separate province. It is physical, biological and historic knowledge placed in a human context where it will illuminate and guide the activities of men.
—John Dewey (1922a, 296)

Let me begin this chapter by summarizing the argument so far. I have argued that (1) standards are to be found everywhere, (2) they are all about power (in addition to being about technical, economic, political, social, and other concerns), (3) the Enlightenment project involved standardizing the world, (4) this spawned a response using standards to differentiate, propelled in part by changes in transportation, communications, packaging, and the neoliberal reforms, and (5) differentiation was and is dealt with by developing organizations that certify and accredit nearly everything and everyone. Furthermore, I have argued that, under certain circumstances, such certifications and accreditations may do violence to persons.

But if standards are all about power in society, then they must also be about ethics and justice. Indeed, as Dewey (1922a, 75) argued, all forms of conduct—customs, habits, standards, laws—have moral implications since "they are active demands for certain ways of acting." While all standards have ethical import, in most instances the ethical issues for which they have import have been largely settled. There is rarely a need, for example, to raise issues about which side of the road to drive on, the standard height of doors, or the standard for body temperature.

That said, it is surprisingly commonplace even for ethicists to see science as largely settled and ethics as the subject of endless debate. A few moments' reflection will demonstrate that this is not so. As Bruno Latour (1987) has noted, science in action is quite different from science that has been completed. The former is the subject of experiment and heated debate, while the latter is taken for granted, accepted, no longer the subject of discussion except in the education of a new generation of scientists. The same is true of ethics. Ethics in action is contentious, subject to considerable experimentation and revision, a place where competing theories and practices clash, but ethics completed is taken for granted, accepted, no longer the subject of discussion. In both instances there are those who still adhere to what others would consider outmoded views. Moreover, someone may come along and—if she can muster sufficient resources, or if the situation changes sufficiently—reopen the debates, but the more tightly linked to other settings the scientific or ethical knowledge is, the less likely it is that route will bear fruit.

In this chapter I explore the ethical import of standards, focusing not on the ethical aspects of standards that are now taken for granted but instead on ethics in action, on how ethics may illuminate the ways we use standards. I begin by summarizing a sixty-year-old debate in analytic philosophy over the ethical import of standards. Then I briefly examine how the three major questions posed by ethicists—questions about consequences, rights, and virtues—each provide insight into standards. From there I attempt to turn those questions around by asking how various standards are justified. Next, I examine how standards relate to the three ways of distributing goods and the good: markets, need, and dessert. The following section notes how standards can contribute to injustice. I conclude by borrowing from business the concept of value chains and show how with sufficient reshaping, it might help us grapple with the complexities of some ethical conundrums.

From Standards to Ethics

There is a great difference between judging the merits of the apple grower as an apple grower and grading his produce. There is an even greater difference between grading his produce and judging him as a man.
—John Llewelyn Evans (1962, 32)

From such factual premises as "He gets a better yield for this crop per acre than any farmer in the district," "He has the most effective programme of soil renewal yet known," and "His dairy herd wins all the first prizes at the agricultural shows," the evaluative conclusion validly follows that "He is a good farmer."
—Alisdair MacIntyre (1984, 58)

These two seemingly opposed comments by philosophers epitomize the debate with respect to the relationship between standards and ethics. Yet surprisingly, despite what appears to me to be quite obvious—namely, that standards are and must be about ethics (among other things)—philosophers have been generally rather silent on the issue.[1] Many apparently cavalierly dismiss standards, especially standards for things, as having no ethical content.

But one particular philosopher did write an important, although much debated, paper on standards, James Opie Urmson (1915–). Urmson is a philosopher in the analytic tradition, focused largely on language. In discussing grading, the example that Urmson chose, likely because of its mundane character, was that of grading physical objects: apples. Urmson argued that "grading, like speaking English . . . is something which you cannot in a full sense do without understanding what you are doing" (Urmson 1950, 147). This is the case because there is a set of principles that is employed to engage in grading, and these principles must be knowingly employed. To grade apples you don't need to have ever tasted one, but you must understand the criteria by which they are to be graded.

This in itself would be insufficient to link the practice of grading to ethics, but Urmson goes on to argue that the terms "good" and "bad" are merely general grading labels used by philosophers and laypersons in sorting persons, objects, processes, and practices. Of course, some criteria are taken to be more important than others by those who are doing the grading. Moreover, the criteria themselves are quite obviously different for apples than for people. However, for Urmson, in both instances a judgment is called for, and that judgment is an ethical one.

Numerous critics have challenged various aspects of Urmson's view. For example, Karl Britton (1951, 528) argued that whereas the technical criteria for grading apples were essentially fixed and public, the criteria for judging that someone was good did not bring closure to the issue. As he put it, "Moral conduct is not a particular kind of activity and virtue is not a particular art." Another critic made a similar argument, asserting, "We should think a person who dismissed a movie because it failed to fulfil some set standard a most insensitive person and no fit judge of art" (Browning 1960, 239).

But this assertion appears a bit too strong. As Paul Taylor (1962) argues, not all criteria can be discerned via simple empirical tests, nor can all criteria be equally easily formalized. (It is probably easier to formalize what should not be included as criteria for a good film: the company from which the film was bought, the color of the can in which it was stored, and so forth.) The standards for Harlequin novels and novelists described earlier illustrate this point, but standards they are. In addition, standards are never fixed. They are always subject to interpretation, although some standards are more or more frequently disputed than others. Hence, identifying clear criteria by which to grade films on their artistic value is far less likely to be successful than is identifying criteria for grading apples (although the criteria for grading the suitability of films for children are authoritative). That said, it seems reasonable to argue, all other things being equal, that one should spend more time pondering whether someone is a good person than whether a film is a good film, and more time considering whether a film is a good film than whether a given apple is a good apple.

Another challenge to Urmson sought to distinguish between beings before they are graded and after the fact. In that article, M. J. Baker (1951) argued that Urmson's criteria apply only before the actual process of grading begins. In contrast, once the being has been graded in a formal manner, a label (what I called an indicator earlier) is attached to that being (e.g., an apple is identified as grade A, a person is identified as a high school graduate). The label purports to tell us something about that being, or about beings of the class to which that being belongs. Thus, to say that Jill has a high school diploma is to say something about her, and about the class of persons having high school diplomas. Baker suggests that the problem arises when the grading is ethical in character, that is, a ranking of good or bad. In those instances, "then the ethical problem is one of discovering what is the peculiar merit indicated by the use of moral grading labels, and what are the various ways in which this merit is manifested" (Baker 1951, 535). In short, Baker distinguishes two kinds of grading—one in which ethical judgment is called for and one in which it is not.

However, Baker is unable to tell us how or when to distinguish one type of grading from the other. David Miller (1965) resolves this problem for us. He notes that grading requires moral judgment only if one knows the *ranking* of the categories to which one grades; if no ranking is employed, then one merely has a set of nominal (and hence nonethical) classifications, what I earlier referred to as divisions. Divisions—whether of apple varieties or of eye color—have no moral import. In contrast, ranks, filters, and Olympics all involve the determination that some being is "better" than some other being. In short, Baker and Miller amend Urmson's assertion by noting that it is relevant for beings that are graded based on three of the four extant types of standards.

That said, divisions can be turned into ranks or filters with little difficulty, as the widespread presence of racism based on skin color—a seemingly nominal category—clearly illustrates. Both the postbellum American South and the apartheid South African regime did this with rather disastrous and clear ethical consequences (Bowker and Star 1999). Even today, racists are happy to transform the nominal classification of skin color into a ranking system. Hence, one might say that "pure" divisions are a limiting case; in practice, such classifications are rarely fully nominal.

C. A. Baylis (1958) challenges Urmson from another angle. He notes that Urmson is unable to tell us why people might agree on certain criteria as the relevant ones for grading. He argues, citing Dewey, that there are no intrinsic goods, as nothing is solely an end. But this critique appears merely to point to the limits of his analysis rather than in any way undermining Urmson's basic premise. Put differently, Urmson never asked why particular criteria were chosen; he showed only that the employment of criteria (in the case of ranks, filters, and Olympics, but not divisions) was necessarily an ethical ranking. Asking why a given set of criteria is employed is a means of grappling with the ethical content of a particular situation. As David Miller (1965, 113) writes, "Only by means of qualitative linear ordering can we say meaningfully of items that they are significant or insignificant, divine or corrupt, good or bad, beautiful or ugly." In other words, the use of some sort of system of ordering is necessary to and indicative of moral judgment.

But perhaps the most serious challenge to Urmson's view came from John Llewelyn Evans (1962). Evans notes, following Aristotle, that there are many uses of the word "good." He goes on to argue that philosophers have either (1) tried to identify a common element that unites them all, or (2) argued that the various uses were radically different and merely led to confusion. Evans is willing to concede that Urmson's position appeared to provide a means by which one could move away from seeing evaluations of good or bad as merely subjective. After all, the very fact that formal standards require criteria by which they are employed provides at least some level of objectivity. But he argues that there are two forms of objectivity involved in standards and grading. The first involves "the relation of an individual judgment to the criteria employed," while the second indicates "the status of the criteria themselves" (Evans 1962, 29). Put differently, the person who is involved in grading must employ judgment in using the *received* criteria: wormy apples should be separated from sound ones. Graders may do this well or poorly. The second aspect of objectivity involves asking whether the criteria used, such as worminess, are in fact valid and objective criteria. Evans argues that grading is objective only in the first sense, and that this applies to all beings that are graded.

So far, so good. But then Evans introduces a particularly important point: in order to grade a being, that being must be sufficiently like other members of the class to afford comparison according to whatever criteria are used. In other words, one can grade apples by color, size, and shape because they are all apples. We could, and in certain very unusual instances we do, grade all sorts of otherwise unlike things—apples, dogs, rocks—based on color, size, or shape or some other criterion. But in most ordinary situations we do not. Thus, in the vast majority of situations, likeness of the members of a class must be determined before the grading process can begin.

However, Evans argues that people cannot be morally graded, since they are not members of the same class in the way that apples are. As he puts it, "I should maintain that you never, or very rarely, get two men sufficiently alike so that we could say that one good man is just like any other good man, nor is one good man good in precisely the same respects in which any other good man is good" (Evans 1962, 30). In short, each of us is unique with respect to goodness and badness. Certainly the goodness expressed by Gandhi was quite different from that of Lincoln, so Evans's point would appear valid.

That said, his position would appear to be contradicted by our everyday experience. After all, we grade people all the time. We do it every time a test is administered in school, every time our insurance rates go up after an automobile accident, every time someone is fined for returning a library book late, every time someone is diagnosed with a disease, and every time that someone wins a gold medal at the Olympics. How does Evans resolve this apparent contradiction?

Evans asserts that we make two fundamentally different kinds of evaluations of persons. The first kind involves standards, criteria, and grading, "where we take nothing into account in arriving at our estimations except the actual performance of the person or persons to be judged" (Evans 1962, 31). The second kind involves including the context, the person's "endowment," or the extenuating circumstances into our judgment. Thus, as he suggests, when a rich man steals a loaf of bread for the fun of it, the situation is hardly the same as when a poor man steals a loaf in order to eat. But the very example provided illustrates how problematic his position is. If both were to be caught and brought to court, both would be found guilty of stealing the loaf. A judge might well mete out a harsher punishment to the rich man, but the judgment would likely be the same in both instances, since the criteria for arriving at that decision in the trial would have been the same.

Evans also argues that each person is unique from the vantage point of morality, and that the same argument can be applied to works of art and literature. According to Evans, in each of these instances there are two obstacles to grading people: On the one hand, there is no clear set of criteria that can be developed that allows us to grade people and their artistic and literary creations. On the other hand, people are responsible for their artistic and literary creations, as well as for their other deeds; therefore they can be and are judged.

I really want to sympathize with Evans. But it appears that we are in a double quandary. First, whether we like it or not, in our contemporary world we are subject to more and more instances in which formal standards are employed and we and our creations are graded. Second, as Alfred Schutz ([1932] 1967) observed, we only know people as types. People we know well we know as many types, while those we hardly know, we know as a few types. In short, there do not appear to be two separate domains, one in which standards and objective criteria apply and another in which they do not. Instead, it appears there is something more akin to a (messy) continuum running from purely objective, robust, indeed even mechanized grading of persons and things (think of fingerprint readers and automated color sorters, respectively) to highly situated judgments. But Evans is surely right on one point: As I noted in the previous chapter, to the extent that we exclude considered, situated judgment and judge persons on formal criteria alone, we may well do violence to some persons.

Finally, Jerry Clegg (1966, 138) challenges Urmson by arguing that "grading is a physical operation and a label is a physical, material artifact, usually printed paper, but decidedly not a mere spoken or thought-of word." Such processes, he asserts, must necessarily involve some official authority, while mere consideration of merit does not. Put differently, whether we are talking of apples or of persons, some notions of good and bad are backed up by some formal organization such as a firm, a national government, or a private voluntary organization, while others are matters of individual judgment. Thus, I may argue that Nelson is brilliant even though he failed to get a high school diploma. I may argue that a particular apple is excellent although it failed the grading process. In doing so, I must reject the grading system established by an authority. But the very fact that I have no authority to change that grading system demonstrates that what I have done is somehow different from what those employing the formal grading system do. Moreover, my inability to influence the official grading reduces the impact of my downgrading it. Thus, without saying so, Clegg raises the question of power as well as distinguishing (as I do in this book) between formal standards and informal ones.

Here again, this hardly invalidates Urmson's position, although it points to some of its limitations. However, here too, perhaps the counterclaim is too strong. Certain standards and the criteria for their application are and are likely to remain informal or rather vague. These are usually called customs, norms, traditions, mores. Many may be violated without sanction, and some are honored only in the breach. Yet we can all think of instances of unauthorized groups of people taking matters into their own hands, ignoring the standard, developing other criteria, and even replacing the official standards with those they favored.

What can we learn from this debate about grading about the kinds of standards of concern in this book? First, it appears that there is an irreducible moral element to standards whether in the form of ranks, filters, Olympics, or divisions (classifications), although divisions in their "pure" form appear to be of little or no moral import. Second, this irreducible moral element appears to apply to nonhumans as well as to humans, since standards for nonhumans—even for apples—necessarily implicate humans. We are the ones who order things. And to paraphrase MacIntyre, the farmer who produces good apples is a good apple farmer. Third, some standards are far easier to formalize, even quantify, than others; formalization and quantification enhance the objectivity of the grading (but not necessarily of the criteria) but do not eliminate the need for moral judgment. Fourth, formal standards can and perhaps must do violence to some persons by insisting that judgment be made without regard for their unique circumstances. Finally, formal standards are necessarily linked to some institutionalized form of authority that provides them with (more or less) force, although that force is sometimes overcome by other equally or better organized persons with different notions as to what the standards should be.

But if Urmson and his supporters and critics provide us with solid claims as to the ethical import of standards, they do not take the "empirical turn"; that is, they tell us relatively little about how ordinary people develop and evaluate standards. It is to that issue that I now turn.

A Rapid Tour through Ethics (with Apologies to Ethicists)

Virtually all theories of ethics can be divided into three major approaches for analysis: consequences, rights, and virtues. Each has something to tell us about standards, and standards reflect differently on each of the three approaches. Unfortunately, most of the existing work takes a consequentialist position, and more usually a cost–benefit or risk–benefit view. A bit of reflection reveals why this is the case. First, those standards that require scientific and engineering investigations (likely a growing proportion of the total of formal standards) draw on those disciplines' concerns for costs and risks. Second, economics, which includes the vast majority of studies of standards, is dominated by utilitarian theories (one but hardly the only consequentialist perspective) that emphasize trade-offs between costs and benefits. Yet rights and virtues approaches display standards in a different light. Let us consider each of these distinct approaches.

Consequences

The consequences of standards are of considerable importance to us all. Standards lead us down certain paths and not others. Food safety standards are generally used to avoid undesired consequences—death and illness. Moreover, standards not only have costs and benefits associated with them, the costs and benefits of both creation and compliance, they often distribute the costs to some and the benefits to others—or, more precisely, the distribution of costs and the distribution of benefits are not the same. Some standards also incorporate estimates of risk and either implicitly or explicitly distribute those risks among various persons and things. Thus, for many standards, asking questions about consequences is quite reasonable. I return to this distributive question later in this chapter.

Rights

Standards are also bound up with our notion of rights. Here it is useful to distinguish between rights that allow us the freedom to do something (positive rights) and those that provide us with freedom from something (negative rights). Standards are implicated in both instances. Positive rights are often inscribed in constitutions as well as other legal documents. They are also found in documents such as patients' bills of rights or in a contract giving a customer the right to return a product that does not work as claimed, and are themselves standards. Thus, in democratic societies, we value (among many other things) the rights to speak freely, to practice a religion, and to have a fair trial.

Similarly, negative rights may be inscribed in constitutions or laws as standards. Again, in democratic societies we value the rights to privacy, to protection from unreasonable search and seizure, and to protection from cruel and unusual punishment. We also create standards in which we inscribe rights: freedoms from unsafe food and water, from quack physicians, from incompetent teachers.

But standards may also abrogate or enhance rights in less obvious ways. Several examples will help clarify this point. Let us first look at standards for bridges. As a child, my grandfather would walk with me across the George Washington Bridge from New York to New Jersey. When the bridge was constructed, standards were such that it was expected there would be a sidewalk for pedestrians. Today, on many newer bridges in the United States, sidewalks are no longer provided. Indeed, walking across those bridges is quite dangerous and may be prohibited by law.

Similarly, Langdon Winner (1986) tells the story of the construction of the New York City parkway system. He argues convincingly that Robert Moses, then Parks commissioner, deliberately insisted that overpasses on the parkways be rather low to prevent buses from using those roads. As a result, low-income, mostly African American residents of the city were largely barred from the new public beaches to which some of the parkways led. In short, in the former case the rights of pedestrians to cross bridges, and in the latter case the rights of poorer citizens who do not own cars, were limited by the standards. Indeed, it can be argued that the redesign of American cities around automobile transportation has had numerous undesirable effects. We now have the right to choose among hundreds, perhaps thousands, of models of automobiles (providing that we can afford them), but for the residents of many cities we have lost the right to choose our means of transportation. Moreover, even those of us who do not own or use motor vehicles are subject to their ill effects, including urban pollution and an increased risk of death or injury on crossing the street.

At the same time, standards can extend rights. Consider the ways in which new standards for curb cuts at corners, wheelchair ramps on buses, TTY (teletypewriter) phones for the deaf, and specially equipped toilets have extended the rights of handicapped persons. Similarly, consider how U.S. civil rights legislation eliminated many of the most egregious forms of racial discrimination.

Virtues

Standards may also be understood from the vantage point of virtues. Standards may promote, or at least remain neutral with respect to, virtues such as honesty, fairness, trust, integrity, generosity, patience, tolerance, and respect. They may also dilute or even denigrate these and other virtues. At their worst, standards may even promote vices.

Greetings are an obvious example of standards promoting virtuous behavior. In general, in ordinary circumstances people greet each other in standardized ways that promote civility. Hence, when I walk into an office and meet someone whom I have never met before, we exchange greetings, perhaps shake hands, or engage in other behavior that mutually demonstrates and establishes—performs—civility. The standard that says one should take one's turn (at the doctor's office, waiting to see a city official, using the toilet) also promotes civil behavior.

But the same is also true of more formal standards. The honor system as used in some educational institutions is designed to promote honesty and integrity. While persons are not carefully supervised, violations of the standards are treated as unacceptable and are often subject to sanction.

Laws are often designed to promote virtuous behavior and to punish bad behavior (at least as defined by those making the laws). Hence, laws requiring persons to pay their debts, as well as bankruptcy laws, are designed to promote fair exchange and to forgive debts that cannot possibly be repaid, respectively.

Standards may also erode certain virtues, either deliberately or unwittingly. Standards that are inadequately enforced may serve to encourage contempt for the standards. Widespread corruption, for example, likely encourages those persons who would otherwise be honest to engage in corrupt behavior. For example, university admissions standards that are widely ignored will encourage the submission of false documents or the bribing of university officials.

In sum, each of the three major philosophical positions on ethics offers insights into the subject of standards. They each offer a perspective from which to view and to query standards. This can be especially useful when designing new standards. But they tell us little about how people substantively justify the use of standards in actual everyday situations. In short, it is insufficient merely to say that a given standard costs less, is more virtuous, or promotes certain rights in some abstract sense. We also need to examine how that standard is justified.

Justifying Standards

We not only ask ethical questions in the three general ways noted above, we employ different substantive justifications in different parts of our lives. As Alfred Schutz and Thomas Luckmann (1973) argue, we carve up our lives into provinces of meaning. The cognitive framework I use in discussing my work with my boss is quite different from the one I use in discussing family issues with my wife. The obverse of this is that we all have multiple social selves—selves that are displayed as we interpret and reinterpret social situations. Thus, the self I display to my wife is different from that which I display to my children, which is in turn different from the one I display to my friends.

It should be emphasized that there is nothing at all unusual about this. Indeed, what would be unusual would be if we inhabited the same province of meaning or displayed the same self to everyone we met. Thus, for example, I would discuss my sex life not only with my wife but with co-workers and children. It would be equally unusual if we were to hermetically seal off these different selves from each other. This would be an instance of what psychiatrists call multiple personality disorder. Were this to be enforced by some authority, it would be a grave injustice. As Michael Walzer (1994, 38) notes, "A just society . . . makes for complicated life plans, in which the self distributes itself, as it were, among the spheres, figuring simultaneously as a loving parent, a qualified worker, a committed citizen, and apt student, a discerning critic, a faithful member of the church, a helpful neighbor."

Furthermore, each of these selves is social. That is to say, as George Herbert Mead ([1934] 1962), Alfred Schutz ([1932] 1967), and others have noted, the self is itself a social creation. Each of us is socialized in multiple settings and learns (more or less well) how to behave in each of those settings. Infants who grow up in environments in which they fail to interact with others have incomplete selves and tend to remain social isolates.

The multiplicity of selves that each of us displays means that we must employ different justifications when the legitimacy of our actions is challenged in different settings. Virtually no one—including philosophers—goes around contemplating the ethical significance of various aspects of the world all the time. For most of us, such contemplation occurs only when we have a conundrum, when we are confronted with a dilemma, when what we are doing is challenged. When this occurs, and we reject the use of violence, we must then justify our actions in order to explain how they are legitimate. But which justification? Michael Walzer (1983) calls the various justifications available to us "spheres of justice," while Luc Boltanski and Laurent Thévenot ([1991] 2006), coming out of a quite different tradition, call them "worlds."

Boltanski and Thévenot argue there are several different worlds in which different justifications apply. Thus, they go beyond those philosophical notions of ethics that attempt to demonstrate ethical universals that apply at all times and in all places and make the perhaps somewhat startling claim that different justifications have more or less weight in different social settings. By a careful reading of how-to manuals, Boltanski and Thévenot identified six different worlds of justice, which they call the civic, merchant, industrial, inspirational, domestic, and opinion worlds, to which Lafaye and Thévenot (1993) have recently added the environmental world.[2] Boltanski and Thévenot leave open the possibility that other worlds of justice might well exist, might have existed in the past, or might come into being at some later date. They argue that justifications are quite different in each of these worlds but that all of us inhabit multiple worlds simultaneously.[3]

While in some instances merely reciting the justification is sufficient to produce agreement, more often it is necessary to construct tests and trials that demonstrate the worth of the person or thing within a particular world. Hence, the quality of a particular brand of automobile might be tested by its winning an award for styling (or not). The quality of a Catholic theologian might be tested by that individual's rising to sainthood (or not). Moreover, in some rare instances, new worlds are created—think here of the rise and widespread acceptance of environmental justifications in the past several decades—although the precise paths by which these justifications arise and become legitimate is still unclear.

To Boltanski and Thévenot's argument we might add that each of these worlds must have its own appropriate standards—standards that state the criteria required of persons and things that have (or lack) standing in this or that world. Standards are not the same as tests or indicators, although they are usually tightly linked together. Moreover, standards can and must be different for different worlds. Furthermore, these standards can be appealed to when the pursuit of a particular justification is in doubt. But this is all rather abstract, so let us examine the standards for several worlds and provide several examples to help to bring this all into focus.

Merchant Standards

Consider the standards used in the marketplace. A great salesperson is usually viewed as someone who has extraordinary skill in selling, as demonstrated by a lengthy and impressive record of sales. Standards frequently used in the marketplace are volume of sales, profits made, throughput, and the like. Debates can and do exist as to how to measure all these things, but nearly everyone—even those who find the marketplace abhorrent—agrees that they are central to the merchant world.

Determination of value in the market is always the result of a sum of values. Thus, as economists rightly point out, the value of an object is equal to the sum of the value added at each step in the production process. That sum may include positive values ("Look at the high quality of the workmanship," or "Nancy has excellent sales skills") or negative values ("Note that the object is scratched," or "Edgar did not make the sales goals for this quarter"). Moreover, values in the market are always formed in binary terms. One buys or sells, accepts the price or rejects it, considers something or someone to be worth the cost or not. Although haggling may go on, in principle and usually in practice, most bargaining is about price. Put differently, a central feature of the market world is that in principle, everything has its price; everything can be made commensurable. The key is that the price must reflect the quality of the good or service to be sold. Hence, market standards tend to be focused on consequences; the bottom line is the bottom line.

Standards in the merchant world are largely of two sorts. The first consists of those designed to create, promote, lubricate, or enhance the market itself—literally, to create a moral order of the market to which a moral being is expected to conform. Thus, standards applying to contract enforcement, the prohibition of dangerous products, standard weights and measures, standard currency, and the clear labeling of products are designed to smooth transactions in the marketplace. These standards say little about the specific nature of any particular item that might be bought or sold in the marketplace. A second kind of market standard is designed to make myriad incommensurable things—including physical objects, animals, and human labor—into items that can be bought and sold in the market. This would include standards of identity that qualify things for the market. For example, in order to qualify for the market, things or people's labor must be for sale.

Indeed, Appadurai (1986a) argues that objects go in and out of the commodity state (much as human beings display different selves?). Thus, when I drive an automobile before paying for it, I do so as part of the test of the value of that vehicle; its commodityness is unchanged by my so doing. In contrast, once I take possession of the vehicle, it is withdrawn from the world of commodities. It may become personalized, by virtue of bumper stickers, objects in the glove compartment, or wear and tear on the tires. Should I decide to sell it at a later date, I will likely remove the symbols that singularize it, have it "detailed," and return it to the state of a commodity. The same point can be made of persons. When seeking a job, people often adopt a slightly different persona. They may dress a bit differently, act a bit more politely and deferentially, be more prompt than usual. After obtaining the job, that behavior will likely change, in large part since one is no longer "on the job market."

Tackling this problem from a slightly different perspective, Callon (1998a; 2002) argues that goods must be qualified to enter the marketplace. They must be made comparable to other goods and they must be made calculable. This requires that standards be invoked, that measurements be made. Qualification is in part today the work of marketing professionals. They position the object (on a shelf, in an ad, in a store, in what they hope is a desirable setting to potential buyers) so as to encourage consumers to attach themselves to it.

Note also that market standards consist of both legal frameworks and various far less sanctioned customs and standards. For example, only certain persons have the right to sell certain things. In most instances this person would be the owner of the thing for sale or the owner's agent, to whom the power to sell has been delegated. It is for this reason that I cannot (legally) try to sell you the Brooklyn Bridge. But far less formal standards are also invoked by the market. Thus, such things as how change is given—generally into one's hand in the United States, and into a tray (where it can be seen by both parties) in France—are part of the market exchange process but are rarely written down.

Finally, money allows the expression of relations in the same unit: "It provides the currency, the standard, the common language which enables us to reduce heterogeneity, to construct an equivalence and to create a translation between a few molecules of a chemical substance and human lives. Money comes in last in a process of quantification and production of figures, measurements and correlations of all kinds" (Callon 1998a, 21–22).

It is usually the case that objects sold in the merchant world need only be saleable or desirable to confer greatness. Put differently, their quality needs to be equilibrated with their price (Eymard-Duvernay 1995). They need not have any efficacy or utility. Consider, for example, the pet rocks that were sold in the mid-1970s. Their creator, Gary Dahl, was a marketing genius, able to get people to purchase ordinary rocks boxed with instructions for their care. Doubtless, standards for success in the merchant world are generally ranks. But the merchant world also uses divisions to differentiate among products, thereby enhancing the wealth of persons.

Civic Standards

Standards in the civic world are considerably different. Civic standards do not take the viewpoint of buyer or seller but focus on creating and measuring the qualities of *public* goods. Thus, standards for clean drinking water, the width of roadways, defining which side of the road on which one should drive, how much education and health care should be provided by public authorities, and what kind of street furniture (lampposts, traffic signals, etc.) should be found in public spaces are usually the subject of civic standards. They often have clear legal sanctions attached. However, such things as the placement of street furniture, the order in which streets should be cleaned, and the placing of police are generally far less codified. Furthermore, standards for many civic activities are enforced (more or less) by the populace as a whole. These include standards for appropriate queuing behavior, the appropriate distance between strangers passing on the street, the politeness of drivers of vehicles, taking turns in public places, and myriad other displays of civic behavior (e.g., Goffman 1971).

Standards in the civic world are not based on the talents of any particular individual but on the virtuousness of those involved (Boltanski and Thévenot [1991] 2006). Moreover, standards in the civic world are often, perhaps always, collective. The great nation, the great city, cannot be understood except as a collectivity with certain features. And, the standards by which persons are judged in the civic world attempt to capture their contribution to these collective goods.

Industrial Standards

In the industrial world standards are designed to promote industrial values such as speed, efficiency, precision, accuracy, and the production of needed things. Hence, the standards of the industrial world are nearly always precise, accurate, mathematical, replicable. These standards are designed to standardize, to ensure that everything is the same. But the calculations involved in industrial standards are not designed so as to sell some good as much as to subject persons and things to the industrial discipline. Rather than the spontaneity of a sale, industrial standards are designed to produce repeatable events, to make persons or natural objects perform according to predefined plans and targets. Thus, industrial standards such as the control charts initially developed by Walter Shewhart (see chapter 2) are often wrapped up in efficiency. The scientific instrument that allows the rapid and accurate testing of thousands of blood samples and the machine that turns out millions of identical potato chips embody industrial standards. Similarly, the metrological devices and reference materials produced by the National Institute of Standards and Technology are typical of industrial standards.

Not surprisingly, those who excel at the creation of industrial goods, such as scientists, engineers, and industrialists, are those who are fêted in the industrial world. A steady stream of biographies shows such persons as the exemplars against which others are judged. Hence, scientists such as Pasteur and Einstein, engineers such as the Wright brothers, and industrialists such as Henry Ford and Andrew Carnegie are often held up as exemplars against whom others should be judged. In general, proponents of the industrial world emphasize filter standards, sorting out those things that make the grade from those that do not. In fewer cases, they emphasize Olympian standards, for example by giving awards for industrial design.

Inspirational Standards

In the world of inspiration, standards are quite different. A great artist is not revered based on how many paintings she has sold or what they brought on the market. (Of course, those who sell art are very concerned about sales; however, they are art merchants.) Instead, a great artist strives to meet the usually ambiguous standards for producing great art, which include both the facility with which materials are worked and the inspiration manifest in the painting.

Standards in the world of inspiration promote the universal through the unique. Thus, the brilliance of Leonardo as an artist, Pasteur as a scientist, Michelangelo as a sculptor, Plato as a philosopher, or Mohammed as a prophet cannot be reduced to a set of codified standards. True, each of them underwent a set of tests and trials that demonstrated his or her inspiration. Yet for each of them, and for any others we might care to name, the tests and trials were themselves different and unique. Indeed, what is common to standards in the world of inspiration is the impossibility of codifying them at all. Here in the world of inspiration, claims to greatness are simultaneously the result of unique attributes and universal recognition. Indeed, the very act of codifying would tend to undermine the standard by intimating that all that is required to meet the standard is a set of clearly defined actions. In contrast, standards in the world of inspiration are always accompanied by a sense of mystery and awe. Thus, most are Olympian in nature. Standards for art are nearly always of this type. They consist in holding up exemplars of extraordinary inspiration and comparing others to them: a painting worthy of Monet, a play worthy of Shakespeare, a symphony worthy of Beethoven. No one, expect perhaps a social scientist (e.g., Sorokin 1937–41), would think of ranking paintings or plays or symphonies based on some objective criteria, even though they might well point to the brush strokes in a Van Gogh, the chords in a Beethoven symphony, the elegance of a turn of phrase in Shakespeare.

Thus, we can not tell an aspiring artist merely to emulate Leonardo or Picasso so as to achieve greatness. We cannot tell an aspiring scientist merely to emulate Pasteur. We cannot tell an aspiring philosopher merely to emulate Plato. Even were they to excel in this regard they would merely have succeeded in producing an excellent *copy* of the original inspired work. It is precisely the uniqueness of the contributions of persons who are inspired that allows others to judge them as inspired.

Domestic Standards

The domestic world is the world of personal relations, not only those of the family but those of all sorts of personalized interactions. Like the world of inspiration, the domestic world abhors precise standards (thus blurring the distinction between standards, customs, traditions, and the like). Indeed, it emphasizes traditions, hierarchy, position. Consider the traditions of gift exchange. In most Western societies the standards for reciprocity with respect to gifts are those of rough equivalence in value. Calculating the actual monetary value of a gift is considered extremely rude and undermines the very nature of the gift exchange.

Moreover, standards in the domestic world tend to emphasize hierarchy. For example, when my children come for dinner I may serve hot dogs, provide margarine for the bread, and more than likely will take the cooking pots directly from the stove to the table. Conversely, when my boss comes I will serve steak, provide butter for the bread, and put everything into serving dishes before bringing the food to the table. In short, the standards of the domestic world largely take the form of ranks; each standard displays where in a hierarchy of people and things I and my guests stand.

Standards in the domestic world specifically rule out distinctions between public and private. Instead, they emphasize the means by which positions in various hierarchies are to be measured. A relatively common assignment in introductory sociology classes has students go home during fall or spring vacations and treat their family as if the student were a guest in the home. The purpose of the assignment is to demonstrate the invisible rules that govern domestic behavior and that are constantly performed by the actions of family members. Thus, asking one's mother or father if it is okay to remove a food item from the refrigerator would usually be considered at best strange behavior. Indeed, it violates the unwritten but nevertheless very real standards that govern domestic behavior by shifting one's place in the household hierarchy from that of son or daughter to that of stranger or distant relative.

Domestic standards also include those of proper comportment in a wide range of circumstances. These standards include both unwritten conventions about behavior in large or small organizations (what subjects I can discuss with my boss or colleagues), as well as much of the information one finds in various employee manuals (e.g., who reports to whom, what benefits are provided to whom) and in organizational charts. The *Warrant of Precedence* (see box 1.1) is a good example of a domestic standard, carefully codifying each person's place in an extraordinarily complex hierarchy.

Finally, the standards of the domestic world include those for loyalty. In the domestic world, one is expected to be loyal to both one's superiors (to support the policies enacted by your boss) and to one's inferiors (to protect those one supervises). This contrasts sharply with the market and opinion standards, where loyalty is irrelevant and a good bargain or a moment of fame is of far greater import.

Opinion Standards

In the world of opinion one's greatness is dependent on what one's contemporaries think of one. Greatness in acting, sports, and pop music depends not on any intrinsic quality as it does in the world of inspiration but on what others think of you. Fame is demonstrated by the degree to which one is recognized as famous, honor by the degree to which one is honored. But unlike in the world of inspiration, recognition in the world of opinion is fickle; what is recognized as great today may be ignored tomorrow. This is perhaps what Andy Warhol meant when he noted in an interview that "in the future everybody will be world famous for fifteen minutes." Hence, standards in the world of opinion tend to be of the Olympian or ranked form. They are often precise, based on opinion polls, but ever-changing.

In the world of opinion, standards are largely based on the opinions of others. Thus, politicians and movie stars fret over their standing in opinion polls. Moreover, certain opinion polls are taken far more seriously than others. An opinion poll taken by the Gallup Organization is far weightier than one taken by the Republican Party.

Similarly, the standards for judging the rankings of movie, music, and sports stars are based on their public visibility. While these opinion standards are generally less precise than those of opinion polls, they nevertheless are standards. Thus, being seen at certain events, being chased by paparazzi, having one's picture displayed frequently on the television screen, having an interview with a well-known columnist or news anchor, and even being the subject of gossip columns all provide gauges for judging the worthiness of persons in the world of opinion.

If this is unconvincing, then consider what the absence of these events means. The movie star who is no longer invited to certain events, no longer chased by paparazzi, no longer interviewed on television, is now passé, a "has-been." And for the most part, these events are not easily manipulable by the person whose career is withering. One can hire agents to stage events, but this is in most instances only marginally useful. And even if one thinks one is a great politician, actor, movie star, or whatever, one's personal view of oneself remains independent of the opinion of others.

It is also worth noting that in the world of opinion, standards are based largely on the alleged talents (or lack of them) of those who are widely known—whether in art, sports, politics, music, or some other domain. Moreover, those talents are to be made public, transparent, available to all who care to know—and the more who know, the more renown that person is said to have. This is in sharp contrast to the standards for the inspired world, which eschew renown, or those of the industrial world, which dismiss it as irrelevant.

Environmental Standards

In recent years we have seen the rapid, even spectacular, rise of justifications based on the environment, nature, ecology (Lafaye and Thévenot 1993). Militant social movements, legal sanctions, and collections of technoscientific data are commonly justified based on their contribution to environmental protection. Standards for the protection of the environment, whether highly precise (emissions standards) or rather vague (standards for aesthetically pleasing landscapes), are an attempt to define the degree to which a given person, organization, process, product, or thing contributes to the protection of nature, which is seen as the common inheritance of humanity. Thus, the promotion of greenness is now commonplace—one can talk of green automobiles, green cities, green products—and is used to counter market, industrial, and even civic justifications. Most environmental standards are filters; those beings that pass through the filters are considered to be environmentally sound, to contribute to the building of a better, more environmentally friendly world.

As Boltanski and Thévenot note, a perfectly reasonable and acceptable justification in one of these worlds might well be rejected out of hand in another. For example, consider the standards used to determine the quality of an automobile. A "good"-quality automobile, one might argue, should have rapid acceleration, get good gas mileage, and provide a comfortable and safe ride for the passengers. Note that all of these claims for a good automobile can be and usually are the subject of industrial standards, that is, they appeal to the industrial world. However, one could—and manufacturers do—make claims that an automobile is lovable (domestic), brilliantly designed (inspirational), a good value (merchant), ideal for urban use (civic), produces little pollution (environmental), and is part of a long tradition of excellence in motoring (domestic). By contrast to the industrial claims to quality, developing standards for these claims (not to mention tests and measures for them) would be quite different. Indeed, standards of this sort might even conflict with those of the industrial world. For example, the most environmentally friendly vehicle would probably not have the most rapid acceleration; the vehicle that is part of a long tradition in motoring may be lacking in certain cutting-edge design features.

Sometimes divergent justifications can lead to conflicts over standards. A case in point is the recent flap over kosher standards, which involve the pursuance of particular slaughtering procedures (Freedman 2007). More than half the kosher slaughterhouses in the United States are certified as such by the Orthodox Union.[4] A certain conservative Rabbi Allen was concerned that the workers in kosher slaughterhouses, often immigrants from Guatemala and Mexico, were poorly paid and unaware of proper safety procedures. Arguing from biblical injunctions against abusing the destitute, he proposed that an expanded kosher standard be developed to include social justice. The conservative Rabbinical Assembly accepted the notion. However, the Orthodox Union rejected the idea, claiming that issues about wages and safety were the proper province of the state and not of the rabbinate. In short, one interpretation is that what the Rabbinical Assembly saw as appropriate for a standard from the domestic world was seen by the Orthodox Union as more appropriately handled by standards from the civic world.

Different Worlds Require Different Kinds of Standards

When we compare the various sorts of standards that are commonplace in each of these worlds, we begin to see a pattern emerging (table 5.1). While no totally unambiguous statements can be made, it appears that different types of standards are often found in different worlds. Thus, when conflicts occur across worlds of justification, the very types of standards considered acceptable might well differ, a point I develop below. But first let us briefly consider how different kinds of standards tend to predominate in different social worlds.

Within the world of inspiration, standards are usually Olympian. An example should clarify this point. Some years ago, I gave a class of social science undergraduates the assignment to interview a faculty member in the biological sciences and write a two-page paper on the faculty member's

Table 5.1
Worlds of justification, examples of standards, and common types

World	Standards	Common type of standards
Merchant world	Volume of sales, profits made, value added, wealth accumulated	R
Civic world	Cities with vibrant economy, social life, sense of collective belonging, effective provision of public services (schools, hospitals, water, sewer, electricity, gas)	F
Inspirational world	Display of brilliance, insight, avoidance of publicity	O
Domestic world	Manners, comportment, loyalty	R, D
Industrial world	Precision, accuracy, usefulness, efficiency, formal qualifications	R, D
Opinion world	Fame, renown, eminence, notoriety	O
Environmental world	Nonpolluting, clean, in harmony with nature	F

Note: O = Olympic, R = ranks, D = divisions, F = filters.

research. I received some twenty-five papers and read through them. Every single paper noted that Professor X was at the cutting edge of her or his field. Disclosing that to the class brought some snickering. But further discussion revealed what I later recognized as a standard. How could anyone engage in the highly disciplined activity known as research—any field of research—unless that person believed he or she was on the cutting edge?

This is not to suggest that one cannot rank researchers. Indeed, that is often done. But arguably no one could undertake serious research unless they were confident that they could do something they believed to be important, vital, groundbreaking, central to the topic at hand.

Olympian standards are also commonplace within the opinion world. Hence, there is one Miss America, one Best Actor of the Year, one winner of the World Series (even if it is limited to the United States and Canada), and one best wine of the year. Here again, one could employ ranks, but being second best often doesn't quite make the grade.

In the civic and environmental worlds, filter standards are far more common. There are good reasons for this. Consider the case of city water systems. The key to a successful system is to meet the requirements for one. This might include meeting safety requirements, being palatable, not leaking significantly, and so on. Undoubtedly, some water system somewhere is the safest, most palatable, and has the fewest leaks, but clearly, not all water systems can meet that standard, nor would we expect them to do so.

The merchant, domestic, and industrial worlds appear most likely to use ranking standards. In the merchant world, sales often depend on ranking a bit higher than other products or services (e.g., higher quality or lower price). While Avis may be No. 2, they try harder, as their ads suggest. Where they are in the ranking is important, as it is for all other auto rental companies. In the domestic world, where hierarchies are important, rankings are commonplace forms for standards, too. Similarly, in the industrial world the entire thrust of the programs for enhancing quality—whatever names they might have—are all about improving the company's rank in a complex hierarchy.

These tendencies become problematic when standards are being developed that cross several worlds. For example, an environmental standard for auto emissions that required the very lowest emissions possible (an Olympian standard) would meet resistance from manufacturers on the ground that it would be so costly as to price virtually all vehicles out of the market. A filter standard that demanded that vehicles produce no more than a given amount of emissions would be far more acceptable. Similarly, a standard for the inspirational world that was in the form of a filter would likely be unusable.

Problems may also arise when those committed to the standards for one world attempt to argue that their standards should be applied to *all* worlds. Thus, those committed to the market sometimes believe that the standards for the market should be applied in all worlds. Alternatively, those committed to the civic world may wish to apply its standards to all other worlds as well. Or those committed to the world of inspiration might argue that its standards should applied everywhere.

Examples of such a desire can, alas, be found everywhere. Orthodox Marxists who believed that the state should be the sole dispenser of all justice founded the now-defunct Soviet Union nearly a century ago. The laissez-faire politicians of the mid-nineteenth century were so convinced of the rightness of market-based decisions that when the potato crop failed, they allowed millions of Irish peasants to starve or forcibly migrate rather than provide famine relief. At least some contemporary neoliberals appear to share this boundless faith in the market. And members of certain Christian, Muslim, Jewish, and Hindu sects today are so convinced that their particular standards for the world of inspiration must be applied to all human endeavors everywhere that they are willing to kill innocent bystanders and die themselves to support their beliefs.

There is a great irony in all of this, since it is well-nigh impossible to construct any social world without appealing to many worlds of justification. Soon after the creation of the Soviet Union, for example, black markets in all sorts of goods appeared and remained until the demise of the Soviet state in 1989. The family garden plots provided to peasants on collective farms provided much of the fruits and vegetables eaten at that time. If asked to justify such actions, Soviet citizens would doubtless have argued that without private plots there would have been no fruits and vegetables. And, of course, even the Soviets had to pay some attention to the Russian Orthodox Church (even as they tried to suppress it), to public opinion, and to industrial design.

Attempts to found modern nations built entirely on religious doctrines have been equally unstable and disastrous. The contemporary Iranian state provides a good example. Its economy is nearly in ruins. Corruption reigns in both state and private affairs. Ironically, although it is one of the world's major producers of oil, it must import oil to satisfy the needs of its citizens. In short, the attempt to trump all other justifications with those of the world of inspiration turns out not to be particularly inspiring to a vast number of Iranians.

Given the failures described above, I have little doubt that the neoliberal project of a market society—one in which the state acts only to construct markets and enforce market rules—will turn out to be as devastating as the reality produced by the Soviet state. If markets are to replace most state functions, the fate of democracy appears grim. And the neoliberals may discover that, to paraphrase one of its champions, F. A. Hayek, there is more than one fatal conceit, more than one road to serfdom. As Dewey ([1915] 1976, 40) once remarked, "Only when the 'good' is resolved into simple and unalterable units, in terms of which old situations can be equated to new ones on the basis of the number of units contained, can an unambiguous standard be found." In the realm of ethics, such unalterable units are rarely found.

How Standards Distribute Goods

As Michael Walzer (1983) notes, there are three ways in which goods can be distributed: First, goods can be distributed by market exchange—they can be bought and sold in the marketplace. Second, people can receive goods based on what they deserve. Thus, one can receive an Olympic gold medal or ten years in prison because one deserves it. Finally, goods can be distributed based on need. Thus, food, clothing, or shelter can be provided to the poor, hospital care can be provided to an accident victim, formal education can be provided to the young, because they need these goods. This would seem to apply to each of the various worlds identified by Boltanski and Thévenot, although more to some than to others. Of course, the merchant world tends to distribute things almost entirely based on market exchange, while the world of inspiration tends to distribute its rewards based mostly on dessert. We expect that artists, theologians, musicians, and others who are inspired will get their just desserts; distribution of these goods based on market exchange or need seems vaguely repugnant. The civic world makes room mostly for need (education, health care, provision for the poor) and dessert (court decisions, civic medals, awards), and under certain generally limited circumstances for market exchange (e.g., fees for park admissions, hunting permits, driver's licenses).

Consider the case of religion. Religions are instances of the world of inspiration. They focus on some sort of transcendence of the world of everyday. They put forth a set of standards for practitioners which are more or less easily measured. Of course, the very practice of religion requires engaging in certain kinds of market exchanges to purchase vestments and religious artifacts, build houses of worship, and so on. People who provide these goods and services must be compensated for them. But practitioners of most religions would be horrified if all religious standards could be met by simply paying a sufficient sum of money to the church. Indeed, the purchasing of indulgences was one of the main criticisms leveled against the Catholic Church by the leaders of the Reformation.

Much of the twentieth century can be seen as a battle between two extreme views, the view holding that the state was and should be the final arbiter of all ethical and moral standards and the view that the market should fill this role. The former held that social goods should be distributed based on dessert and sometimes need, while the latter held that social goods should be distributed based on market exchange.

Injustice

Just as there is no single standard for the good, there is no single standard for the bad. Each world has its own means for violating or denigrating its standards. Table 5.2 lists some behaviors that violate the central principles of various worlds. Some of these examples of bad behavior are largely

Table 5.2
Examples of behavior that violates standards of particular worlds

Civic world	Buying votes, bribing judges or other government officials, using one's office to extract favors
Merchant world	Deliberately selling mislabeled goods, bribing buyers' agents to make sales
Inspirational world	Engaging in sinful behavior, plagiarism
Industrial world	Producing inconsistent or shoddy goods
Domestic world	Ignoring obligations to one's family, friends, superiors
Opinion world	Engaging in secretive, anonymous behavior

confined to a particular world. Others, by contrast, involve transgressing the boundaries between worlds. For example, ignoring obligations to one's family can hardly be seen as problematic in the industrial world, but using one's family ties to excel in the industrial world is problematic. Similarly, buying votes is a breach of the central principles of the civic world, but buying stock in order to vote at corporate stockholder meetings is quite acceptable.

Moreover, when the merchant world is invoked in education and health care, new forms of inappropriate or even illegal behavior can emerge. Hence, surgeons might perform unnecessary surgery and claim unusual expenses in order to enhance the bottom line of a hospital (Klaidman 2007). Similarly, when the merchant world extends too far into education, new forms of inappropriate behavior emerge there.

Value Chains

So far, in discussing justice, justifications, and injustices I have made it appear as if each judgment, each valuation, each justification was distinct from all others. But such is usually not the case. If I excel in my academic work, my family celebrates; more important, if I am dismissed from my job for incompetence, my family life is affected as well. The social networks of which I am a part are all linked together, sometimes loosely, sometimes more tightly. The notion of value chains, although usually applied solely to the production, distribution, and sale of various things, may be usefully applied to persons here as well.

At each step in the process of producing goods, from the raw materials all the way to final consumption, standards of all sorts are applied. Of course, value chains do not exist as empirical entities; rather, they are a convenient conceptual apparatus for extracting a particular set of relations from the social fabric.[5] However, by extending the notion of value to include not merely monetary but all values, the same concept can be applied to people as well. Each of us as we make our way through life must submit to standards measured by tests and trials. These tests and trials determine our qualities as much as other tests are designed to determine the qualities of objects. And just as the standards demanded at each stage of the value chain are somewhat different, so the standards demanded of us, both individually and as participants in various organizations, vary throughout our lives. These are what Arnold van Gennep (1960) meant when he wrote of "rites of passage."

As infants we are measured against a variety of standards for normality of behavior, crankiness, rate of growth, rapidity of language acquisition, and so on. Our parents, relatives, and neighbors define us as lovable, clever, gullible, outrageous. Later, as schoolchildren, we are labeled overachievers, average, or underachievers in terms of dozens of tests, some more standardized than others. Some of us will find ourselves in special classes for the gifted and others in special classes for remedial work. Most of us will find ourselves in an ill-defined middle. Some of us may be tracked toward college while others are encouraged to prepare for manual jobs.

As adults, we find ourselves tested as we enter the workforce, form our own families, and participate in a wide variety of community activities to one degree or another. Moreover, throughout our lives some of us may encounter extreme tests demanding extraordinary displays of courage, love, bravery, compassion, honor, and other values. Regardless of how well we do, we will doubtless pass some of these tests and fail others.

The same applies to organized groups of humans such as families, educational institutions, hospitals, large corporations, small businesses, the military, sports teams, and even nation-states. A seemingly endless array of standards, tests, and trials defines all our social institutions. In short, just as objects are tested and tried as they go through their life courses, so are human beings, both as individuals and as members of institutions. Standards for some of these tests are quite specific, requiring passing written exams, demonstrating particular skills, or performing tasks in certain ways. Others remain and are likely to remain somewhat vague and indistinct. Yet one may argue that those standards that remain vague and indistinct are often the ones we hold most dear. When we evaluate others in terms of their love, courage, bravery, kindness, or similar value, it may be difficult or impossible to define precisely the criteria by which such judgments are made. Nevertheless, we make such judgments all the time, and we place great stock in them.

Furthermore, a small industry exists in all complex societies of persons whose job it is to develop and administer the standards, the tests, and the trials. Thus, military officers may be cited for bravery, scholars for their insights, inventors for their inventions, religious leaders for their saintliness, artists for the art they produce, and baseball players for the number of home runs they score. On a more mundane level, we find ourselves measured frequently as we apply for driver's licenses, register to vote, purchase alcohol or tobacco, or are arrested and charged with crimes. In most instances the persons who design and administer the tests are perceived as legitimate by those who are tested; in other instances not only the tests and the standards but the persons who administer them are challenged.

Both the value chains for people and those for things are constantly contested because success on the tests and trials along the value chain demand the display of *different* values—values that are sometimes complementary but more often contradictory. As Boltanski and Thévenot ([1991] 2006) have shown in a somewhat different context, we do not live in a fully ordered world in which there is one good but rather in a partially ordered world in which there are multiple goods. Moreover, success in achieving one good may restrict success in meeting other goods. And even the goods themselves may be highly contested. An example will help clarify these points.

Some years ago some colleagues and I conducted a study of the canola value chain (Busch et al. 1994). Canola is the name that was given to the rapeseed plant (*Brassica napus* and *Brassica rapa*) after two compounds were removed from it through conventional plant breeding. Erucic acid was removed from the oil as at the time it was believed to be potentially toxic. Glucosinolates were removed from the meal as they cause goiters to develop when fed to farm animals. We examined the entire value chain from the production of seed to the retail sale of cooking oil and margarine. What we quickly discovered was that each human actor in the value chain—seed producers, canola farmers, grain elevator operators, crushers, food processors, meal manufacturers, retailers, livestock farmers, and final consumers—had a somewhat different notion as to what characteristics were desirable ones for canola. For example, seed producers wanted seed of uniformly high germination rate. Canola farmers wanted varieties that produced high yields, even stands, and did not lodge (fall over) during windstorms. Elevator operators wanted canola that was easily handled without being squashed or otherwise damaged. Crushers wanted canola that easily flaked (the process by which the oil is separated from the meal). Food processors wanted canola that produced a clear, light-colored oil with a pleasant odor, or that could be easily processed into margarine. Feed manufacturers wanted meal that was high in protein to spur animal growth. Retailers wanted products with a long shelf life so that the product would not go stale on the store shelves. Finally, consumers wanted an oil that tasted good and was low in saturated fat.

What I described above for canola is equally true for every product in every value chain, from automobiles to computers. That is, different actors along the chain demand that the product conform to different standards, as measured by its passing different kinds of tests and trials displaying different kinds of values. This would be of little interest were these tests merely additive. But much of the time they are not. For example, farmers' desire to maximize yield could easily lead them to plant varieties that proved unacceptable to food processors. Elevator operators' desire to have varieties that are not easily squashed might pose problems for oilseed crushers. Several issues emerge from this set of problems.

First, the actors along the chain usually must *compromise* if they are to achieve their ends.[6] Clearly, farmer Jones may insist on growing the highest-yielding variety of canola even though it produces overly soft seeds containing oil with a pungent odor. However, she will soon find that no one wants to buy it. Alternatively, she may find that someone will buy it, but only for a very low price, perhaps lower than the cost of production. Thus, farmer Jones finds that she must compromise, growing what is perhaps not the highest-yielding variety but one that is of acceptable quality to those further along the value chain and that they are willing and perhaps eager to purchase. The same is true for the production of other things. For example, steel manufacturers who produce abundant steel that is of inadequate tensile strength for buyers' uses will find themselves forced to compromise or depart the market.

Second, extraordinary success in the maximization of one value will be downgraded if other values are not at least present to a sufficient degree as demonstrated by other tests and trials. This follows from the first point. Put differently, if farmer Jones persists, even as her neighbors praise her for attaining such a high yield, others will denigrate her achievement precisely because it fails to satisfy the needs of the chain as a whole even in the most minimal way. Again, the same is true for all other objects. The steel manufacturers described above will be considered foolish by others in the industry if they insist on producing steel of inadequate strength.

Third, what has just been noted for the relations among actors in a given value chain applies as well to those not part of the chain but affected by it. For example, a highly efficient producer of gasoline that, by virtue of its efficiency, is also a large-scale polluter will have its status downgraded below one of lesser efficiency that is environmentally responsible. A retailer that maximizes its profits by paying its workers substandard wages and benefits will be downgraded in the eyes of many members of the general public. A clothing manufacturer who produces beautiful, stylish, long-wearing clothes through sweatshop labor will be devalued by many, especially when compared to one that is more socially responsible.

Fourth, like the animals in George Orwell's allegorical *Animal Farm*, while all actors in value chains are equal, some are more equal than others. That is to say, although compromise is a necessary feature of all value chains, often some actors are better able to impose their views on others (and, by definition, some actors have the views of others imposed on them). This is particularly so when one or a few powerful actors confront the many, as is often the case in supply chain management. For example, in contemporary food value chains there are millions of farmers and billions of consumers, but a much smaller number of large retailers and processors.[7] As a result, the processors and especially the retailers are often able to impose their desires on other actors. However, they are limited in their ability to do this since they too depend on the existence of the entire chain for their success.

Conclusion

Even as standards are technical rules, they are also compromises among diverse values, themselves drawn from different worlds or orders of worth. Standards are attempts to fix values, to embed them in particular products, processes, persons, practices, and organizations. Standards construct publics since they implicate various persons up and down the value chain, as well as those in networks that surround it. At the same time, standards determine how power and other social goods will be distributed among persons and things. But if standards are and must be about the distribution of social goods, then they are also about the means of governance. This brings us to the age-old tension between democracy and expertise. That is the subject of the next chapter.

6   Standards and Democracy

Standards creation appears to pose some particularly difficult challenges for democracy. Many standards appear to require some form of professional or technical or scientific expertise if they are to be effective. So it appears we are left with a conundrum: either we let the experts decide for us or we decide ourselves. If we let the experts decide, then democracy must be largely abandoned. If we decide, we are likely to make fools of ourselves— or worse. And to make matters more frustrating, some standards (e.g., for decision making, for the provision of necessary public services) are essential to any form of democracy; indeed, without the stable realities that they provide, we would be hard-pressed to have democracies at all. We seem to be in a situation much like that of the ancient Greeks, who tried to square the circle—to draw a square with the same area as a circle using only a compass and ruler. This inevitably leads us to a very old debate about expertise and democracy. The main positions in that debate seem to be immutable.

The Case for Expertise

Plato ([360 BC] 1994) in *The Republic* argued that the business of government was far too complex and required far too many skills to leave to the average person. And he knew all too well that his teacher, Socrates, had been forced to drink the hemlock by the members of a democratic jury. Democracy was tantamount to mob rule for Plato. Hence, he proposed the creation of a class of guardians, men and women who would be raised and educated to lead a republic. Plato believed that if the material wants of such persons were fulfilled, if they were properly educated in the arts of governing, and if they were relieved of the mundane tasks of daily life, they would produce a society quite superior to the monarchies, oligarchies, and democracies of his day.

Nearly two thousand years later Francis Bacon ([1605/1626] 1974), in his posthumously published utopian novel *The New Atlantis*, imagined a world run by a council of men whom we would today call scientists and engineers, organized into what he called the House of Salomon. The state would be subordinate to the House of Salomon; the experts would call the shots. Bacon was particularly distrustful of politics and believed that doing away with it would considerably improve the lot of humankind. He reasoned that just as one could read the Bible to learn of the written works of God, one could read the "Book of Nature" to discern God's plan for nature. Experts trained in the use of the scientific method would learn how God organized nature, and would design policies in conformity to that plan. This in turn would allow us to avoid politics altogether. More recently the eminent scientist E. O. Wilson in *Consilience* (1998) picked up the Baconian banner—even citing him favorably in the epigraph—and has argued that improvements in our understanding of biology will allow us to make better ethical judgments.

In the early twentieth century Walter Lippmann, in a widely circulated book, *Public Opinion* (1922), argued that the public could not possibly understand the complex technical issues facing a modern society. As Lippmann (31) put it, "I argued that representative government, either in what is ordinarily called politics, or in industry, cannot be worked successfully, no matter what the basis of election, unless there is an independent, expert organization for making the unseen facts intelligible to those who have to make the decisions." Lippmann went on to argue that nearly everyone, including those with a strong interest in politics, was very much taken up by the everyday activities of their lives such that they scarcely had time to briefly examine—let alone carefully consider—the myriad complex and often technical decisions that had to be made all the time. Hence, he argued, although we might have opinions on whether the military was adequately prepared, what policies should be enacted to improve transportation, how much smoke steel mills should be permitted to emit, how schools should be improved, and a virtually endless list of other contentious issues, these were not—*nor could they be*—the result of careful reflection. Indeed, even those who carefully reflected had to depend on second- or third-hand information filtered through the media, media that were neither sufficiently reliable nor sufficiently expert to provide the appropriate information on which to base decisions.

Furthermore, in order to aggregate, to sum individual opinion into public opinion, the countless variations among persons had to be simplified into a few straightforward positions. Thus, the ill-considered opinions of particular individuals, based on hearsay and rumor, deception and inaccuracy, had to be aggregated into something even more vague and error-prone called public opinion. This meant, for Lippmann (1922, 248–249),

> It is no longer possible . . . to believe in the original dogma of democracy; that the knowledge needed for the management of human affairs comes up spontaneously from the human heart. Where we act on that theory we expose ourselves to self-deception, and to forms of persuasion that we cannot verify. It has been demonstrated that we cannot rely upon intuition, conscience, or the accidents of casual opinion if we are to deal with a world beyond our reach.

It takes little reflection to realize that this more than two-millennia-old problem is still with us. The creation, modification, and enforcement of many standards appears to require considerable expertise—whether in science, engineering, law, medicine, finance, or other fields of expert endeavor. At the same time, acquiring this expertise is often a long and arduous process.

Indeed, today virtually every government of every industrialized nation has a plethora of regulatory agencies staffed with experts of various kinds, many of whom are charged with creating, modifying, adjusting, enforcing, and checking on an enormous array of extremely complex, often arcane, standards. In the Food and Drug Administration one finds chemists, biologists, medical doctors, toxicologists. In the Treasury Department one finds financial specialists, economists, tax accountants, and experts in protecting against counterfeiting. In the Interior Department one finds land management specialists and persons knowledgeable in numerous subfields of biology and geology. And the National Institute of Standards and Technology employs some 2,500 staff, most of whom hold doctoral degrees, solely for the purpose of creating, adjusting, and maintaining various reference standards. The private and nonprofit sectors have an even more diverse range of experts and specialists busily making, modifying, and interpreting standards. Clearly, no one could possibly claim to have even a casual acquaintance with all these highly specialized fields of endeavor. Alas, it would seem that democracy is merely an outmoded, idealistic dream to which many of us continue to adhere.

The Case for Democracy

While defenders of expertise have been quick to challenge democracy, until very recently democratic theory has been largely silent on the question of expertise. There are a variety of reasons for this. First, when conceived of by the ancient Greeks, democracy was about the affairs of the polis, which

for the most part did not include questions about the technoscientific
world. Indeed, in ancient Greece, craft work was largely the province of
slaves, while those who were free men—women were excluded—were to
be occupied by the far more important affairs of state. Even Greek science
and mathematics, brilliant as it was at times, avoided the practical affairs
of daily life (*technē*), focusing far more on something akin to what we
would today call the theoretical world (*epistemē*).

Second, although we are often loathe to admit it, our modern notion
of representative democracy originated not so much with the ancient
Greeks as with the aristocracy of Western Europe. It was the English aristocracy's desire to control the actions of King John that led in 1215 to the
writing of the Magna Carta, an iconic document in the history of modern
democracies. Even then, it took another four hundred years for these liberties to be taken seriously. In any case, the immediate concern with curbing
the powers of the king took precedence over any concern with expertise.
Such experts as did exist were those who understood political processes,
narrowly conceived. Moreover, they were experts because of their broad
experience in political affairs, not because of any specialized knowledge
they were thought to possess. When in November 1660 twelve illustrious
followers of Bacon created the Royal Society, their own version of the
House of Salomon, it was clearly subordinate to the king, not the other
way around.

Third, liberal political theory, as initially propounded by Hobbes and
especially by Locke, specifically excluded expertise from the domain of
politics. For Hobbes, monarchy was to be preferred as a form of government because it was most effective in producing order, in enforcing the
social contract among men, and in creating a commonwealth. But Hobbes
was clear that a good sovereign would provide defense against foreign
enemies, ensure peace at home, promote the enrichment of subjects, and
allow subjects freedom to enjoy "harmless liberties" (Hobbes [1658/1642]
1991). Expertise clearly fitted into this last category. It was not the province
of the state.

Similarly, one looks in vain for guidance from other classical liberal
philosophers. Expertise simply did not occupy much space in their models
of the social order. Hence, John Locke, despite his training in medicine
and familiarity with "mechanical philosophy" (Milton 2001), is largely
silent on the matter of expertise. Similarly, Adam Smith, ever anxious to
develop a market society, says little about the nature of the expertise
required by those who would craft those goods beyond noting the consequences of the minute division of labor.

There are several likely explanations for this seeming oversight. First, during the Middle Ages, technoscientific change occurred, but it was rather slow (but see White 1962) and was deliberately limited precisely because those in power understood that rapid change might undermine their control. Second, Bacon and other early proponents of mechanical or experimental philosophy claimed they were merely revealing the material works of God. Moreover, much of what they initially accomplished was more likely to be displayed in the foyers of European courts or demonstrated before awed audiences than used in the transformation of the larger society. Third, until a few centuries ago, the consequences of expert knowledge for political life were rather limited. It was only in the late nineteenth century that industrialized nation-states had sufficient confidence in the knowledge of experts in the sense that we understand the term today to begin to delegate authority to them (Stanziani 2005).

But it is helpful to begin an examination of the democratic critique of expertise with the British satirist Jonathan Swift, who nearly three hundred years ago poked fun at the very idea that experts could run a society. That was exactly what they did in the mythical kingdom of Lagado:

In these colleges the professors contrive new rules and methods of agriculture and building, and new instruments and tools for all trades and manufactures, whereby, as they undertake, one man shall do the work of ten. . . . The only inconvenience is that none of these projects are yet brought to perfection, and in the meantime the whole country lies miserably in waste, the houses in ruins, and the people without food or clothes. By all which, instead of being discouraged, they are fifty times more violently bent on prosecuting their schemes, driven equally by hope and despair. (Swift [1726] 1947, 227)

Perhaps Swift was being somewhat unfair in his critique of the very beginnings of modern science in his day. Yet we must also concede that Swift has a point. At the same time as there are clear successes that derive from modern technoscience—consider antibiotics, indoor plumbing, and electricity, among many others—history is also littered with a sufficient number of broken promises to encourage a degree of skepticism. The *Titanic* was unsinkable. The *Columbia* space shuttle was perfectly safe to fly. Nuclear energy was going to make electricity so cheap there would be no need to monitor it. Building superhighways in and between our cities was going to eliminate congestion. The widespread, nearly indiscriminate use of pesticides led to a litany of problems (Carson 1962). Countless other incidents can be recounted as well.

It was only during the twentieth century that proponents of democracy began to grapple directly with the problems posed by proponents of expertise. Perhaps no one was a greater defender of democracy in the face of expertise than the American pragmatist philosopher, John Dewey (1859–1952). Dewey took Lippmann's critique very seriously, so much so that he responded first with a book review (Dewey 1922b), then with a set of lectures published under the title *The Public and Its Problems* (Dewey 1927). Dewey agreed with Lippmann on the importance of *both* democratic decision making and scientific expertise (Marres 2007), but he argued that an educated public could and must make decisions about public issues.

In the book review, Dewey seemed initially to agree with Lippmann about the state of representative democracy, and even to go one step farther, arguing for the inclusion of social scientists among the experts: "The union of social science, access to facts, and the art of literary presentation is not an easy thing to achieve. But its attainment seems to me the only genuine solution of the problem of an intelligent direction of social life" (Dewey 1922b, 288). But in his conclusion he argued that the education of experts was insufficient.

In his book, Dewey developed his response into a full-fledged reply. First he dissected the very notion of "the public." He concluded that "Transactions between singular persons and groups bring a public into being when their indirect consequences—their effects beyond those immediately engaged in them—are of importance" (Dewey 1927, 64). In short, for Dewey a public was created only in those instances in which some apparently private, local, individual, or small-group action had important consequences for "bystanders." In more contemporary terms we might say that issues arise when there are threats to people's "livelihood" (Marres 2007). Thus, if I sell you a tomato at an agreed-upon price, there is no public involved. Similarly, if I develop a more effective means for producing tomatoes on my farm by varying the spaces among the plants and planting marigolds to reduce the insect populations, there is no public involved. However, if I decide to spray my tomatoes with an airborne insecticide that also blows into your home, then there is an issue. It is precisely this that makes (some) standards the province of public discourse and democracy.

But at the same time, Dewey rejected the received notion developed by many of the classical liberals that democracy was merely the aggregated views of isolated individuals with clearly defined *personal* preferences: "The idea of a natural individual in his isolation possessed of full-fledged wants, of energies to be expended according to his own volition, and of a ready-made faculty of foresight and prudent calculation is as much a fiction in psychology as the doctrine of the individual in possession of antecedent political rights is one in politics" (Dewey 1927, 102). Thus, the problem for Dewey was the profusion of publics, the growing centrifugal tendency of contemporary society.

The solution to the problem of expertise lay in a multifaceted program aimed at direct or deliberative or participatory rather than representative democracy. Participatory democracy was both means and end for Dewey; it would secure liberties and would provide the conditions for self-realization through the creation and re-creation of a vibrant community. It would not merely be a place where the divergent preferences of various individuals would be the subject of negotiation or bargaining; it would be far more like Aristotle's polis, a place where preferences, commitments, and concerns were discovered, (re)created, and transformed (see de Vries 2007).

The model for that participatory democracy appears to have been science itself. For Dewey, science was a kind of participatory democracy. It was one in which all ideas were potentially subject to reflection and critique, even radical dissent,[1] not solely through discourse but, more fundamentally, through experimentation—through practice. It was a democracy in which results were made public and widely disseminated, permitting everyone to examine and interpret findings. And it was a democracy in which scientists chose freely in which publics (i.e., disciplines) they would participate. In short, science provided a model of public decision making.

The key to creating this participatory democracy was education. Perhaps more than any modern philosopher, Dewey was passionately concerned about education. For it was universal public education that would provide students with the skills to apply a version of the scientific method to everyday public issues. Significantly, his version of the scientific method was not the dogmatic one favored by Bacon and Descartes and designed to lead to Truth. In contrast, for Dewey the outcome of scientific research was not the discovery of a set of brute facts about the world but the *creation of a set of transformative actions that might be performed*. Put differently, Dewey saw science as providing society with a model for social experimentation, for both the formation of social habits—standards—and for reflecting on and modifying them in a considered manner.

But perhaps Dewey's view of science was a bit too rosy. Dewey had joined the faculty at the University of Chicago at a time when it was a grand new experiment in interdisciplinarity. By contemporary standards, his colleagues in the sciences and the liberal arts at Chicago were generally broader in their perspective and more open to extradisciplinary challenges than their counterparts at most universities, then or now. One observer notes that Dewey likely developed his model for science from discussions with his colleague and family friend at Chicago, biologist Jacques Loeb (Pauly 1987). In addition, Dewey died before the philosophy of science began its empirical turn. In his day, most philosophers and social scientists were content to examine science from a distance, to exalt science based on its public relations rather than to carefully examine the process of doing science itself. It was still widely believed that science consisted in building the cathedral of knowledge one experiment at a time. Dewey viewed science as it would like to be rather than as it was (and is), warts and all.

In contrast, more recent commentators have been much more critical of expertise in general and science in particular. For example, philosopher Paul Feyerabend noted that instead of debating publicly the case for technoscience, politicians and scientists often assume its excellence as a means of knowing. Moreover, he argued that Western views have prevailed because they have been foisted on other peoples (Feyerabend 1978). Feyerabend's critique has been pursued in more detail by a significant number of contemporary historians and social scientists. For example, Susantha Goonatilake (1982) has convincingly argued that much of what passes for Western science was appropriated from other parts of the world and that the Western powers suppressed science in their colonies, seeing it as a threat to the status quo. Ironically, the inventions that Bacon used as exemplars of what a new science might do—printing, gunpowder, and the compass—were all imports from China.

Geologist David Sarewitz (1996) has tackled the issue from a different angle. He has examined the myths that are perpetrated by scientists and politicians alike. The myth of infinite benefit says that if some science is good, more must be better. But the evidence to support this assertion is simply not there. As Sarewitz notes, despite vast increases in funding for medical research, U.S. citizens' health has not significantly improved over that of other industrial nations. In fact, nations that spend far less now surpass the United States on key measures of health care.

Sarewitz also notes what he calls the myth of authoritative knowledge. As he explains:

When scientists ridicule public opinion, they implicitly argue that wise decisions about the potential societal impacts of research can only be made by experts. But if, as researchers say, the results of their work are intrinsically unpredictable, then experts are no more qualified to assess the potential for negative social consequences than any other reasonably informed person. If, on the other hand, such consequences can be generally foreseen, then the public has a right and an obligation to be involved in deciding what sorts of actions should be considered; technical knowledge can guide but not dictate these decisions. Either way, the question of technical expertise is secondary. Of primary importance are the underlying social and moral dilemmas that drive public concern. (Sarewitz 1996, 65)

He goes on to note that questions can only be answered authoritatively when there are no controversies. If there is a political controversy that involves scientific expertise (or any other kind), then the science is simply drawn into it.

The myth of accountability is that scientists have been fully accountable for their actions if they have convinced their colleagues of the value of their work. Hence, accountability is limited to convincing the cognoscenti of physics or genomics or electrical engineering or even sociology of the value of the work accomplished. But if such work is paid for by the public purse, then at the very least it would appear that (some representatives of) the public should be permitted to evaluate it as well. This is not to advocate for "no-nothingism," a dismissal of the value of intellectual knowledge. But it is to argue that, if scientists are to use public funds, they have an obligation to explain the importance of their results to a larger public, and members of that public have to be able to ask (and have answered) questions about both the assumptions underlying such research and its consequences.

Sarewitz also describes what he calls the myth of the endless frontier. Borrowing from the title of a 1945 government report by Vannevar Bush (1945), he challenges the idealized view in which all knowledge flows from the scientific laboratory to the engineering laboratory, where it is scaled up or down, thence to the factory, where it is turned into a product for sale. In point of fact, numerous empirical studies show entirely different paths. For example, one British nineteenth-century scientist opined that "the science of thermodynamics may be said to be the result of the steam-engine" (Osborne Reynolds, quoted in Cardwell 1971, 186). In other instances science is driven by the problems encountered in attempts to improve new technologies. Leonard S. Reich (1985) recounts the ways in which science was implicated in the competition between General Electric and Bell over light bulbs and vacuum tubes a century ago. In other instances, invention is the result of trial and error and has no scientific theory behind it. Thomas Edison is said to have tried hundreds of compounds before developing a satisfactory filament for the electric lamp. Other technologies, such as the bicycle, are largely the result of effective tinkering over an extended period of time (Bijker 1995). Finally, many of the products of technoscience are the result of negotiated agreements about standards, as I hope I have made clear in this book. Such standards are never solely the product of pure science (whatever that might be) but always involve compromises among the industrial, the political, and the economic worlds.

Finally, Sarewitz describes the myth of unfettered research. This is the notion that any line of basic research is as likely to lead to social betterment as any other. Since one cannot know in advance where basic research will lead, one cannot attempt to direct it in any way. From this perspective, the best approach is to let research go on unfettered, led by the conviction that it will doubtless lead to socially desirable outcomes. Unfortunately, this myth ignores the fact that nearly all scientists work within an institutional and organizational framework that both limits and directs what they do. This limiting and directing need not be heavy-handed—a good case can be made that it should not be so—but it exists nevertheless. It is necessary if for no other reason than that there are groups in society that financially support, argue and lobby for and against, see potential for profit-making or losses, gain or lose prestige, for nearly all research. Such groups include not only the countless manufacturers, the consumers who might purchase the products, and the government officials that stand to gain in stature from new scientific discoveries but also the vast array of suppliers of scientific equipment and instrumentation, university administrators strapped for cash, and the scientists themselves, who doubtless want the funding to get on with their preferred work. Taken together, this complex network of more or less well-represented publics belies the myth of unfettered research.

The myths that Sarewitz describes show some of the limits of expertise. The work of experts does not automatically lead to infinite benefits. Expert knowledge is authoritative only when there is no controversy. Everything on the public agenda is controversial; that is why it is there. If experts are to spend other people's money, their accountability must extend (at least) to those who pay the bills. Nor can experts claim they are at the sharp end of a cornucopia from which all knowledge flows: knowledge can come from tinkering as much as from grand theories. And finally, there is no unfettered research, nor should there be: research is a social product, not the result of the genius of an isolate.

But while Sarewitz's concerns build on Dewey's call for democracy, they remain incomplete. To them we need to add several other points to complete the case for democracy. In particular, supporters of the position that expertise trumps democracy have invented two similar but distinct and important responses: cost–benefit analysis (CBA) and risk analysis. Both of these may be seen as belonging to standard-setting exercises, although they are usually auxiliary standards and not those for particular products, processes, persons, or practices. Let us examine each of these in turn.

Cost–Benefit Analysis

Everyone can agree that things have costs and benefits associated with them. If I buy an automobile, it has a certain cost (both the initial price and the cost of operation and maintenance), as well as a certain set of benefits (point-to-point transport, comfort, speed, etc.). It may be easier to sum up the costs in monetary terms than the benefits, but clearly it can be accomplished. However, CBA is not generally used for such relatively straightforward calculations.

CBA owes its origins to the U.S. Army Corps of Engineers, where it was initially developed to hold off powerful interests opposed to the Corps's projects (Porter 1995). By the 1950s it had become institutionalized. Today it is commonly used as a means to determine whether or not a given *important* project should be pursued. Hence, environmental impact statements, overviews of the effectiveness of medical procedures, and even investments in education may be subject to CBA. Given that CBA requires expert knowledge and is expensive and time-consuming to conduct, it is generally not used for trivial projects, those deemed unimportant, or those deemed essential. This is true regardless of cost. One does not perform CBA in determining which brand of canned peas to purchase at the supermarket, nor does one use it in determining whether to repair a national monument.

Like all tools, CBA is, and must be, based on a set of assumptions. In general, in CBA "Facts are considered as relevant only if they fit into the preconceived scheme and relate to the predetermined goal. And, to be considered, they must be in quantitative form, either by nature or by arbitrary assignment of a price" (Hoos 1972, 72). Put differently, users of CBA—by virtue of its very nature as an analytic tool—must assume that (1) only certain kinds of things count as costs and benefits and (2) these costs and benefits are amenable to quantification.

Furthermore, CBA assumes (3) that the issue before us, whatever it might be, is best expressed in the form of an *economic* calculation in which the most efficient solution (i.e., the greatest benefit for the lowest cost) is the desired outcome. As such, "Cost-benefit analysis does not, because it cannot, judge opinions and beliefs on their merits but asks instead how much might be paid for them, as if a conflict of views could be settled in the same way as a conflict of interests" (Sagoff 1988, 38).

But even if we accept the goal of efficiency, it may well be the case that the most efficient economic solution is not the most efficient in terms of, for example, minimizing energy use, unemployment, pollution, or some other equally reasonable goal (Melman 1981). Proponents of CBA might object, arguing that these other outcomes can be transformed into prices and included in the CBA equation. That is undoubtedly the case. Yet to do that, one must make the rather heroic assumption that current prices (or some other price using a different set of heroic assumptions) are the right ones to enter into the equation. For example, a CBA involving the use of considerable quantities of petroleum would be considerably different under 2011 pricing compared with 1965 pricing.

Furthermore, CBA must necessarily exclude distributive questions. This is true for both costs and benefits. After all, the costs will be the same regardless of who pays them and the benefits will be the same regardless of who receives them. To include distributive questions would clearly involve supplementing the rule of efficiency with some other rule from outside the CBA framework.

In addition, CBA almost always involves comparing the costs now with the benefits at some point in the future. Since the future is at best seen through a glass darkly, those expected benefits to persons in the future may be inflated or deflated, depending on the (more or less well-examined) predilections of the analyst. In practice, as Langdon Winner (1986) notes, the tendency is to insist on a scientific rationale for determining the costs of the new technologies even as it accepts a much broader definition of the (promised) benefits. Furthermore, it avoids grappling with a central policy question for any society: What kind of future do we want? As Daniel Bromley (1998b) puts it, instead of beginning with a desired future and working back to the present, CBA merely extrapolates the present into the future.

In sum, although it is shot through with questionable assumptions, CBA tends to shift decision making from the public to an expert elite with the wherewithal to gather the "hard data" needed to arrive at a decision. Moreover, this process is often conflated with consent, although it may well be that the most cost-effective policy to pursue is unacceptable to the vast majority of the public.[2]

Risk Analysis

Much of what I noted above for CBA applies as well to risk analysis. Risk analysis is usually seen as consisting of four parts. Risk identification involves the determination that something is a risk. Risk assessment involves the quantification of the risk as a statistical probability of harm. Risk management involves the development of strategies to mitigate the risk. Finally, risk communication involves communicating the risk to the affected public such that groups and individuals can change their behavior accordingly. This model of risk has been challenged elsewhere, which need not concern us here (Abraham and Reed 2002; Busch 2002; Jasanoff et al. 2005; Kysar 2006). What is important to us here are (1) how it defines the problems at hand and (2) how it stratifies participants in the debate over a given risk into several different categories.

As the name implies, risk analysis frames controversies as situations in which there may be risks to be avoided. "Framing orders experience, eases confusion, and creates the possibility of control. Problems that have been framed are capable, in principle, of being managed or solved" (Jasanoff 2000, 69). This is sometimes quite appropriate, even necessary, but in other instances framing the situation, deliberately or unintentionally, as one of risk would exclude concerns relevant to one or another group, thereby distorting the debate.

For example, consider the case of toxic compounds. If the toxicity is framed as a natural phenomenon, risk mitigation is likely to be considered the province of the state. If it is framed as an industrial effluent, it is likely to be considered the province of industry. If it is framed as genetic suscep-tibility to the compound in question, it is likely to be considered the province of a particular worker (Salter 1988). Each framing has nontrivial distributive consequences.

The debate over genetically modified organisms (GMOs), especially plants and animals, provides another case in point. Supporters of GMOs generally argue that the modified plants and animals now on the market have been shown to pose no risk to humans, and likely none to the envi-ronment, either. Even if we put aside the validity of this expert claim, numerous other concerns have arisen about GMOs. These concerns include the growing concentration of control over the world's seed supply by a small number of companies, the inability of farmers to legally replant the seeds from genetically modified plants, religious objections to the violation of species lines or to the insertion of genes from "unclean" organisms into those fit for human consumption, consumers' right to choose what they eat, and what some have referred to as the "yuck" factor, a generalized revolting feeling that some persons claim to have when confronted with GMOs. What is important to note here is that, regardless of the validity of these counterclaims, they cannot be fitted into the risk analysis frame-work. They are simply and summarily ruled out of court. Hence although legislative debates, litigation, and the WTO dispute settlement process are centered on risk, the matters of greatest concern to some have little or nothing to do with risk.

In addition, risk assessment is defined as the province of experts—generally in the sciences and engineering, but occasionally in other fields such as economics. In contrast, risk management and risk communication are generally assigned to administrators. This division has the convenient effect of "purifying" the science that is undertaken by assessors and segmenting it from the necessary intercourse with one or more publics. As Jensen and Sandøe (2002, 247) suggest, "The general picture painted by these definitions is this: risk assessment is an exclusively scientific process, and in the process of risk analysis it comes first. Risk management, on the other hand, is a political/administrative process, which comes after, and is based on, the relevant risk assessment."

Occasionally members of the general public might be included in risk identification, but even then they are usually held at bay. In particular, if publics cannot make their case in scientific terms, in the language of a statistical probability of harm, they are likely to be dismissed as cranks. Yet, since the very definition of what risk is frames the analysis, it is hardly a trivial issue. Any real public participation tends to occur as a reaction to the decisions made by others (Krimsky 1984). Yet risk analysis is seldom if ever undertaken as a purely scientific project. Its entire intent is to inform policy (Rosa 1998).

Moreover, risk assessment asks assessors to make two quite different kinds of decisions. On the one hand, they are asked to answer a largely technical question: What is the statistical probability of harm associated with X? In the best of circumstances this question can be given a reasonably clear and reliable answer; in some instances the data are contradictory or missing. In still other cases there is great uncertainty over the results, although that uncertainty is often unreported. On the other hand, assessors are asked to determine whether the risk is one worth taking. This is not a technical question at all; it is a normative question, and one that assessors are poorly prepared to answer in their role as assessors.

An example is useful here. Like many older persons, I have been diagnosed with leg cramps of unknown origin. I used to take an off-label medication (i.e., one not approved specifically for this condition), quinine sulfate, as a preventative. However, the Food and Drug Administration banned the off-label sale of the drug as it appears to increase the risk of cardiac dysfunction. Patients were asked to check with their physician as to what to do. No equally effective drugs are available. Note what has happened here. First, assessors determined an elevated level of risk of cardiac dysfunction associated with taking this drug. Then they had to determine that the risk was sufficiently high to warrant a ban on its off-label sale. The latter decision is one for which the assessors can claim no special expertise. Indeed, the FDA advice—talk with your physician—is indicative of the discomfort of the assessors in banning the drug.

Another issue concerns the arcane character of risk assessments. All but the most highly industrialized nations are ill-equipped to carry out such assessments. However, this hardly means there is widespread agreement over the details of risk assessment among scientists in industrialized nations. As the GMO issue illustrates, U.S. scientists have tended to focus on risks associated with products, while UK scientists have tended to focus on those related to processes (Jasanoff 2000). European nations tend to give more official weight to the "precautionary principle" than does the United States (but see also Busch 2001; Kysar 2006). That said, the less industrialized nations usually lack the trained personnel, instrumentation, and software needed. This puts them at a disadvantage relative to nations that can execute risk assessments. In particular, such nations lack the means to challenge risk assessment decisions they believe are inaccurate or unfair.

Yet another limitation of risk analysis is the tendency on the part of assessors to presume that things will be done as expected, that standards of good (manufacturing, agricultural, medical, etc.) practice will be observed. But as Charles Perrow (1984) observed a quarter of a century ago, accidents are normal. As Robert Burns wrote, all "the best laid plans of mice and men, often go awry." Those most aware of where and when "normal accidents" are likely to happen are not usually the risk assessors but those persons who each day use the technologies in question, who each day are subjected to the standards, and who, often with great effort, attempt to maintain them. Hence, even if the assessors get it right, what they get right is often desituated, all about the average or typical case. Hence, one might well argue, as does Brian Wynne (1989) in an analysis of the relations among sheep farmers and scientists after the Chernobyl disaster, that scientists can be as parochial in their views as anyone else.

Finally, it is essential to recognize, as has Ulrich Beck (1992, 21, emphasis in original), that "*Risk* may be defined as *a systematic way of dealing with hazards and insecurities induced and introduced by modernization itself.*" In short, there is a great irony associated with RA: it exists largely as a result of attempts by scientists and engineers to grapple with the unintended consequences of technoscience itself. That irony is doubled by the peculiar tendency to treat all risks to our health and our environment as if the actions of people were not part of the equation, as if further scientific study of the problem were sufficient to resolve it. Yet as Wynne et al. (2007, 21) note, all "innovations are really socio-technical innovations, because organisational competencies, business-to-business linkages, and value chains and industry structures more broadly have to be renewed as well."

Nevertheless, many scientists appear incapable of grappling with the painful fact that technoscience is itself the source of many of the most serious problems facing the world today—climate change, nuclear war, air pollution, chemical waste dumps. These problems are all too easily dismissed as side effects, mistakes, the misuse of science. "Laboratory science is systematically more or less blind to the consequences which accompany and threaten its success" (Beck 1994, 30). Instead, like the scientists of Swift's ([1726] 1947, 227) Lagado, "instead of being discouraged, they are fifty times more violently bent on prosecuting their schemes, driven equally by hope and despair."

Furthermore, as Beck goes on to note, one cannot experience risk—even in the way in which one can experience a cost or a benefit. One may experience illness from contact with a toxic compound, but one cannot experience the risk itself. Risk is an artifact, a probability, part of a world transformed by technoscience. Yet even as one cannot experience it, one can—and one does—attempt to avoid it. Hence, in much of the industrialized world, and to a lesser extent elsewhere, there is an endless procession of risk-avoiding behavior. We avoid transfats, foods treated with pesticides, toys with lead paint, noxious fumes from internal combustion engines, toxic waste sites. This is not to suggest that we should treat these risks cavalierly; to the contrary, some of them are likely to harm some of us— some proportion, that is.

Risk analysis also promotes the standardized differentiation described in chapter 3. Risks are constantly subdivided; groups are reduced in size, but since risks are *probabilities* of harm, they are never (contra Beck) fully individualized. As François Ewald (1991, 203) notes, "Strictly speaking there is no such thing as an individual risk; otherwise insurance [and all other risk estimates] would be no more than a wager." Hence, I may discover that, given my smoking one pack of cigarettes a day, being male, being sixty-five years old, eating a diet that is too high in fat, and working in a profession that does not require much exercise, I (and all those in this thin slice of humanity) have a high probability of experiencing a heart attack. But in a seemingly absurd way, the probability can never apply only to me.

In sum, both cost–benefit analysis and risk analysis are attempts to forecast, to predict some future state of events. At best they are somewhat heroic attempts to forecast the future, sometimes shedding light on real and important problems and opportunities. At worst they are pseudoscience, means for obfuscating what is at stake and of concentrating decision making in the hands of a technoscientific elite that falsely claims to have the answers. Proponents of both approaches tend to forget that "[e]very single moment always hides endless contingencies—which, if we look at them carefully, are likely not simply to be contingent. That means that most elements relevant to making or unmaking the *goods of life* involved in making a decision escape the moment of that decision" (Mol 2002, 170, emphasis in original).

Yet none of this is to argue that costs or risks should be ignored. We need to have estimates of what things will cost. We need to know what risks are involved, and we need to know the benefits that might follow from the costs or risks. Therefore, the expert opinions of cost–benefit analysis and risk analysis must be inputs into a democratic decision-making process in which contingencies are revealed, and not prepackaged outputs that obviate democracy. Perhaps Lewis Carroll put it best in *Alice in Wonderland*:

Alice Would you tell me, please, which way I ought to go from here?
Cheshire Cat That depends a good deal on where you want to get to.

Moreover, as noted in chapter 2, even those who acquire expertise in one field lack it in others. Molecular biologists know little about materials science. In a recent editorial, the editor of *Science*, Donald Kennedy (2007, 715), noted that "we're all laypeople these days: Each specialty has focused in to a point at which even the occupants of neighboring fields have trouble understanding each others' papers." This sorely limits the role that expertise can play, since the experts might well misdefine the situations that need to be subjected to standards.

Thus, while setting standards often requires expertise, that is not all it requires. Standards are also about ethics. Standards are all wrapped up in questions about who we are, how we want to live, and what is the right thing to do. The answers to these sorts of questions may incorporate expert knowledge, since that knowledge can offer new opportunities for action. But an expert claiming to have clear answers to these questions should be viewed with considerable suspicion.

Of course, some philosophers, ethicists, theologians, and other thoughtful individuals have studied these questions far more than the average person. Doubtless they can provide guidance, help us clarify our views, point out contradictions and dilemmas, and otherwise shed light on our concerns. But as with other experts, we only let them make decisions for us at our peril.

Conclusion

At a conference some years ago, French sociologist Bruno Latour jokingly suggested that scientists should be considered like plumbers. When you or I have a problem with our plumbing and call in a plumber, we point to the problem and ask the plumber to fix it. We don't call in a plumber and ask him or her to look around our home and see if anything is in need of fixing. Indeed, doing so would likely not only prove quite costly but also cause considerable disruption to our lives.

This comment appears to me to point the way to a meaningful resolution of the apparent conflict between standards and democracy. Our current regulatory bodies and standards agencies, as in the story of Latour's plumber, give free rein to the experts. Our regulatory bodies and standards agencies appear to be designed based on the models proposed by Plato and Bacon, if not that proposed by Lippmann. Yet by recognizing that standards are both technical *and* moral projects, we can begin to design institutions that grant both the experts and the general public their due.

This notion, as Winston Churchill is said to have expressed it during World War II, is as follows: experts should be on tap, not on top. Since that time, other commentators have used the expression in a number of ways. For Jacob Viner (1940) in his 1939 presidential address to the American Economic Association, it meant that expert advisers to government should yield to the demands of public officials. More recently, an EU expert cited in a report argued, "Science comes first but it is followed by a consideration of other legitimate factors" (Wolf, Ibarreta, and Sørup 2004, 24).

I find both of these approaches somewhat lacking. The former limits public oversight of experts to elected officials; this seems to me too narrow a view of democratic governance. The latter puts science (and presumably expertise) ahead of public concerns. The Families and Democracy project appears to be a better way to integrate concerns of experts and ordinary citizens.[3] This particular project focuses on the relations between family counselors and families: "The model emphasizes the importance of a public life to a full human life. It rejects the notion of private life cut off from life in the 'commons,' and posits that the privatization of contemporary life leads to the unhealthy dominance of the market and the state over human affairs" (Doherty and Carroll 2002, 584). A key feature of the approach is that families and professionals are active contributors and designers of a collectively desired future. Experts are on tap, not on top.

Moreover, this idea can be set within a more robust notion of what democracy is and might be. In the classical liberal notion of democracy, autonomous individuals come together to govern collectively. But the points I have made earlier stand against it: (1) we each learn who we are and how to be in the world through the acquisition of habits, mores, customs, and traditions given to us by others; (2) as individuals, our knowledge of the world and even of ourselves is incomplete; (3) cognition is distributed in that we—whether experts or laypersons—can only validate our knowledge through coordination and interaction with others; and (4) we all live in multiple worlds of justification. In short, we are "between autonomy and sociality" (Rasmussen 1973).

Thus, any robust notion of democracy must lie between the myth of total autonomy and the countermyth of complete community. Indeed, to the extent that standards development by experts involves deliberation and debate (as well it should [Jasanoff 2003]), it lends support to the four points noted above. But if the experts can only claim expertise with respect to the *technical* aspects of their projects, then the standards they develop are likely to be inadequate. Only through democratic processes can this problem be remedied. But what might those processes look like? Clearly, for the kinds of endeavors discussed in this book, we cannot return to the direct democracy of the ancients.

It is a commonplace notion today that there is a complex division of labor in nearly all human affairs. Moreover, that division of labor has increased in complexity over time. That increase in the division of labor has brought with it a more fragmented distribution of cognition, posing extraordinarily complex problems of distribution and coordination. The Soviet response took the route to sociality; it put the state in charge; it provided solutions to some problems—there is some justification for the "*Ostalgia*" found in the former East Germany—but it ultimately failed. The neoliberal response takes the route to individuality. It attempts to use the market and market-like mechanisms in each instance to enhance certain forms of individual freedom. As Michael Walzer (1994, 23) suggests, in this neoliberal state, "Hierarchy is replaced by universal striving and contention, romantically celebrated as the seed-bed of individuality and self-realization." Like the Soviet model, it resolves some problems but has led to others, coercing all of us to be entrepreneurial at all times, widening the gap in the distribution of income and wealth, and hollowing out democracy into merely another form of market. Both approaches bring a rapid end to deliberation.

However, the division of labor—from the family to the workplace—is both the source of my individuality *and* of my sociality. Through the division of labor I learn about my individual uniqueness. I develop certain habits, skills, ways of being and acting. I even learn about those aspects of myself that are never revealed to others (Thévenot 2006). But I also learn about what I have in common with others. I learn both what we value and how I am valued (or not) by some larger group. Moreover, as Marx and de Tocqueville both understood, the minute division of labor leads to a mind-numbing narrowing of cognition. Standards, for both people and things, determine in large part how that division of labor is to be organized, how society will be performed, in what reality we shall dwell. Hence, the formation of standards is central to the (re)structuring of society. As such, the formation of standards is central to the success or failure of contemporary democratic governance as well.

It is precisely because our knowledge is always incomplete and fragmented, because cognition is distributed, because we may do violence to others, that we need to have both experts and laypersons involved in most standards setting activities. Deciding what standards we want necessarily involves deliberation, negotiation, and compromise among diverse persons and organizations, as well as in the design and use of myriad objects. No single best solution is possible. Society is neither a logical problem nor a mathematical equation to be solved but a commons to be improved. Appeals to efficiency, to technological superiority, or to the market as the value that trumps all others must necessarily be based on idealized models that are never realized in practice. As such, they merely avoid the difficulties involved in the process of standards setting, with no guarantee of better outcomes. Experts—including those who proclaim the virtues of efficiency or the market above all other values—should be on tap, not on top.

Conclusions: Another Road to Serfdom?

The great antinomies of human life are never solved by grasping one polarity and forgetting the other. Our problem is not to get rid of control in any absolute sense but to find a new kind of control that will allow a wider freedom.
—Robert Bellah (1975, 84)

Austrian-born economist F. A. Hayek ([1944] 2007) argued more than half a century ago in his widely cited book, *The Road to Serfdom,* against the concentration and abuses of state power known variously as fascism, Soviet communism, and Nazism. In Britain when Austria was annexed by the Nazis, Hayek decided not to return, and became a British citizen. The book became widely known in the United States when it was serialized by *Reader's Digest.* Pivotal to that work and much of his later work was a strong revulsion for central planning and an equally strong belief in the market.

But if Hayek rightly warned us against the serfdom posed by the over-reliance on central planning that robbed people of the ability to make their own decisions, he was largely blind to the potential for serfdom to be found in overreliance on the market or other idealized worlds. Hayek (1945, 519) rightly criticized his more orthodox colleagues for ignoring a central feature of the human condition, namely, that all knowledge is partial and incomplete. As he put it, "knowledge of the circumstances of which we must make use never exists in concentrated or integrated form, but solely as the dispersed bits of incomplete and frequently contradictory knowledge which all the separate individuals possess." However, Hayek and his colleagues substituted a formal logical model of the competitive market for the "superb eye of power" of the state. In short, even as Hayek decried attempts at state planning, even as he emphasized the incomplete-ness of knowledge, he believed he had found an Archimedean point—the kind of certainty that comes from logic or mathematics—on which to build a better society. It was this competitive price system that provided a solution to the lack of complete knowledge. It was not a perfect solution but the best one available: "Fundamentally, in a system where the knowledge of the relevant facts is dispersed among many people, prices can act to coordinate the separate actions of different people in the same way as subjective values help the individual to coordinate the parts of his plan" (Hayek 1945, 526).

While Hayek was doubtless right about prices as a convenient and often desirable coordinating mechanism for a wide variety of exchanges, he failed to explain why a society should reorganize its legal system to meet the specifications of a logical model, even if that model did demonstrate that competitive markets did not lead to monopolies and did increase wealth most efficiently. How could all the myriad concerns, values, interests, ills of a society be even partially captured by such a model? How could all human institutions be best served by basing them on prices?

Moreover, he failed to recognize that attempts to *positively* create a market society could be as oppressive as attempts to create a planned society. As Michael Walzer (1983, 119) notes, "A radically *laissez-faire* economy would be like a totalitarian state, invading every other sphere, dominating every other distributive process." Finally, Hayek failed to realize that even if markets functioned nearly perfectly, even if all the conditions of the model were met, slight deviations from the model would produce erratic and even unpredictable effects (Ormerod 1994). Indeed, Joseph Stiglitz won the Nobel Prize in Economics in 2001 (with George Akerlof and Michael Spence) for demonstrating that the very fact of asymmetric information—a commonplace in all market situations—made the ideal world of the model unattainable. Put differently, efficiency is, and is likely to remain, a highly elusive goal. Moreover, it is far from clear that it should trump all others.

But it was not solely because of its conformity to the market model that Hayek saw a price system as desirable. It was also because it "advances [civilization] by *extending the number of important operations which we can perform without thinking about them*" (Hayek 1945, 528, emphasis added). In this single statement he simultaneously emphasized the importance of customs, traditions, and standards and swept them under the rug.

In his later work, Hayek (1973, 1975, 1976, 1979) extolled the virtues of custom, tradition, and common law even as he decried the mischief accomplished by statutory law. For him, habits that are formed unconsciously were considerably more desirable than those formed consciously. What he failed to see was that this merely enshrines the existing order as somehow always more desirable than any alternative imagined future.

While it exalts the nearly random experimentation performed by custom and tradition, it denigrates the experimentation performed more carefully by standards and laws.

As a result, curiously missing from the work of Hayek and other neoliberals is much in the way of discussion of standards, certifications, or accreditations.[1] As Hayek (1979, 62) briefly notes, "It can hardly be denied that the choice of the consumer will be greatly facilitated, and the working of the market improved, if the possession of certain qualities of things or capacities by those who offer services is made recognizable for the inexpert." Hence, in this rather short passage, he opens the door to private standards. He then lists some things that might be the subject of standards and certifications, including building codes, food safety laws, and the competence of physicians. He concludes by noting, "All that is required for the preservation of the rule of law and of a functioning market order is that everybody who satisfies the prescribed standards has a legal claim to the required certification, which means that the control of admissions authorities must not be used to regulate supply" (Hayek 1979, 62).

This is the crux of the matter. In his zeal to radically restrict and reshape the role of the state, Hayek paid little attention to how one would avoid "regulating supply" in a world of *private* standards. Moreover, since he firmly believed in the superiority of the market over all other institutions, he hardly noticed that the withdrawal of the state would create for firms, NGOs, and others the *horror vacui* to be filled as rapidly as possible in order to stabilize and organize markets.[2]

There is thus an irony in all of this: the very attempt to perform the logical model, to create the perfectly efficient and competitive economy, creates new avenues of resistance by those appalled by (the consequences of) the model *and* those largely supportive of it. As a result, NGOs begin to create standards as an attempt to resist the neoliberal juggernaut, while companies and industry associations create standards to protect themselves from the neoliberal market they otherwise embrace. The very attempt to restructure the legal framework so as to allow the model to exhibit its truth claims opens up new spaces both for self-interest and for the creation of islands in which neoliberalism may be limited if not avoided.

Furthermore, the proliferation of standards, certifications, and accreditations, which may be understood as the fate of virtue under the conditions of neoliberalism, opens new possibilities for action not considered by its proponents. Specifically, neoliberalism requires for its success that the laws and standards that protect the model and ensure that reality (nearly) mirrors its functioning remain largely *invisible* to actors in the market society that is constantly growing in scope. This was already recognized by Adam Smith in his discussion of the invisible hand, but to Smith, the workings of the economy were indeed largely opaque. However, the very proliferation of standards not under the control of the state, as well as the need to harmonize legal frameworks across nation-states so as to extend markets globally, brings the issues associated with the instantiation of standards and laws to the foreground. Hence, the Achilles' heel of neoliberalism is to be found in bringing to consciousness the implications of standards and laws, in creating new locales of debate and dialog precisely in the places where the rules of the market(s) are made and the neoliberal market society is constantly defended. Moreover, just as Gödel showed that arithmetic cannot be explained solely by reference to itself, so one cannot employ the market as a means to shape and formulate the standards and laws that form its framework for action. Alas, even the defenders of neoliberalism find themselves forced to engage in such dialog.

Thus persons, states, and organizations use standards (and certifications and accreditations) strategically to gain an advantage in the marketplace. Some firms were already aware of the strategic use of standards as a result of the enhanced role of trade embodied in the several rounds of the General Agreement on Tariffs and Trade. But in 1994 the neoliberal approach to standards was incorporated into international law. In particular, Article 2 of the World Trade Organization's Technical Barriers to Trade Agreement, "Preparation, Adoption and Application of Technical Regulations by Central Government Bodies," severely limits the powers of governments to impose standards, especially those not clearly discernible in the final product or service.

The widespread use of market standards, certifications, and accreditations to differentiate has led to another, seemingly oxymoronic pair of extremes not envisioned by neoliberals: a world in which everyone is constantly confronted by a vast—indeed, overwhelming—range of decisions, many of which are life-changing, *and* where everyone is subject to incessant checking and audit, certification and accreditation.

Choices and Devoirs

The widespread use of standardized differentiation has plunged us into a world of obsessive choice and decision. Yet, as the UK's Food Ethics Council (n.d.) notes in its *Ethics Toolkit*, the choice card is always a joker. Sometimes the choice has longstanding positive or negative consequences; at other times it hardly matters. Either way, the choice often conceals the myriad decisions taken elsewhere, as well as the situatedness, the framing, the setting that has already taken place but is invisible. Decisions and habitual action are not opposite ends of a spectrum but two sides of the same coin. The plethora of decisions makes their situatedness invisible.

Of course, many trivial decisions are easily avoided by simply ignoring the vast range of choice most of the time and developing habits of choosing in much the same way that one has chosen in the past. Hence, confronted by twenty-five different kinds of ketchup, I no longer ponder over which brand and size to buy but rapidly reach for the one I bought last time. However, the consequential decisions are time-consuming and require considerable skill—skill that the middle class might have, the upper class can buy, the lower class is rarely able to pursue, and the "underclass" cannot pursue at all.

The market and marketlike arrangements currently in vogue as solutions to all sorts of social welfare problems create an incessant need for each of us to make decisions. That, in itself becomes a burden, a yoke, a duty, an encumbrance, a *devoir*[3] for which we are ill-prepared. The neoliberal argument goes something as follows: A particular good A is currently provided by the government (or is currently creating a number of externalities). Therefore, good A should be better subjected to the discipline of the market such that its price better reflects its "real" costs. But a perverse effect of this kind of disciplining is that it turns all sorts of habitual actions to which we pay little attention into conscious ostensibly "rational" decisions to which we are forced (lest we suddenly discover that we are penniless, left out, excluded) to pay attention. This includes differential tolls on highways, decisions as to which schools to send our children to, when a medical expense is or is not reimbursable, etc. A list of some of them can be found in table 7.1. To this we may add the burden, accepted by some and rejected by others, of "ethical" consumption. Should I buy the fair trade coffee? The bird-friendly coffee? The one with the lowest carbon footprint? Or should I buy only locally produced food, whatever that might mean? And so on.

All of these personal or family decisions share certain features in common. First, each is not a single decision but a rather large set of decisions that need to be made, often within a short period of time. Some people may relish making their way through some of these decisions, but others will find them impenetrable. Second, many, perhaps most, of these decisions are not trivial ones, such as whether to have vanilla or chocolate ice cream, but are potentially life-changing. Failure to make the "right" decision can have tragic consequences for you, your family, poor farmers

Table 7.1
The growing list of *devoirs* required of citizens

To which elementary or secondary school will you send your children? Is it safe? How does it rank in various ranking schemes? How far away is it? Will it provide your child with a "solid foundation"?

Which health care plan is best for you and your family?

Which package of college loans will best keep costs under control?

In which retirement plan should you invest?

What type of mortgage should you take out on your home?

What kind and value of various kinds of (health, life, home, auto, liability) insurance do you and your family need, and what can you afford?

What kinds of expenditures, donations, and the like will minimize your income tax bill?

Which mobile phone plan will minimize costs while providing the desired features?

What time should you leave for work and return so as to take advantage of differential pricing of tolls, train, and rapid transit fares?

Should I engage in genetic counseling before deciding to have a child? If the tests suggest some problem, what steps should I take among the several offered to "remedy" that problem?

Which foods should I buy to feed myself and family? Should I buy only organic foods to minimize exposure to pesticides? Should I avoid refined sugars? Too many carbohydrates? Do I get sufficient vitamins? Too much cholesterol?

Which brand of auto should I buy? What kind of warranty does it have? Will I be safe in it? Will it reflect my status adequately? Will it pollute the environment?

Should I recycle my trash? Should I limit my purchases as much as possible to items that are easily recycled? Should I avoid purchasing certain items from certain companies because they are known polluters?

somewhere, the environment. Third, such decisions take a very significant—and growing—portion of one's time. From an economic perspective, these are added transaction costs, costs that literally overwhelm some persons and groups. From a more general perspective, these decisions intrude on other activities, from getting the work at hand done to enjoying one's friends and family. Fourth, they often require considerable skill in matters that are rather arcane. In short, the multiplication of these kinds of decisions increases the burden on nearly everyone while providing a significant advantage over government services for only a few.[4] Fifth, all these decisions are posed as economic decisions, usually involving the purchase of some good or service (or not). While doubtless one needs goods and services to get on with one's life, all those things not easily translated into market prices are downgraded, made invisible. One need not be an incurable romantic to believe that love, friendship, family, friends, and compassion, as well as trust and honesty, are necessary components of a meaningful life. But they are square pegs that do not fit easily into the round holes provided by the market.

Moreover, those at the top of the economic ladder can and often do hire professionals to help with the decisions they face as individuals. Accountants, financial consultants, wedding planners, and a host of other professional advisers are ready to help you if you can afford it. In contrast, those in the middle class make better or worse decisions depending on their skills; sometimes their choices are limited (in ways that may be positive or negative) by organizations with which they interact, especially their employers. Those who are poorly educated, who have low-paying jobs, who work extremely long hours, or who are subject to frequent unemployment often make choices based on erroneous and inadequate information.

Those at the very bottom of the economic ladder—the so-called underclass—tend to be excluded altogether. Indeed, they are the residual, the leftover: "They are . . . the walking symbols of the disasters that await fallen consumers, and of the ultimate destiny of anyone failing to acquit herself or himself of the consumer's duties" (Bauman 2007, 31–32). They are "other." They are different from you and me; they are the box on the questionnaire that one checks when the named categories do not fit. Hence, when President George W. Bush vetoed a measure to provide health coverage to lower- and lower-middle-income children, his justification in terms of consumer choice failed on two counts: First, a significant portion of those children received no health care coverage at all. Second, it maintained a substantial burden for millions of low and middle-income Americans.

For the vast majority of us, time is always in very short supply. While the "official" workweek has remained rather stable at about forty hours for quite a while, in recent years many of us are working more than ever before. This is especially true for professional and managerial workers, precisely those for whom overtime pay is not required (Rones, Gardner, and Ilg 1997). It is also true—although with more dire consequences—of those at the very bottom of the economic ladder, who sometimes work two full-time jobs to make ends meet. In addition, as more women have entered the paid workforce, the time available for family activities has declined. Furthermore, new technologies—cell phones, the Internet, email—have had the effect of blending work and leisure in a manner previously impossible. This can hardly be considered the creative work called for by critics but rather yet another set of *devoirs* that threaten to envelop us.

In short, if the last century was that of the welfare state, in which we would be cared for from cradle to grave, we are now entering an era in which each of us is expected to engage in complex quasi-mathematical calculations of costs, risks, and benefits of all the myriad actions of daily life. And if the critics of the welfare state noted its stifling effects on human freedom, the emerging state of choice is likely to be far more restrictive. Rather than a paternal state telling us what to do, we are now to have inside each of us an insatiable need to make choices, to endlessly calculate our life chances, to view life as a large bazaar in which we must make our fortunes before we perish. As the cynical bumper sticker says, "He who dies with the most toys, wins."

Living the Certified Life

At the same time as we are faced with the burden of choice as consumers and in the ever receding world of citizens, we are also faced with the impersonal but ever present world of checking. Not only are things to be checked but we are to be checked. As children we are to be tested incessantly to ensure that we and our teachers and our schools are certified as having met the standards. As workers we are to be audited to ensure that we did not lie on our job applications, that we are doing our jobs in the approved manner, that we are not engaging in unhealthy practices (e.g., smoking, drinking, drug use) that might impinge on our work practice or cause our employer's insurance premiums to rise. As applicants for various social programs we are checked, monitored, cajoled, to ensure that we respond within the established standards. Each of these—and many more— "shallow rituals of verification" (Power 1997) binds us to our assigned tasks as surely as medieval serfs were bound to the land. Audit, certification, accreditation, even as they are ever more needed, threaten to replace the interpersonal virtues of trust, honesty, integrity, and honor with a far more odious means of holding society together.

There is yet another way to consider the rapid growth of standards, certifications, and accreditations. Let me suggest an analogy: Catholicism relies on confession; protestant sects rely on internalization of certain norms of behavior. Marx ([1844] 2009) once remarked that Luther "turned priests into laymen because he turned laymen into priests." Analogously, the neoliberal policies recently enacted have the goal of turning economists into laymen by turning laymen into economists. Standards, certifications, and accreditations demand that we become economists in the sense that we weigh the costs and benefits, the efficiency of each decision before we make it—and we do so as largely autonomous individuals. Constant's (1814) "superb eye of power" and Foucault's (1977) panopticon require constant surveillance of persons and things. But in the new standards economy such surveillance is unnecessary since, at least as the project is envisioned by its neoliberal proponents, conformity to the standards, although hardly noticed, must become part and parcel of the socialization process. Thus, the market-based standards economy of neoliberalism may be interpreted as the protestant—no, Puritan—answer to the Catholicism of the neoclassical school. Such a world, I submit, would be insufferable. It would be one in which the burden of calculation and choice, as well as the burden of incessant audit, would be placed squarely on our shoulders every waking moment. It would be a world in which the tyranny of the state would be replaced by the tyranny of the market.

This is not to suggest that the civic and market orders of worth are the only roads to serfdom. Alas, there are many others. The world of industry can also be a road to serfdom. Marx knew what he was talking about when he described the Satanic mills. Dickens's novels provided another view of the appalling conditions found in Britain in the mid-nineteenth century.

The world of inspiration can be a source of serfdom as well. While all of us need to act in inspired ways at times—bursting into song, jumping for joy, creating an object, pondering the wonders of the world—living full-time in the world of inspiration leads to its own form of serfdom. Inspired states of whatever persuasion have a long and tragic history of catastrophe. This was evident in the Crusades, in which thousands went off to their deaths in a bizarre attempt to save the Holy Land for Christendom. It was also evident in the tragedy at Jonestown, Guyana, in 1978, when the followers of preacher Jim Jones all died in a mass suicide and murder. More recently, members of Islamic fundamentalist sects have been inspired to commit suicide—and to kill others along with them—in their efforts to demonstrate their zeal for a peculiar, distorted form of Islam.

Even environmentalism can be a road to serfdom, as Luc Ferry (1995) warned recently. As he convincingly argues, the Nazis had their form of environmentalism—one that extolled the virtues of hiking in the forests, but also supported the extermination of Jews, Romani, homosexuals, the disabled, and other "pollutants." Even today, while hardly connected to the Nazi atrocities, there are proponents of environmentalism who mistakenly put the interests of an imagined earth-organism ahead of the needs and desires of real human beings.

The Limits of Private Standards?

All this suggests that markets and standards are no more undiluted goods than are other human institutions and organizations. They are useful and necessary, but they must be limited in scope and subject to critical revision. Indeed, there are several major problems confronting the world today that will test the market world. Significantly, all of them are problems associated with standards. Among them:

• The recent "food crisis" appears to be the result of (1) the growing use of corn (maize) as ethanol has raised the price of corn on world markets, (2) rising soybean prices, as large areas in soybeans in the United States have been diverted to the more lucrative corn production, (3) the coincidence of poor wheat harvests in both Argentina and Australia, and (4) speculation in grain and oilseed markets, leading to rapid price increases. Together, these created a "perfect storm," leading to rapid increases in food costs, especially in poorer nations. One result is that several nations moved to block exports. This may actually have made the situation worse. Regardless of how the food crisis is ultimately resolved, it has destroyed the myth that food prices can be left entirely to the workings of the world market. The situation is likely to lead to new calls for state regulation in light of the failure of private standards to prevent dislocation and hunger.

• The financial crisis of 2008–2009 was apparently the result of what was in itself a brilliant set of innovations in housing finance. Specifically, banks and mortgage companies, with a great deal of help from the changed policies of nation-states, developed a global market in "securitized" mortgages. Whereas as little as several decades ago mortgage markets were quite local in character, requiring intimate knowledge of the borrower and of local conditions, today they reach through most of the industrial world and into parts of the rest of the world as well. As one observer explains, "Securitization implies the transformation of illiquid financial assets into liquid capital market securities and has been the critical financial innovation that has allowed private and public actors to finance local property development and housing in the national and international capital markets" (Gotham 2006, 257).

This transformation was the result of the invention of a variety of new forms of investments based on very elegant and precise mathematical models of risk (Derman and Wilmott 2008). These could and did travel across governmental jurisdictions. The idea itself was rather simple. The default rates on mortgages were well known. Thus, by aggregating large numbers of mortgages and securitizing them—that is, by making them available as securities—one could sell a low-risk product on capital markets around the world. The problem with this idea, as we know all too well now, was that it was easy to game the system. Thousands of persons whose credit was shaky could and did obtain mortgages, sometimes under fraudulent conditions. All went well until the default rate started to rise, at which point many persons and institutions began to realize that what they had thought were very safe investments were actually very risky—or, to put it more precisely, what they thought were known risks were now mere uncertainties, in which risks were no longer calculable. Once this was recognized, banks began to fail and liquidity vanished. In short, a technical innovation and changed regulations allowed new standards for investments to be established, but without the necessary controls to ensure that the system could not be gamed. As the leader of the UK's Conservative Party David Cameron put it, "We need to make sure we don't have people inventing financial instruments, profiting hugely from their creation, but not understanding the contagion they can spread" (quoted in Bartiromo 2009, 14).

• Global climate change poses a series of new dilemmas for our increasingly global society. Although a few doubters remain, it appears that climate change is upon us and that we are likely contributing to it to a significant degree. Climate change will require considerable new capital investment and will test the existing trading networks to their very limits, putting greater pressure on land and water resources, shifting cropping patterns, and creating a wide range of somewhat unpredictable social dislocations, including floods, droughts, and extremely hot or cold weather. It is doubtful that private standards will be sufficient to cope with these problems. They will likely lack both the needed sanctions and the consistency that state or international standards would have.

It would take little effort to add to this list of crises (e.g., declining support for education, declining oil reserves, new epidemics). But taken together, these three global crises pose a considerable threat to political stability and to the standards on which the contemporary market order was created over the last half century. There appears to be only a small number of likely paths for improvement. First, it is conceivable, although improbable, that these problems (and perhaps others only now on the horizon) will be addressed through the existing market mechanisms and global market institutions such as the World Trade Organization and the International Monetary Fund. Second, it is possible, although highly unlikely, that individual action—the formation of new habits of reuse, recycling, and conservation, the consumption of things produced close to home, living in more modest homes—will improve the situation. Third, it is possible that, at least for some time, the pendulum will swing back toward state regulation of the economy, and private standards will be markedly reduced in scope and once again limited to a great degree by state authority. Indeed, the U.S. and UK bank bailouts are essentially nationalizations, putting governments in control of a significant portion of the world's financial assets. Fourth, there may be a significant expansion of *international* state regulation of society. In other words, it is possible that pressure will build to create other institutions of global governance comparable in scope to the WTO. In particular, state institutions might be developed that govern the environment, regulating carbon emissions and other forms of pollution, or that govern the supply of grains and oilseeds, guaranteeing food security. Fifth, regional blocs might emerge that attempt to grapple with some of these issues in a regional context and try to fence off the rest of the world. Finally, it is possible that some entirely new approach to the increasingly global problems facing humanity will be developed.

Which of these come to pass will depend largely on the future actions of myriad actors. But regardless of the path eventually taken, the problems associated with standards will remain. Thus, the challenge is not to eliminate standards, to return to some mythical past during which standards were of trivial importance. Instead, it is to develop standards that allow us to compromise among worlds of justification and to ensure that seemingly benign standards do not lead to gross injustices. It is to this issue that I now turn.

Building Standards That Are Fair, Equitable, and Effective

Given the enormous range of persons, processes, products, and practices that are the subject of standards, any attempt to address this topic must necessarily be either general and somewhat abstract or specific but limited in scope. But before beginning to develop standards I believe one must ask one central question, make one central choice: Are standards the most appropriate form of governance in this particular instance? In other words, among the many potential classes of recipes for realities, including laws, regulations, customs, and habits, are standards likely to produce better results? If not, then one or more of those other routes should be pursued.

Assuming that standards are the preferred route to governance, and since neither space nor time permits an exposition on the myriad standards to which we find ourselves and our lives subjected, what I will attempt to do here instead is to provide a set of guidelines—standards for standards—that may be helpful in determining whether standards achieve their ethical objectives. (I leave the technical, economic, legal, and other objectives to others, realizing full well that any workable standards must involve compromises.) I submit that better standards ought to have certain qualities, qualities that are at once ideals but are also realizable in some recognizable form. They are listed here in no particular order:

1. *Delegate to subsidiary bodies when possible.* The notion of subsidiarity is of recent origin. Proponents of subsidiarity trace its roots to early twentieth-century Catholic doctrine. In recent years it has been enshrined in law by the EU. The idea behind the principle is fairly straightforward: "The principle seeks to allocate responsibilities for policy formation and implementation to the lowest level of government at which the objectives of that policy can be successfully achieved" (Inman and Rubinfeld 1998, 1). In the context of standards setting and enforcement, this means that it is presumed that local knowledge (and customs, traditions, norms, mores, etc.) takes precedence over that to be placed in force over some larger geographic area unless there is a compelling reason why this should not be the case.

2. *Use precaution.* Since standards shape both human and nonhuman behavior, they should not be taken lightly. Precaution may be manifested in proceeding carefully (literally, with care) when someone or something might be at risk, or experimenting first with a small population before extending the standard over a larger area or number of beings. While the "precautionary principle" is often associated with Europe and rejected by U.S. authorities, in point of fact it is widely used nearly everywhere (see Busch 2001). Given our enormous technoscientific capacities and always incomplete knowledge, acknowledging and promoting its use appears only prudent.

3. *Do minimal violence.* Standards can and often do do violence to persons, through their substance or through their enforcement. This may well be partially unavoidable. However, standards are likely to do far less violence to persons if that question is on the table when the standard itself is being designed. Hence, one must ask: Are the rights of certain persons or groups abridged by the standards? Do the standards encourage virtuous and discourage unethical behavior? Are the consequences of the standard such that the benefits far outweigh the social and economic costs? Are the consequences distributed in a fair and equitable manner?

Violence to persons can also be minimized by, whenever possible, focusing on the outcomes rather than on the process. Hence, even though it may be important to certify that individuals have met the standards allowing them to practice a particular profession, in general, one should leave most professional decision making to the professionals; standardizing the process is likely to lead to as many problems as it claims to resolve. For example, there are good reasons to certify teachers, physicians, nurses, mechanics, plumbers, farmers, perhaps even judges. Once that is done, however, further attempts to impose standards do violence by undercutting the exercise of their professional judgment. On the other hand, standards for the outcomes of their work may be appropriate and necessary, provided that such standards consider the situatedness of their work. For example, one may develop standards to compare the death rates of patients across hospitals or the success rates of surgeons, but to avoid doing violence, such standards must include the differences in the initial patient populations, as well as other intervening variables (e.g., the quality of tools and devices available).

We can even go farther. We can and should ask whether the standard in question abridges what Elizabeth S. Anderson (1999, 320) calls democratic equality. As she suggests, "What citizens [in a democratic society] ultimately owe one another is the social conditions of the freedoms people need to function as equal citizens." If a standard changes social conditions in a manner that restricts those freedoms, then the standard in question is unacceptable, no matter what other benefits it might claim to bestow on society.

4. *Make actionable standards.* Standards that are not actionable can be as consequential as those that are. They may suggest that standards are arbitrary and to be ignored, or conversely, they may create anxiety. The color coding standards issued by the Bush White House as Homeland Security Presidential Directive 3 (from green for low alert, through blue, yellow, and orange, to red for high alert) are an excellent example of a nonactionable standard. This is certainly true with respect to the general public; it is likely true for most government agencies as well. When I go to an airport and am informed that the code is orange, just what is it that I am supposed to do differently? How would that compare to a code of yellow? According to the directive, "This system is intended to create a common vocabulary, context, and structure for an ongoing national discussion about the nature of the threats that confront the homeland and the appropriate measures that should be taken in response. It seeks to inform and facilitate decisions appropriate to different levels of government and to private citizens at home and at work."

The general response has been neither to create a "national discussion" nor to inform decisions "at home and at work." Rather, it has been to ignore the standards as so much mumbo-jumbo, or to frighten travelers. Standards of this sort are ill-conceived in that they provide little or no guidance as to implementation. In consequence, they undermine the credibility of the very organization that has issued them, in this case the U.S. government.

5. *Encourage the voices of publics through participation.* Participation is critical both for designing standards that are acceptable to all affected parties as well as for providing legitimacy to the standards themselves. Most standards setting is not analogous to legislative deliberations. Rather than setting broad guidelines, standards setting is usually about the details of products, persons, processes, and practices. It is only secondarily territorial in its applications. It eschews the adversarial character of legislative deliberations since something approaching consensus is generally desired. Participants are generally not elected. Hence, they are encouraged not to act as representatives of particular organizations or constituencies. Thus, developing a truly participatory and deliberative approach to standards setting can be both difficult and frustrating (see Dixon 1978). The key, as Dewey might have said, is to identify the publics involved with the standard. Are they all able to participate? If a representative approach is employed, do the representatives actually represent the publics of concern? Are all affected publics equally empowered in the standards design process?

6. *Use the most appropriate form of standard.* In this book I have pointed to four basic forms of standards: Olympian, ranks, filters, and divisions. Certain forms are best suited for some standards. At the very least, one must ask which form best serves the purpose at hand and the publics involved. One must also ask whether the boundaries among the grades or classes are the right ones. For example, setting a filter standard too high or too low can either allow unacceptable beings to pass through it or deny admission to acceptable beings. Employing an Olympian standard where it is unnecessary can create a winner-take-all society (Frank and Cook 1995). Moreover, standards that are nominally divisions should not be used to conceal what are in fact ranks. Doing so is likely to lead to considerable conflict after the standards are established.

7. *Ask about path dependence.* Since standards manifest path dependence, careful thought should be given to the economic and social costs of introducing a new standard or changing an old one. For example, through an

ill-conceived reduction in the tax on gasoline made from ethanol, farmers were encouraged not only to grow maize for ethanol production but to invest in ethanol conversion facilities throughout the U.S. Midwest. Although it is now apparent that this mode of producing biofuels creates as many environmental problems as it solves (e.g., Hill et al. 2006; Pimentel and Patzek 2005), it will be very difficult to dissuade farmers from the ethanol habit.

Another way of addressing the issue of path dependence is to consider it in terms of the burdens we place on future generations. Every time we start down a given path by implementing standards we risk handing a burden to our children and grandchildren, and those to whom we hand that burden are hardly in a position to complain. The burden is twofold. First, we may burden our children with the consequences of our sometimes irreversible interventions. The accounting standards that make oil cheap by basing the raw material price solely on the cost of extraction not only encourage global warming; they burden future generations with finding other energy sources. Second, the very standards themselves may become a burden, as they may be ill-conceived and require frequent and painful revision. As Douglas Kysar (2006, 31) recently remarked, "Like children, future generations are part of the 'Achilles' Heel of liberalism'—that vulnerable location for interest holders who are imperfectly situated to identify and assert their rights or interests in the manner that liberalism demands of them."

8. *Design appropriate tests*. Standards can only be effectively implemented if there are tests available to measure compliance with the standard. (The tests need not always be precise or even mathematical in form.) However, tests may fail to capture the most salient dimensions of the standards or may be easily gamed, or both. Building better standards means giving greater attention to whether the tests (1) measure what they claim to measure in a sufficiently robust manner, (2) provide sufficient hurdles to prevent exaggerated claims or even outright fraud, and (3) have neither too great nor too little precision and accuracy for the purposes at hand. Overly precise tests make meaningful distinctions less "real." Moreover, false precision is costly. It increases transaction costs.

In addition, tests must be developed carefully in order to avoid looping (per Hacking 1999), and to measure sufficiently robustly that which is of concern. Proxies are likely to lead to troubles; skin color is a problematic proxy for intelligence. When tests for people are necessary, they should be robust. Although no test can take into account the myriad differences in skills and abilities among persons, in general, tests that include a wider range of *relevant* skills and abilities are preferred to those that are narrower. Finally, as much attention should be given to what is *not* tested as to what is.

9. *Open new avenues to thinking and acting by making routine things habitual.* Better standards are those that relieve people of dull, boring, routine work by eliminating the necessity of thinking about them all the time. At the same time, standards should not relieve people from the necessity of making life-changing judgments; all standards have an irreducible hermeneutic element. The key question is how much judgment is embedded in the text or object and its accompanying tests and how much is interpreted by living persons. Some standards must be spelled out in exquisite detail, others are best left to those who must implement them. But at the same time, standards open up new opportunities for thought and action. Improvements in standards should accomplish both these tasks, and they should not do it by merely redistributing the tasks at hand.

10. *Review standards, tests, and indicators frequently.* Standards, tests, and indicators cannot be fixed permanently since as the world changes they become obsolete. In particular, tests that were once valid may no longer be so. For example, auto emissions tests were initially designed to measure emissions while the vehicle was stationary. This works fine with conventional internal combustion engines but is meaningless when applied to hybrid vehicles. Such vehicles emit nothing while stationary but do have emissions when moving. Similarly, the skills required to be an engineer, a lawyer, a physician, or a construction worker are quite different today from what they were fifty years ago.

11. *Use law experimentally.* We should thank the neoliberals for their insistence on the importance of law. In democratic societies law is a means— perhaps the most important one—for constructing our common future. But once we jettison the neoliberals' insistence that law be used to make society conform to the premises of an abstract mathematical model, we can begin to use law in ways quite different from those currently in vogue. Specifically, we can use law simultaneously to promote trade, investment, and wealth creation *and* to limit market processes where they are ineffectual, unjust, or environmentally destructive. We can begin to see laws as what they are: social experiments in need of frequent revision in light of their consequences, of our always incomplete knowledge, and of our own changing notions of what constitutes a better future, rather than as closer and closer approximations to some fixed model.

Thus, instead of either/or, instead of *either* the world of security, territory, population, described by Foucault, *or* a world of community, justice, and equality, instead of the domination of one form of justification (e.g., by the market or by the state), we can instead begin to think of *both and*!

As Dewey (1929) suggested, experiment is all about opening up new spaces for action. If an experiment is a controlled experience, one in which the outcome is replicable such that the phenomenon of interest may be made to appear again and again whenever the conditions for its appearance are made to be present, then an experiment is the prologue to each standard. But at the same time, every time the standard is invoked, it involves repeating the experiment and verifying if the expected outcome is made real.[5] Like recipes, standards and laws do not simply appear as fixed sets of ingredients to be measured and ordered in a particular manner. To the contrary, they must be based on experiment, much as a cook might experiment before determining how to prepare a given dish. And just as an experienced cook might still fail to produce the desired dish, so even a frequently used standard may fail.

Furthermore, just as nothing compels us to consume any particular dish, or to consume a range of dishes in a particular order, nothing compels us, other than our enchantment by the amplification brought by the new, to insist on the necessity of turning any or every path revealed by experiment into a standard or law by which the same phenomenon might be made to appear again and again.

Also, experiments are and must be performed. Even if they are modeled on aspects of nature, they must be performed. Hence, when our ancestors invented the seasons, they did so by developing means for counting the days, measuring the length of days, noting the location of the rising or setting sun, engaging in rituals, and otherwise performing the seasons with more or less success. Similarly, today's geneticists, who use (standardized) scientific instruments to make particular genes reveal themselves, must also perform their experiments.

But standards, however correct, serving in the manner we desire them to serve, do not reveal the Truth (Heidegger 1956). Truth as a final state of certitude and completeness is always beyond reach. There are always other aspects of the world to be revealed, other experiments to be performed. Furthermore, once we begin to see experiments as performances, we may conclude that certain performances, even while they are correct, do not lead us to truth, justice, or other valued ways of being and therefore should not be transformed into standards or laws.

Moreover, since business firms, corporations, foundations, and private voluntary associations are creatures of the state, that is, of the law, we can use law to reshape these entities. Why should corporations have the same or even more rights than people? Why should they be immortal? What kinds of standards, certification, and accreditations should corporations be allowed to require of suppliers? Should large corporations, large foundations, and PVOs be broken up into smaller ones or forced to disband after a fixed period of time, or when certain conditions are met? Why should corporations be run solely for the benefit of shareholders? There is nothing whatever that is necessarily fixed about these entities.

Admittedly, this is a rather abstract and perhaps not very satisfactory set of imperatives for producing good standards. Therefore, let us consider the case of evidence-based medicine (EBM), which appears to embody most if not all of them. As Timmermans and Berg (2003) argue, EBM makes certain practices handy or habitual. It also has the virtue of enhancing the professional space occupied by physicians and other health care providers. It does so by insisting on the careful examination of evidence before engaging in treatment, as opposed to a rule-of-thumb approach. But it in no way tells practitioners which evidence to examine, how to interpret that evidence, or whether it applies to a particular patient. Those matters remain confined to professional judgment, although admittedly, that judgment is perhaps subject to more careful scrutiny by others. At the same time, it is hard to argue that EBM brings along with it any major injustices. In short, EBM appears to meet the criteria outlined above: It makes certain practices routine, it opens up new avenues for creative thought and action by requiring practitioners to ask some rather profound questions, and it appears to promote (or at least not reduce) justice among those involved in health care.

In contrast, the national educational standards embodied in the No Child Left Behind Act in the United States and similar testing in the United Kingdom fail to incorporate most of these imperatives in their practices. Clearly, the standards make (some) knowledge about educational performance handy, habitual, routine. However, they are far more problematic when the other guidelines are considered. Indeed, it is precisely because NCLB provides such clear indicators for students, teachers, and schools that I must judge it negatively on the other aspects. First, the act drastically reduces the professional space available for teachers, local school boards, and school administrators in determining what shall count as knowledge. Second, it rewards those students who are successful in memorizing the particular kinds of information that are included in the test, as well as those students who have developed test-taking skills. Finally, in an ironic twist, the insistence on uniform testing punishes those students, teachers, and schools that are the weakest. Not surprisingly, those schools are over-whelmingly located in poor neighborhoods.

In short, NCLB reduces the professional space occupied by teachers and school administrators. Similarly, it sends the message to students that only certain kinds of skills and competencies are worth developing—those measured on the tests. In the worst instances it substitutes coerced hours of mindless drill for careful learning. Finally, it produces outcomes that are less just than those it was ostensibly designed to replace. It erroneously assumes that poor educational outcomes are solely the result of poor teaching as measured by standardized tests. But hundreds of studies show the strong correlation between family income, time spent in the classroom, and success in school, while hundreds of other studies show the huge variation in the physical infrastructure, teacher preparation, and teacher salaries across schools and school districts. EBM may not produce cookie-cutter medicine, but NCLB surely appears to produce cookie-cutter education.

While I believe that these are two illustrative examples, they should be considered just that—illustrations. It is conceivable that a more thorough examination and debate might lead to somewhat different conclusions. What is essential is to put those issues on the table for debate by the relevant publics, rather than having them hidden away from view.

Toward a Trustworthy Society

As Hayek had hoped, our contemporary world is one in which the role of the nation-state has been transformed and restricted, especially in certain aspects of social life. The monetary policies initially advocated by Milton Friedman have also been largely institutionalized as well. It is not too much of an exaggeration to say that in international affairs, it is the international market and financial institutions, the WTO and the International Monetary Fund in particular, that trump all others and limit the scope of many individual state actions. Thus, many state actions designed to protect national industries or to conserve certain aspects of local cultures or to protect the environment are either prohibited or far more difficult to accomplish.

One effect—perhaps even an intended one—of the creation of a truly global society has been the tight coupling and integration of all sorts of persons and things. Much of the world today is highly dependent on the rapid, largely unimpeded movement of people and things around the world. Hence, even if the crises noted above are quickly resolved, there will likely be others to replace them. Moreover, each of these crises is likely to increase demands by the vast majority of persons for actions on the part of politicians. One possible response would be to attempt to disconnect these newly created links with the world economy. But far more likely is that a consequence of these crises will be to illustrate to everyone the increasing need for global governance. Ironically, such a move would turn the tables on the neoliberal approach, creating a new hybrid of state and market, and perhaps rekindling demands for social justice that can only be met on a global scale. And just perhaps, those demands for social justice might allow us to move beyond the shallow trust of a certified and accredited world toward one in which nearly everyone, far more secure in their places in the world, would be more trustworthy.

So the next time that you tie your shoe, drive your car, go to the hospital, drink a cup of coffee, take a class, or engage in any of the activities that together make up everyday life, think of the myriad standards that are involved in those activities. Think of the power that standards have and must have over all of us. Ask yourself who established those standards and what justifications they used in establishing them. Think of who wins and who loses as a result of standards. Think of what virtues and vices are made manifest through standards. Ask yourself whose rights are supported and whose rights are abridged as a result of standards. And, perhaps most important, ask yourself how standards might be used, modified, or transformed to produce a more just and caring world.

Notes

Introduction

1. This is not to suggest that there is anything wrong with social scientists, marketing researchers, economists, financial analysts, and others focusing their work on either habits or choices and ignoring the other category of practice. However, those standards for research have their pitfalls as well.

2. The NIST produces reference materials (standardized samples of various things) and sells them worldwide. The standardized samples include oyster tissue, human hair, nickel isotopes, and lead silica glass. Each sample is designed to be compared with those used in industrial processes of various kinds (National Institute of Standards and Technology 2007).

3. By 1940 the Soviet Union had more standards than any other nation, but enforcement problems plagued the nation until well after World War II (Berliner 1957).

Chapter 1

1. Other contemporary accounts provide similar imagery (e.g., Dutton and Freeland 2005; Oberg 1965).

2. A colleague has argued that there is nevertheless an asymmetry here: people make standards for both people and things. Yet even this asymmetry appears to me to be overstretched. Things made to standards change our behavior. Once in place, standards for things change the behavior of people even as we change the behavior of things.

3. Moreover, there are numerous instances in which standards replace law, as is the case apparently in the cotton (Bernstein 2001) and diamond industries (Mercuro and Medema 2006).

4. Knut Blind (2001) notes, using the case of Switzerland, the positive correlation between the number of standards a nation develops and its export trade. Standards makers clearly have an advantage over standards takers.

5. Heidegger goes on to argue that this leaves us with an inauthentic world, leaves us homeless, devoid of any bearings. Perhaps so, but I will let others debate this question.

6. Were I to curse the lock or kick at the snow, others might find it amusing, somewhat analogous to Don Quixote's tilting at a windmill. The person blocking my way acts so as to display the power relations he has with me; I cannot so easily discern the power relations embedded in the lock.

7. Wendy Nelson Espeland and Mitchell L. Stevens (1998) note that commensuration has been given short shrift by social scientists.

8. As Theodore M. Porter (1995) notes, "the bureaucratic imposition of uniform standards and measures has been indispensable for the metamorphosis of local skills into generally valid scientific knowledge." Hence, both commensurability and incommensurability are in large part created by standards.

9. Alas, some people do see these lower-grade products. In an effort to get by on an inadequate income, some consume pet food; others glean harvested fields, finding vegetables discarded or overlooked by harvesting equipment.

10. Divisions are standards by which we divide up the world, irrespective of the origin of the criteria used to apply the standard. Apple varieties are as much the result of human endeavor as they are of nature's bounty. Nimbus clouds are not formed by us. But in both instances the divisions are ours; they are classifications we care about.

11. For a detailed account of the protracted conflicts involved in establishing genetic classifications for organisms, see Hull (1988).

12. This does not rule out the possibility that some extreme nationalist group might define citizenship in a very different way, for instance by bloodlines, in effect establishing a kind of counterstandard.

13. As Witold Kula (1986, 81) notes, "the Jews kept their standards in the Temple, the Romans in the Capitol, and Justinian ordained that they must be kept inside the [basilica of] Hagia Sofia." Standards have always been kept in special sacred places under special conditions.

14. This was not always the case. Prior to the establishment of grain standards and the treatment of grain as a liquid, each sack had to be inspected and tested—a cumbersome and lengthy process. See Cronon (1991).

15. This is so much the case that the bibliographies of written standards often cite only other standards.

16. Local practices, however messy they may appear to us, also require considerable investment of time, money, organization, and method. David Turnbull (1993) pro-

vides the example of the construction of Gothic cathedrals: such knowledge had to be discarded or devalued in order for modern standards of architecture to become the dominant form.

17. There have now been several false starts with respect to electric vehicles. It remains to be seen if the current efforts will be successful.

18. Economists have been arguing for some time as to whether the QWERTY keyboard is the most efficient (David 1985; Liebowitz and Margolis 1990). Some see QWERTY as an example of market failure, while others argue that it is an example of market success. I remain agnostic on this issue.

19. It should be remembered that gauge, the distance between the rails, is only one of a wide range of things that need be standardized in order to successfully link multiple railroads. See, for example, Rehm (1910).

20. See also the discussion of factories in chapter 2.

21. I admit that I spent an inordinate amount of time in the lavatory doing this fieldwork. My belated apologies to the other passengers.

Chapter 2

1. Whereas in most Western societies evenness is valued, in China odd numbers have magical significance.

2. Denise Schmandt-Besserat (1980) argues that tokens used to mark goods appeared in the Middle East ca. 8500 BC.

3. Type fonts pose another entirely different problem. Even the revered pica and point (1 pica = 12 points) vary in size from one nation to the next. None are easily converted to metric measure (Kuhn 2008).

4. According to the ancient Jewish historian Flavius Josephus ([94] 2009, I:II, 2), Cain "was the author of measures and weights. And whereas they lived innocently and generously while they knew nothing of such arts, he changed the world into cunning craftiness." Perhaps this explains the strong moralizing in the biblical passages quoted here.

5. Robert D. Romanyshyn (1989) takes this back one step farther, arguing that Alberti's invention of linear perspective drawing in 1435 promulgated the notion that a single way of viewing the world was the correct one.

6. Quantum physicists partially abandon these rules: They argue that the quantum level is the smallest available to human knowers, that the set of physical properties at the quantum level is not manifested at larger scales, and that uncertainty is central to quantum phenomena. In a very different way, much qualitative sociology also rejects these Cartesian premises.

7. This is not to suggest that scientists are uninterested in the results of their work. To the contrary, most scientists are passionately committed to the hypotheses they test. But they also understand that the more significant their findings, the more the scientific community will work hard to demonstrate their limitations. Hence, even as hyperbole is de rigeur in planning new research programs, there is considerable incentive to avoid it in reporting research results.

8. The situation has changed today, but far less than might appear to be the case at first glance. Today, knowledge of English has become mandatory as nearly all international scientific journals are published in English. Computers use the same software for analyses. Moreover, there is a global market for highly sophisticated scientific instruments, as well as for less expensive items such as glassware and microscopes. As Scott Lash (2001, 114) notes, "The laboratory is a generic space."

9. Such an approach is likely to lead to correctness, but may actually stifle the quest for truth. See Heidegger (1956).

10. While much effort was and is expended on standardizing laboratory rats, their use as proxies for human beings remains highly contested. See, for example, Gold, Manley, and Ames (1992).

11. Despite standardization, the interpretation of radiographs and scans remains complex and at times ambiguous. As with all standards, some level of interpretive flexibility remains.

12. Stephanie Coontz (1992) notes that after the war, women were systematically squeezed out of jobs they had held during the war. In some instances women were even subjected to electroshock treatment to force them to accept domestic roles.

13. For a contemporary list of national standards bodies, see World Standards Services Network (2009).

14. This was not as far-fetched as it might seem. In the newly created Soviet Union, engineers played a pivotal role. Allied to the capitalists, they were a threat to the new regime; allied to the proletariat, they were seen as capable of rapid industrialization of the nation. See, for example, Kendall E. Bailes (1974, 1978). Even into the 1930s this was the prevailing view in the USSR. See, for example, Prokofyev (1933) and Stalin (1932).

15. Many years earlier Thomas Carlyle ([1829] 1878, 188) had decried this, noting, "Not the external and physical alone is now managed by machinery, but the internal and spiritual also."

16. Readers not familiar with the history of standards making should be aware that the U.S. trajectory differs considerably from that of most other nations in that ANSI does not receive government funds for its activities. In contrast, in most other

nations, national standards bodies are at least quasi-governmental bodies. See Office of Technology Assessment (1992) for an overview.

17. It appears that a few factory-like units existed in ancient Rome and ancient China. However, for our purposes here, the factory is best understood as a modern innovation.

18. George Pullman built his company by making passenger railcars and leasing them to railroads. He also built a town, Pullman, Illinois, just south of the city limits of Chicago in 1880 where workers were required to live. After a disastrous strike in 1894, the courts required Pullman to sell off the town, and it was annexed to the city of Chicago.

19. This approach works only when the product in question can be controlled. In contrast, in the production of many agricultural commodities where physical qualities are important, farmers have only limited control over the quality of their output. Hence, growers of berries and nuts today often use high-speed sorters that check everything that comes off the line, rather than a sample.

20. It is important to note that the array of different standards for filing is apparent only to one who travels from place to place and to large multinational firms that need to employ the same standards everywhere. When initially introduced, such filing standards needed only to be "universal" locally.

21. John W. Meyer and colleagues (1997) make a similar case for nation-states.

22. For a much more detailed analysis of both Chicago law and economics and public choice theory, the reader is referred to Mercuro and Medema (2006). This analysis is heavily based on that work.

23. This is not to suggest that all interpretation is done away with under this principle. Posner (1987) was right in noting that he is not a potted plant. Interpretation and judgment are necessary to the application of all laws and standards.

24. Ronald Coase (1960, 43) is far more circumspect than contemporary practitioners of CLE. He argues that "it is, of course, desirable that the choice between different social arrangements for the solution of economic problems should be carried out in broader terms than this and that the total effect of these arrangements in all spheres of life should be taken into account."

25. One might object that two mathematicians might have a "conversation" that consisted entirely of successive attempts to solve an equation. However, such a conversation could not take place unless there was already a tacit agreement on the way each operator worked and that they were in fact trying to solve an equation.

26. A similar view can be found in the computer science literature. See, for example, Richard J. Boland and colleagues (1992).

Chapter 3

1. Of course, since standards often combine multiple measures of humans or non-humans, there are some that are both universalizing and particularizing at the same time.

2. Only the most recent edition notes the importance of government intervention in the grain market. None discusses how standards create the homogeneous commodities known as grains.

3. To be sure, larger chains have an advantage over smaller ones in that they can experiment with several stores and see what effect a given change has on the bottom line for those stores.

4. In point of fact, quality is not merely high or low but varies in multiple ways. Hence the informational asymmetry is more complex than it might otherwise appear.

5. Ronald Coase (1988) has argued that all exchanges incur transaction costs. While some of the costs noted above are transaction costs in that they are clearly linked to particular transactions, others are better understood as the costs of maintaining modern markets. These include the costs of maintaining bureaus of weights and measures, health and safety laws, civil courts, etc. Without these, modern markets simply cannot exist. No wonder Aristotle thought goods exchanged in the market to be incommensurable in all other ways!

6. This section draws heavily on the careful work of Marc Levinson (2006).

7. Internet-related communications such as text messaging and email can be either broadcast or narrowcast.

8. Of course, no two beings are ever exactly the same. What is of concern here is that beings may be "the same" for all intents and purposes, in the sense that the differences do not matter.

9. However, even in today's world very few of the have-nots have lifestyles.

10. The creation of a "positive program" for neoliberalism did not spring fully grown from the head of Zeus. Earlier works had already warned of impending doom, setting the problem up as one of good versus evil or as either collectivism or individualism. See, for example, Belloc (1912) and von Mises ([1927] 1985). Other observers trace it to interwar Germany and the works of Alexander Rüstow and Walter Eucken (Megay 1970).

11. This is precisely the reverse position of that taken in the postwar years by critics of capitalism: "While Adorno, Horkheimer and other Critical theorists insist that there is a causal connection between capitalism and fascism, the neo-liberals consider the Third Reich not to be the product of liberalism but instead the result of an absence of liberalism" (Lemke 2001, 193).

12. Foucault (2008) distinguishes between American and German (*ordoliberal*) variants of neoliberalism. However, given their interpenetration, for our purposes here, we can neglect that distinction.

13. Doubtless, as social beings, we tend to spontaneously create a wide range of social forms, including friendships, kinship groups, markets, and organizations. But Hayek claims that the spontaneous order of the market is somehow unitary and efficient. Several problems emerge here. First, as noted in table 3.3, for modern markets to exist, quality, quantity, currency, and price must be standardized. Yet there is considerable debate over the best way to do this, with different persons and groups arguing for the use of different standards. Second, there is nothing necessarily efficient about any particular spontaneous order; they are simply the ones that have evolved historically (Sugden 1989). Finally, the freedom claims for the spontaneous order of the market are undermined by virtue of the fact that participation is rarely optional (Bromley 1998a).

14. Hayek (1975, 437) clarifies: "I regard it in fact as the great advantage of the mathematical technique that it allows us to describe, by means of algebraic equations, the general character of a pattern even where we are ignorant of the numerical values which will determine its particular manifestation. We could scarcely have achieved that comprehensive picture of the mutual interdependencies of the different events in a market without this algebraic technique."

15. In a largely unexpected way, the neoliberals resurrect Hobbes's rejection of empirical knowledge and his plea for certain knowledge based on mathematics (see Shapin and Schaffer 1985) even as they critique positivist and neoclassical economists for trying to quantify everything in their search for truth. While for Hobbes it was geometry, for the neoliberals it is a kind of formal logic and mathematics; it is the certainty that comes from knowing that what is on the right side of the equation is equal to what is on the left side, *quod erat demonstrandum.*

16. Personalized medicine is at best an exaggeration, at worst utter nonsense. In order for any medical or pharmaceutical intervention to be tested, one must have a population on which to test it. The result is a *probability* that a given population will be effectively treated by the intervention. How a given person will respond is knowable only after the fact.

17. See also GenomeWeb (2008). The widespread character of the problem is also illustrated by the attempt to develop international guidelines for quality assurance (Organisation for Economic Co-operation and Development 2007).

18. Susan S. Silbey and Patricia Ewick (2003) note that the U.S. 1990 Occupational Safety and Health Act similarly required that each laboratory produce a safety plan without specifying the nature of the plan. The governmental mandates for Hazard Analysis and Critical Control Points plans in the food industry follow a similar path.

19. These are the criteria used by the U.S. government in encouraging federal participation in standards deliberations. See Office of Management and Budget (1998).

20. Note that taking on a new nationality as an immigrant is quite different from that of the nationality in one's nation of origin. The new nationality involves some concessions to the destination society, but it also involves invocation of long-forgotten practices from the country of origin, abandonment of others, and the creation of new alliances across ethnic lines that those in the country of origin might well find abhorrent.

Chapter 4

1. In any contractual relationship some irreducible level of trust is necessary since contracts can never be complete (Williamson 1994).

2. Gregory Tassey (2000) argues that supply chains are even becoming the most important level of policy analysis.

3. It should be noted in passing that this argument can quite easily be flipped on its head; one could argue that all markets are the result of organizational failures. Neither position seems particularly plausible to me. In fairness to Williamson, he has now abandoned that rather extreme view.

4. From the old French word *franchise*, meaning freedom. Most franchisees would likely find this problematic.

5. This is not to suggest that my trust in my wife or in anyone else is unbounded. There are always limits on trust in others, largely due to differences in skills and knowledge.

6. I thank John Stone for suggesting this particular term. It is developed further in Loconto and Busch (2010). A TSR might include what are usually called conformity assessment activities, but it might also include aspects of the social world such as education and medical care.

7. Courville, Parker, and Watchirs (2003) are certainly correct in arguing that there is nothing necessary about the violence done by audits. They suggest that violence may be avoided or mitigated by ensuring that multiple stakeholder groups are represented in the audit.

8. For a careful comparison between high energy physics and molecular biology, see Knorr Cetina (2003).

9. In contrast, ecologists have much greater difficulty in data sharing because they find it harder to generate standardized data (Millerand and Bowker 2009; Zimmerman 2008).

10. One industry magazine featured an article titled "Growers Beware: Adopt GAPs or Else" (Stier and Nagle 2001). The article warned participants against falling into the trap of believing that certification was only about record keeping.

11. Note that the potential for violence is realized only when a standard is enforced; hence, it is in the audit process (whether formal or informal) that violence may be done.

12. Technically, adoption of the standards embodied in NCLB is not required. However, failure to do so eliminates eligibility for federal funding.

13. The General Accounting Office reported in 1995 that $112 billion in repairs were needed to America's schools. Doubtless, that number has not declined since.

Chapter 5

1. Alternatively, one might argue that philosophers have instead been concerned with the development of metastandards. For example, Jeremy Bentham's famous "greatest good for the greatest number" and Immanuel Kant's "categorical imperative" may be seen as metastandards. I thank Paul Thompson for pointing this out to me.

2. This is supported by a more recent article noting that national activities to protect the environment are on the rise (Frank, Hironaka, and Schofer 2000).

3. It is worth emphasizing that it is all too easy to reify these worlds of justification. Another study, using different empirical data, might well identify a somewhat different set. What is important is not the definitions of the worlds themselves but the existence of multiple worlds, and the apparent impossibility of producing a single unified world.

4. U.S. Jews are generally divided into orthodox, conservative, and reform congregations.

5. As Ray Goldberg (1968) noted forty years ago, value chains are abstractions from reality that are convenient for certain sorts of analyses. More recently, others have suggested that the study of chains and networks can be usefully combined as "netchains." See Lazzarini, Chaddad, and Cook (2001).

6. Jürgen Habermas (1998) and John Rawls (1971) each suggest some procedural standards by which compromises about standards for persons and things might be reached.

7. Jan-Willem Grievink (2003) estimates that as few as seventy retailer "buying desks" control nearly the entire food supply in Europe. Doubtless, the number for the United States is similarly small.

Chapter 6

1. As M. Whipple (2005) notes, Dewey spoke out against what he saw as the excesses of the New Deal, which, he argued, tended toward state capitalism and even fascism.

2. Even critics have tended to frame their remarks in its terms. An important recent exception is the report of the World Commission on Dams (2000), which specifically proposed a participatory "rights and risks" analysis before the construction of any new dams.

3. For several contrasting examples applied to agricultural field days, see Carolan (2008).

Conclusion

1. An exception is a brief mention by Simons ([1934] 1948).

2. Another response—a surprise to many neoliberal analysts—has been for firms to actively lobby for new regulations so as to stabilize the firms' environments. See, for example, Lipton and Harris (2007).

3. The word *devoir*, of French origin, describes it best: the making of choices becomes an obligation, a matter of conscience.

4. Richard H. Thaler and Cass R. Sunstein (2008) recently proposed a solution which they provocatively call "libertarian paternalism." They concede that many people will make wrong choices about various things, and that complex technical choices can be a burden. They suggest that that the most welfare-enhancing choice be made the default (e.g., a given pension plan unless one opts for another), thereby increasing welfare without reducing freedom of choice. However, they fail to recognize that defining the default position requires a social choice made through democratic deliberation. Furthermore, they give only lip service to the fact that many critical choices face us not as isolated individuals but as members of a society.

5. This is a particular problem for the production of safety-critical software. "As a rule software systems do not work well until they have been used, and have failed repeatedly, in real applications. Generally, many uses and many failures are required before a product is considered reliable. Software products, including those that have become relatively reliable, behave like other products of evolution-like processes; they often fail, even years after they were built, when the operating conditions change" (Parnas, Schouwen, and Kwan 1990, 636).

References

Abir-Am, Pnina. 1982. The Discourse of Physical Power and Biological Knowledge in the 1930's: A Reappraisal of the Rockefeller Foundation's "Policy" in Molecular Biology. *Social Studies of Science* 12:341–382.

Abraham, John, and Tim Reed. 2002. Progress, Innovation and Regulatory Science in Drug Development: The Politics of International Standard-Setting. *Social Studies of Science* 32 (3): 337–369.

Adam, Barbara. 2006. Time. *Theory, Culture & Society* 23:119–126.

Adams, C. A. 1919. Industrial Standardization. *Annals of the American Academy of Political and Social Science* 82:289–299.

Adler, Ken. 1995. A Revolution to Measure: The Political Economy of the Metric System in France. In *The Values of Precision*, ed. M. N. Wise, 39–71. Princeton, NJ: Princeton University Press.

Adler, Ken. 1997. Innovation and Amnesia: Engineering Rationality and the Fate of Interchangeable Parts Manufacturing in France. *Technology and Culture* 38 (2): 273–311.

Adler, Ken. 1998. Making Things the Same: Representation, Tolerance and the End of the Ancien Régime in France. *Social Studies of Science* 28 (4): 499–545.

Afman, Lydia, and Michael Müller. 2006. Nutrigenomics: From Molecular Nutrition to Prevention of Disease. *Journal of the American Dietetic Association* 106 (4): 569–576.

Agency for Healthcare Research and Quality. 2007. *Advancing Excellence in Health Care*. Washington, DC: Agency for Healthcare Research and Quality.

Agnew, P. G. 1926. A Step toward Industrial Self-Government. *New Republic*, March 17, 92–95.

Agnew, P. G. 1927. Can Industry Make Its Own Law? *Mining Congress Journal* 13:257–259.

Akerlof, George A. 1970. The Market for "Lemons": Quality Uncertainty and the Market Mechanism. *Quarterly Journal of Economics* 84:488–500.

Alsberg, Carl L., and E. P. Griffing. 1928. The Objectives of Wheat Breeding. *Wheat Studies of the Food Research Institute* 4 (7): 269–288.

American Standards Association. 1941. *Guide for Quality Control and Control Chart Method of Analyzing Data*. New York: American Standards Association.

American Standards Association. 1942. *Control Chart Method of Controlling Quality during Production*. New York: American Standards Association.

Anderson, Chris. 2006. *The Long Tail: Why the Future of Business Is Selling Less of More*. New York: Hyperion.

Anderson, Elizabeth S. 1999. What Is the Point of Equality? *Ethics* 109 (2): 287–337.

Ann Arbor News. 2009. Still Playing Games with Your Bra Size? *Ann Arbor News*, March 8, A5.

Anzaldúa, Gloria. 1999. *Borderlands, la frontera: The New Mestiza*. 2nd ed. San Francisco: Aunt Lute Books.

Appadurai, Arjun. 1986a. Introduction: Commodities and the Politics of Value. In *The Social Life of Things: Commodities in Social Perspective*, ed. Arjun Appadurai, 3–63. Cambridge: Cambridge University Press.

Appadurai, Arjun, ed. 1986b. *The Social Life of Things: Commodities in Social Perspective*. Cambridge: Cambridge University Press.

Apple, Michael W. 2004. Creating Difference: Neo-liberalism, Neo-Conservatism and the Politics of Educational Reform. *Educational Policy* 18 (1): 12–44.

Apple, Michael W. 2006. *Educating the "Right" Way: Markets, Standards, God, and Inequality*. New York: Routledge.

Autoridad del Canal de Panamá. 2005. MR notice to shipping no. N-1-2005. Balboa-Ancón: ACP.

Babbage, Charles. 1835. *On the Economy of Machinery and Manufactures*, 4th ed. London: Charles Knight.

Bachelard, Gaston. [1934] 1984. *The New Scientific Spirit*. Boston: Beacon Press.

Bacon, Francis. [1605/1626] 1974. *The Advancement of Learning and the New Atlantis*. Oxford: Clarendon Press.

Bagley, William Chandler. 1907. *Classroom Management: Its Principles and Technique*. New York: Macmillan.

Bailes, Kendall E. 1974. The Politics of Technology: Stalin and Technocratic Thinking among Soviet Engineers. *American Historical Review* 79 (2): 445–469.

Bailes, Kendall E. 1977. Alexei Gastev and the Soviet Controversy over Taylorism. *Soviet Studies* 29 (3): 373–394.

Bailes, Kendall E. 1978. *Technology and Society under Lenin and Stalin: Origins of the Soviet Technical Intelligentsia*. Princeton, NJ: Princeton University Press.

Bain, Carmen. 2010. Structuring the Flexible and Feminized Labor Market: Global-gap Standards for Agricultural Labor in Chile. *Signs: Journal of Women in Culture and Society* 35 (2): 343–370.

Baker, M. J. 1951. Mr. Urmson on Grading. *Mind*, New Series 60 (240): 530–535.

Balzac, Honoré de. [1838] 1993. *The Bureaucrats*. Evanston, IL: Northwestern University Press.

Bartiromo, Maria. 2009. Facetime. *Business Week*, April 18, 13–14.

Basco, Leonardo K. 2007. *Field Application of in vitro Assays for the Sensitivity of Human Malaria Parasites to Antimalarial Drugs*. Geneva: World Health Organization.

Bauman, Zygmunt. 2007. Collateral Casualties of Consumerism. *Journal of Consumer Culture* 7 (1): 25–56.

Baylis, C. A. 1958. Grading, Values, and Choice. *Mind*, New Series 67 (268): 485–501.

Beck, Ulrich. 1992. *Risk Society: Towards a New Modernity*, trans. M. Ritter. London: Sage.

Beck, Ulrich. 1994. The Reinventing of Politics: Toward a Theory of Reflexive Modernization. In *Reflexive Modernization: Politics, Tradition and Aesthetics in the Modern Social Order*, ed. U. Beck, A. Giddens, and S. Lash, 1–55. Cambridge: Polity Press.

Bell, Alex. 2008. "I'm Stuck in a Jam." *Manchester Evening News*, October 30, 6.

Bellah, Robert N. 1975. *The Broken Covenant: American Civil Religion in the Time of Trial*. New York: Seabury Press.

Beller, Sieghard, and Andrea Bender. 2008. The Limits of Counting: Numerical Cognition between Evolution and Culture. *Science* 319:213–215.

Belloc, Hillaire. 1912. *The Servile State*. London: T. N. Foulis.

Beretti, Melanie, and Diana Stuart. 2008. Food Safety and Environmental Quality Impose Conflicting Demands on Central Coast Growers. *California Agriculture* 62 (2): 68–73.

Berliner, Joseph S. 1957. *Factory and Manager in the USSR*. Cambridge, MA: Harvard University Press.

Bernstein, Lisa. 2001. Private Commercial Law in the Cotton Industry: Creating Cooperation through Rules, Norms, and Institutions. *Michigan Law Review* 99 (7): 1724–1790.

Bhattacharjee, Yudhijit. 2007. Science Education: States Urged to Sign Up for a Higher Standard of Learning. *Science* 315:595.

Bijker, Wiebe. 1995. *Of Bicycles, Bakelite, and Bulbs: Toward a Theory of Sociotechnical Change*. Cambridge, MA: MIT Press.

Blaszczyk, Regina Lee. 2007. True Blue: Dupont and the Color Revolution. *Chemical Heritage* 25 (3): 20–25.

Blind, Knut. 2001. The Impacts of Innovations and Standards on Trade of Measurement and Testing Products: Empirical Results of Switzerland's Bilateral Trade Flows with Germany, France and the UK. *Information Economics and Policy* 13:439–460.

Blind, Knut. 2004. *The Economics of Standards: Theory, Evidence, Policy*. Cheltenham, UK: Edward Elgar.

Blomfield, Reginald. 1912. *Architectural Drawing and Draughtsmen*. London: Caswell and Co.

Boero, Natalie. 2007. All the News That's Fat to Print: The American "Obesity Epidemic" and the Media. *Qualitative Sociology* 30 (1): 41–61.

Boland, Richard J. Jr., Anil K. Maheshwari, Dov Te'eni, David G. Schwartz, and Ramkrishnan V. Tenkasi. 1992. Sharing Perspectives in Distributed Decision Making. In *Conference on Computer Supported Cooperative Work 92 Proceedings*, 306–313, Toronto. New York: ACM.

Boltanski, Luc, and Laurent Thévenot. [1991] 2006. *On Justification: Economies of Worth*. Princeton, NJ: Princeton University Press.

Borges, Jorge Luis. 1964. *Labyrinths: Selected Stories and Other Writings*. New York: New Directions.

Borras, Saturnino M. Jr. 2008. La Vía Campesina and Its Global Campaign for Agrarian Reform. *Journal of Agrarian Change* 8 (2–3): 258–289.

Bourdieu, Pierre. 1998. *The Essence of Neoliberalism*. Le Monde Diplomatique (English edition). http://mondediplo.com/1998/12/08bourdieu.

Bowker, Geoffrey C., and Susan Leigh Star. 1999. *Sorting Things Out: Classification and Its Consequences*. Cambridge, MA: MIT Press.

Bringardner, John. 2008. Winning the Lawsuit: Data Miners Dig for Dirt. *Wired*, July, 112.

Brisbane, Albert. [1840] 1969. *The Social Destiny of Mankind.* New York: Augustus M. Kelley.

British Broadcasting Company. 2007. UK to Tackle Bogus Carbon Schemes. http://news.bbc.co.uk/newswatch/ukfs/hi/newsid_3950000/newsid_3955200/3955223.stm.

Britton, Karl. 1951. Mr. Urmson on Grading. *Mind,* New Series 60 (240): 526–529.

Brockway, Lucile H. 1979. *Science and Colonial Expansion: The Role of the British Royal Botanic Gardens.* New York: Academic Press.

Bromley, Daniel W. 1997. Rethinking Markets. *American Journal of Agricultural Economics* 79 (December): 1383–1393.

Bromley, Daniel W. 1998a. Searching for Sustainability: The Poverty of Spontaneous Order. *Ecological Economics* 24 (2–3): 231–240.

Bromley, Daniel W. 1998b. *Transitions to a New Political Economy: Law and Economics Reconsidered.* Cambridge: Cambridge University, Department of Land Economy.

Brooks, David. 2000. *Bobos in Paradise.* New York: Simon & Schuster.

Brown, Joyce. 1979. *Mathematical Instrument-Makers in the Grocers' Company 1688–1800.* London: Science Museum.

Brown, Phil, Stephen Zavestoski, Sabrina McCormick, Brian Mayer, Rachel Morello-Frosch, and Rebecca Gasior Altman. 2004. Embodied Health Movements: New Approaches to Social Movements in Health. *Sociology of Health & Illness* 26 (1): 50–80.

Browning, Douglas. 1960. Sorting and Grading. *Australasian Journal of Philosophy* 38 (December): 234–245.

Brunsson, Nils, and Bengt Jacobsson, eds. 2000. *A World of Standards.* Oxford: Oxford University Press.

Bureau of Standards. 1912. *Conditions in Certain Cities of the State of California Relative to Weights and Measures.* Sacramento, CA: Bureau of State Printing.

Burke, John G. 1966. Bursting Boilers and the Federal Power. *Technology and Culture* 7 (1): 1–23.

Busch, Lawrence. 2001. Témérité américaine et prudence européenne? *La Recherche* 339 (February): 19–23.

Busch, Lawrence. 2002. The Homiletics of Risk. *Journal of Agricultural & Environmental Ethics* 15:17–29.

Busch, Lawrence. 2007. Performing the Economy, Performing Science: From Neoclassical to Supply Chain Models in the Agrifood Sector. *Economy and Society* 36 (3): 439–468.

Busch, Lawrence, Valerie Gunter, Theodore Mentele, Masashi Tachikawa, and Keiko Tanaka. 1994. Socializing Nature: Technoscience and the Transformation of Rapeseed into Canola. *Crop Science* 34 (3): 607–614.

Bush, Vannevar. 1945. *Science, the Endless Frontier*. Washington, DC: U.S. Office of Scientific Research and Development.

Byers, Alice, Daniele Giovannucci, and Pascal Liu. 2008. *Value-adding Standards in the North American Food Market: Trade Opportunities in Certified Products for Developing Countries*. Rome: Food and Agriculture Organization of the United Nations.

Callon, Michel. 1998a. Introduction: The Embeddedness of Economic Markets in Economics. In *The Laws of the Markets*, ed. M. Callon, 1–57. Oxford: Basil Blackwell.

Callon, Michel, ed. 1998b. *The Laws of the Markets*. Oxford: Basil Blackwell.

Callon, Michel, and John Law. 2003. *On Qualculation, Agency and Otherness*. Lancaster, UK: Lancaster University, Centre for Science Studies.

Callon, Michel, Cécile Méadel, and Vololona Rabeharisoa. 2002. The Economy of Qualities. *Economy and Society* 31 (2): 194–217.

Campbell, David, and Robert Lee. 2007. Introduction. In *Environmental Law and Economics*, ed. D. Campbell and R. Lee, xi–xxxiv. Aldershot, UK: Ashgate.

de Candolle, Augustin-Pyrame, and Kurt Polycarp Joachim Sprengel. 1821. *Elements of the Philosophy of Plants*. Edinburgh: Printed for William Blackwell and T. Cadell.

Cardwell, D. S. L. 1971. *From Watt to Clausius*. Ithaca, NY: Cornell University Press.

Cargill, Carl. 1997. *Open Systems Standardization: A Business Approach*. Upper Saddle River, NJ: Prentice-Hall.

Cargill, Carl, and Sherrie Bolin. 2007. Standardization: A Failing Paradigm. In *Standards and Public Policy*, ed. S. Greenstein and V. Stango, 296–328. Cambridge: Cambridge University Press.

Carlyle, Thomas. [1829] 1878. Signs of the Times. In *Critical and Historical Essays*, 187–196. New York: Appleton and Company.

Carolan, Michael S. 2008. Democratizing Knowledge: Sustainable and Conventional Agricultural Field Days as Divergent Democratic Forms. *Science, Technology & Human Values* 33 (4): 508–528.

Carson, Rachel. 1962. *Silent Spring*. Boston: Houghton Mifflin.

Cashore, Benjamin, Graeme Auld, and Deanna Newsom. 2004. *Governing through Markets: Forest Certification and the Emergence of Non-state Authority*. New Haven, CT: Yale University Press.

Cement Age. 1904. Uniform Municipal Building Laws. *Cement Age*, November, 210.

Chadwick, Ruth. 2004. Nutrigenomics, Individualism and Public Health. *Proceedings of the Nutrition Society* 63:161–166.

Clancy, M. 1998. Commodity Chains, Services and Development: Theory and Preliminary Evidence from the Tourism Industry. *Review of International Political Economy* 5 (1): 122–148.

Clarke, Adele E., and Joan H. Fujimura, eds. 1992. *The Right Tools for the Job: At Work in Twentieth-Century Life Sciences*. Princeton, NJ: Princeton University Press.

Clause, Bonnie Tocher. 1993. The Wistar Rat as a Right Choice: Establishing Mammalian Standards and the Ideal of a Standardized Mammal. *Journal of the History of Biology* 26 (2): 329–349.

Clegg, Jerry S. 1966. On Grading Labels. *Mind*, New Series 75 (297): 138–140.

Coase, Ronald H. 1960. The Problem of Social Cost. *Journal of Law & Economics* 3:1–44.

Coase, Ronald H. 1988. *The Firm, the Market and the Law*. Chicago: University of Chicago Press.

Cochoy, Franck. 1998. Another Discipline for the Market Economy: Marketing as a Performative Knowledge and Know-how for Capitalism. In *The Laws of the Markets*, ed. M. Callon, 194–221. Oxford: Basil Blackwell.

Cochoy, Franck. 2002. *Une sociologie du packaging ou l'âne de Buridan face au marché*. Paris: Presses Universitaires de France.

Cochrane, Willard. 1993. *The Development of American Agriculture*, 2nd ed. Minneapolis: University of Minnesota Press.

Coles, Jesse V. 1932. *Standardization of Consumers' Goods: An Aid to Consumer-Buying*. New York: Ronald Press.

Comité Européen de Normalisation. 2008. *Survey of Fora and Consortia, edition 13–september 2007*. Comité Européen de Normalisation, 2007. http://www.cen.eu/cen/pages/default.aspx.

Connor, Patrick E. 1997. Total Quality Management: A Selective Commentary on Its Human Dimensions, with Special Reference to its Downside. *Public Administration Review* 57 (6): 501–509.

Constant, Benjamin. 1814. *De l'esprit de conquête et de l'usurpation: Dans leurs rapports avec la civilisation européenne*. Paris: Le Normand.

Constant, Benjamin. [1815] 1988. *Benjamin Constant: Political Writings*. Cambridge: Cambridge University Press.

Coontz, Stephanie. 1992. *The Way We Never Were: American Families and the Nostalgia Trap*. New York: Basic Books.

Corbett, L. C. 1905. *Tomatoes*. Washington, DC: U.S. Department of Agriculture.

Courville, Sasha, Christine Parker, and Helen Watchirs. 2003. Introduction: Auditing in Regulatory Perspective. *Law & Policy* 25 (3): 179–184.

Cowan, Ruth Schwartz. 1983. *More work for Mother: The Ironies of Household Technology from the Open Hearth to the Microwave*. New York: Basic Books.

Crane, Diana. 1972. *Invisible Colleges: Diffusion of Knowledge in Scientific Communities*. Chicago: University of Chicago Press.

Cronon, William. 1991. *Nature's Metropolis: Chicago and the Great West*. New York: W.W. Norton.

Croom, Simon, Pietro Romano, and Mihalis Giannakis. 2000. Supply Chain Management: An analytical Framework for Critical Literature Review. *European Journal of Purchasing & Supply Management* 6:67–83.

Crosby, Alfred W. 1986. *Ecological Imperialism*. New York: Cambridge University Press.

Dankers, Cora. 2003. *Environmental and Social Standards, Certification and Labelling for Cash Crops*. Rome: Food and Agriculture Organization of the United Nations.

Darwin, C. G. 1945. The Effect of Tolerances on Measurements. *Metal Progress* 47:941–946.

Daumas, Maurice. [1953] 1972. *Scientific Instruments of the Seventeenth and Eighteenth Centuries*. New York: Praeger Publishers.

David, Paul A. 1985. Clio and the Economics of qwerty. *American Economic Review* 75 (2): 332–337.

de Crèvecoeur, J. Hector St. John. [1782] 1904. *Letters from an American Farmer*. New York: Fox, Duffield.

de Tocqueville, Alexis. [1835–40] 1956. *Democracy in America*. New York: Mentor.

de Vries, Gerard. 2007. What Is Political in Sub-politics?: How Aristotle Might Help STS. *Social Studies of Science* 37 (5): 781–809.

Delinder, Jean Van. 2005. Taylorism, Managerial Control Strategies, and the Ballets of Balanchine and Stravinsky. *American Behavioral Scientist* 48 (11): 1439–1452.

Deming, W. Edwards. 1982. *Out of the Crisis*. Cambridge, MA: Massachusetts Institute of Technology, Center for Advanced Engineering Study.

Department of Justice. 2005. *Crime 2004 in the United States: Uniform Crime Reports*. Washington, DC: U.S. Department of Justice, Federal Bureau of Investigation.

Derman, Emanuel, and Paul Wilmott. 2008. Perfect Models, Imperfect World. *Business Week*, 12 January, 59–60.

Derrida, Jacques. 1978. *Writing and Difference*, trans. Alan Bass. Chicago: University of Chicago Press.

Descartes, René. 1901. *The Method, Meditations, and Philosophy*, trans. John Vietch. Washington, DC: M. Walter Dunne.

Dewey, John. 1922a. *Human Nature and Conduct*. New York: Henry Holt.

Dewey, John. 1922b. Review of Public Opinion by Walter Lippmann. *New Republic* 29:285–288.

Dewey, John. 1927. *The Public and Its Problems*. New York: Henry Holt.

Dewey, John. 1929. *The Quest for Certainty*. New York: G.P. Putnam's Sons.

Dewey, John. [1915] 1976. The Logic of Judgments of Practice. In *The Middle Works of John Dewey, 1899–1924*, ed. J. A. Boydston, 14–82. Carbondale: Southern Illinois University Press.

Dewey, John. [1937] 1987. An Inquiry into the Principles of The Good Society, by Walter Lippmann. Boston: Little, Brown and Co., 1937 (book review). In *Later Works of John Dewey*, ed. J. A. Boydston, 489–495. Carbondale: Southern Illinois University Press.

Dietrich, Michael. 1994. *Transaction Cost Economics and Beyond: Toward a New Economics of the Firm*. London: Routledge.

Dixon, Robert G. Jr. 1978. *Standards Development in the Private Sector: Thoughts on Interest Representation and Procedural Fairness*. Boston: National Fire Protection Association.

Dobbin, Frank, and John R. Sutton. 1998. The Strength of a Weak State: The Rights Revolution and the Rise of Human Resources Management Divisions. *American Journal of Sociology* 104 (2): 441–476.

Doherty, William, and Jason S. Carroll. 2002. The Families and Democracy Project. *Family Process* 41 (4): 579–590.

Doleshal, Barbara. 2007. *Simulated Patient Scenarios*. Association for Academic Psychiatry. http://www.hsc.wvu.edu.

Donaldson, John L. 2005. *Directory of National Accreditation Bodies*. Gaithersburg, MD: National Institute of Standards and Technology.

Dragneva, Rilka, and Joop de Kort. 2007. The Legal Regime for Free Trade in the Commonwealth of Independent States. *International and Comparative Law Quarterly* 56 (2): 233–266.

Drake, Stillman. 1978. *Galileo at Work: His Scientific Biography*. Chicago: University of Chicago Press.

Dreger, Alice D. 1998. *Hermaphrodites and the Medical Invention of Sex*. Cambridge, MA: Harvard University Press.

Dutton, M. L., and J. P. Freeland. 2005. *Aelred of Rievaulx: The Historical Works*. Kalamazoo, MI: Cistercian Publications.

Eco, Umberto. 1988. *Foucault's Pendulum*. New York: Random House.

Egyedi, Tineke. 1995. *Shaping Standardization: A Study of Standards Processes and Standards Policies in the Field of Telematic Services*. Delft: Technical University of Delft, Department of Psychology.

Egyedi, Tineke. 2001. Infrastructure Flexibility Created by Standardized Gateways: The Cases of XML and the ISO Container. *Knowledge, Technology, & Policy* 14 (3): 41–54.

Eisenberg, David M., Roger B. Davis, Susan L. Ettner, Scott Appel, Sonja Wilkey, Maria Van Rompay, and Ronald C. Kessler. 1998. Trends in Alternative Medicine Use in the United States, 1990–1997: Results of a Follow-up National Survey. *Journal of the American Medical Association* 280 (18): 1569–1575.

Elgin, Ben. 2007. Another Inconvenient Truth. *Business Week*, March 26, 96–98, 100, 102.

English, James F. 2002. Winning the Culture Game: Prizes, Awards, and the Rules of Art. *New Literary History* 33 (1): 109–135.

Epstein, Steven. 1996. *Impure Science: Aids, Activism, and the Politics of Knowledge*. Berkeley and Los Angeles: University of California Press.

Epstein, Steven. 2007. *Inclusion: The Politics of Difference in Medical Research*. Chicago: University of Chicago Press.

Epstein, Steven. 2009. Beyond the Standard Human. In *Standards and Their Stories: How Quantifying, Classifying, and Formalizing Practices Shape Everyday Life*, ed. M. Lampland and S. L. Star, 35–53. Ithaca, NY: Cornell University Press.

Ericson, Richard, Dean Barry, and Aaron Doyle. 2000. The Moral Hazards of Neoliberalism: Lessons from the Private Insurance Industry. *Economy and Society* 29 (4): 532–558.

Espeland, Wendy Nelson, and Mitchell L. Stevens. 1998. Commensuration as a social process. *Annual Review of Sociology* 24: 313–343.

Evans, John Llewelyn. 1962. Grade Not. *Philosophy* (London) 37 (139): 25–36.

Evans, John, Emma Rich, Brian Davies, and Rachel Allwood. 2008. *Education, Disordered Eating and Obesity Discourse*. Oxon: Routledge.

Evans, Richard J. 1987. *Death in Hamburg: Society and Politics in the Cholera Years*. New York: Penguin.

Ewald, François. 1991. Insurance and Risk. In *The Foucault Effect: Studies in Governmentality*, ed. G. Burchell, C. Gordon, and P. Miller, 197–210. Chicago: University of Chicago Press.

Eymard-Duvernay, François. 1995. La négociation de la qualité. In *Agro-alimentaire: Une economie de la qualité*, ed. F. Nicolas and E. Valceschini, 39–48. Paris: INRA and Economica.

Fallows, James. 2008. Taxis in the Sky. *Atlantic Monthly*, May, 64–66, 68, 70–73.

Fei Shaotong. [1948] 1992. *From the Soil: Foundations of Chinese Society*, trans. G. G. Hamilton and W. Zheng. Berkeley and Los Angeles: University of California Press.

Ferry, Luc. 1995. *The New Ecological Order*, trans. C. Volk. Chicago: University of Chicago Press.

Feyerabend, Paul. 1978. *Science in a Free Society*. London: New Left Books.

Field, Wooster Bard. 1922. *Architectural Drawing*. New York: McGraw-Hill.

Fitzgerald, Deborah. 1996. Blinded by Technology: American Agriculture in the Soviet Union, 1928–1932. *Agricultural History* 70 (3): 459–486.

Flexner, Abraham. 1910. Medical Education in the United States and Canada. New York: Carnegie Foundation for the Advancement of Teaching. http://www.archive.org/details/medicaleducation00flexiala

Food Ethics Council. n.d. *Ethics—A Toolkit for Food Businesses*. http://www.foodethicscouncil.org/node/355.

Foote, Richard J. 1958. *Analytical Tools for Studying Demand and Price Structures*. Washington, DC: U.S. Department of Agriculture.

Fortune. 1943a. Quality Control. *Fortune*, October, 126–127.

Fortune. 1943b. To One-Millionth of an Inch. *Fortune*, October, 128–129, 214.

Foucault, Michel. 1977. *Discipline and Punish: The Birth of the Prison*. New York: Vintage.

Foucault, Michel. 2007. *Security, Territory, Population: Lectures at the Collège de France, 1977–1978*. New York: Palgrave Macmillan.

Foucault, Michel. 2008. *The Birth of Biopolitics: Lectures at the Collège de France, 1978–1979*. New York: Palgrave Macmillan.

Fouilleux, Eve. 2008. Les politiques agricoles et alimentaires en France et en Europe. In *Politiques publiques*, ed. O. Borraz and V. Guiraudon, 113–146. Paris: Presses de Science Politiques.

Franchising World. 2006. *The Profile of Franchising.* International Franchise Association.http://www.franchise.org.

Frank, David John, Ann Hironaka, and Evan Schofer. 2000. The Nation-State and the Natural Environment over the Twentieth Century. *American Sociological Review* 65 (1): 96–116.

Frank, David John, and John W. Meyer. 2002. The Profusion of Individual Roles and Identities in the Postwar Period. *Sociological Theory* 20 (1).

Frank, Robert H., and Philip J. Cook. 1995. *The Winner-Take-All Society.* New York: Penguin Books.

Freedman, Samuel G. 2007. Rabbi's Campaign for Kosher Standards Expands to Include Call for Social Justice. *New York Times,* May 19, A14.

Freidberg, Susanne. 2003. Cleaning Up down South: Supermarkets, Ethical Trade and African Horticulture. *Social & Cultural Geography* 4 (1): 27–43.

Friedland, William H., Amy E. Barton, and Robert J. Thomas. 1981. *Manufacturing Green Gold: Capital, Labor, and Technology in the Lettuce Industry.* Cambridge: Cambridge University Press.

Friedman, Batya, and Helen Nissenbaum. 1996. Bias in Computer Systems. *ACM Transactions on Information Systems* 14 (3): 330–347.

Friedman, Milton. 1962. *Capitalism and Freedom.* Chicago: University of Chicago Press.

Friedman, Milton. 1998. *Parental Choice: Who Wins, Who Loses – A Look at the Future.* Milton and Rose D. Friedman Foundation. http://www.edchoice.org/The-Friedmans/ The-Friedmans-on-School-Choice/Parental-Choice--Who-Wins,-Who-Loses- %E2%80%93-A-Look-at-t.aspx.

Gaillard, John. 1942. Refined Quality Control. *American Machinist* 86:1430–1432.

Gardiner, Bryan. 2008. Format Wars. *NWA World Traveler,* July, 74.

General Accounting Office. 1995. *School Facilities: Condition of America's Schools.* Washington, DC: GAO.

GenomeWeb. 2008. Genentech Files Citizen Petition Urging FDA to Regulate All Lab-Developed Tests. *GenomeWeb Daily News.* http://www.genomeweb.com/ genentech-files-citizen-petition-urging-fda-regulate-all-lab-developed-tests.

GenomeWeb. 2008. Standardization Groups Launch Initiative to Harmonize "Minimum Information" Checklists. *GenomeWeb Daily News.* http://www.genomeweb. com/standardization-groups-launch-initiative-harmonize-minimum-information- checklist.

Gephart, W. F. 1919. Grading and Standardization in Marketing Foods. *Annals of the American Academy of Political and Social Science* 82:263–270.

Gibbon, Peter, and Stefano Ponte. 2008. Global Value Chains: From Governance to Governmentality? *Economy and Society* 37 (3): 365–392.

Gladwell, Malcolm. 2008. *Outliers: The Story of Success*. New York: Little, Brown.

Goetz, Thomas. 2008. Scanning Our Skeletons: Bone Images Show Wear and Tear. *Wired* 16 (7): 117.

Goffman, Erving. 1971. *Relations in Public*. New York: Basic Books.

Goffman, Erving. 1993. On Face-work. In *Social Theory: The Multicultural and Classic Readings*, ed. C. Lemert, 358–363. Boulder, CO: Westview Press.

Gold, Lois Swirsky, Neela B. Manley, and Bruce N. Ames. 1992. Extrapolation of Carcinogenicity between Species: Qualitative and Quantitative Factors. *Risk Analysis* 12 (4): 579–587.

Goldberg, Ray A. 1968. *Agribusiness Coordination: A Systems Approach to the Wheat, Soybeans, and Florida Orange Economies*. Boston: Harvard University, Graduate School of Business Administration.

Goody, Jack. 1977. *The Domestication of the Savage Mind*. Cambridge: Cambridge University Press.

Goonatilake, Susantha. 1982. *Crippled Minds: An Exploration into Colonial Culture*. New Delhi: Vikas Publishing House.

Gordon, Peter. 2004. Numerical Cognition without Words: Evidence from Amazonia. *Science* 306: 496–499.

Gotham, Kevin Fox. 2006. The Secondary Circuit of Capital Reconsidered: Globalization and the U.S. Real Estate Sector. *American Journal of Sociology* 112 (1): 231–275.

Graff, Garrett M. 2008. Predicting the Vote: Pollsters Identify Tiny Voting Blocs. *Wired* 16 (7): 119.

Green, Heather, and Kerry Capell. 2008. Carbon Confusion. *Business Week*, March 17, 52–55.

Greenman, Milton J., and F. Louise Duhring. 1931. *Breeding and Care of the Albino Rat for Research Purposes*. Philadelphia: Wistar Institute of Anatomy and Biology.

Grievink, Jan-Willem. 2003. The Changing Face of the Global Food Industry. OECD conference presentation, the Hague. http://www.cooperatie.nl/download/17-grievink.pdf.

Guthman, Julie. 2007. The Polanyian Way? Voluntary Food Labels as Neoliberal Governance. *Antipode* 39:456–478.

Guthman, Julie. 2008. Thinking inside the Neoliberal Box: The Micro-politics of Agro-food Philanthropy. *Geoforum* 39:1241–1253.

Ha, Marie-Paule. 2003. From "nos ancêtres, les gaulois" to "leur culture ancestrale": Symbolic Violence and the Politics of Colonial Schooling in Indochina. *French Colonial History* 3:101–118.

Habermas, Jürgen. 1998. *Between Facts and Norms: Contributions to a Discourse Theory of Law and Democracy*, trans. W. Rehg. Cambridge, MA: MIT Press.

Hacking, Ian. 1991. How Should We Do the History of Statistics? In *The Foucault Effect: Studies in Governmentality*, ed. G. Burchell, C. Gordon, and P. Miller, 181–195. Chicago: University of Chicago Press.

Hacking, Ian. 1999. *The Social Construction of What?* Cambridge, MA: Harvard University Press.

Hadden, Richard W. 1994. *On the Shoulders of Merchants: Exchange and the Mathematical Conception of Nature in Early Modern Europe*. Albany: State University of New York Press.

Hadley, Arthur Twining. 1911. *Standards of Public Morality*. New York: Macmillan.

Hamilton, David E. 1986. Agricultural Distress in the Midwest Past and Present. In *From New Era to New Deal: American Farm Policy between the Wars*, ed. L. E. Gelf and R. J. Neymeyer, 19–54. Iowa City: University of Iowa Press.

Haraway, Donna J. 1997. *Modest_witness@second_millennium: Femaleman©_meets oncomouse*. New York: Routledge.

Harding, Sandra. 1991. *Whose Science? Whose Knowledge? Thinking from Women's Lives*. Ithaca, NY: Cornell University Press.

Harlequin Enterprises Ltd. 2009. *Writing Guidelines*. Harlequin Enterprises Limited. http://www.eharlequin.com/articlepage.html?articleId=538&chapter=0.

Haug, Wolfgang Fritz. 1986. *Critique of Commodity Aesthetics*. Minneapolis: University of Minnesota Press.

Hayek, Friedrich August. 1945. The Use of Knowledge in Society. *American Economic Review* 35 (4): 519–530.

Hayek, Friedrich August. 1973. *Rules and Order, Law, Legislation and Liberty*. London: Routledge and Kegan Paul.

Hayek, Friedrich August. 1975. The Pretence of Knowledge. *Swedish Journal of Economics* 77 (4): 433–442.

Hayek, Friedrich August. 1976. *The Mirage of Social Justice, Law, Legislation and Liberty*. Chicago: University of Chicago Press.

Hayek, Friedrich August. 1979. *The Political Order of a Free People, Law, Legislation and Liberty*. Chicago: University of Chicago Press.

Hayek, Friedrich August. [1944] 2007. *The Road to Serfdom*. Chicago: University of Chicago Press.

Heidegger, Martin. 1956. *Existence and Being*. London: Vision Press.

Heidegger, Martin. 1977. *The Question Concerning Technology and Other Essays*. New York: Harper and Row.

Heilbroner, Robert L. 1961. *The Worldly Philosophers*. New York: Simon and Schuster.

Heinz. 2009. Heinz: The Nutrition Experts. http://www.heinzbaby.com.

Heller, Michael A., and Rebecca S. Eisenberg. 1998. Can Patents Deter Innovation? The Anticommons in Biomedical Research. *Science* 280 (5364): 698–701.

Henke, Christopher R. 2008. *Cultivating Science, Harvesting Power: Science and Industry in California Agriculture*. Cambridge, MA: MIT Press.

Hermisson, J., G. P. Wagner, L. A. Meyers, H. Bagheri-Chaichian, J. L. Blanchard, L. Chao, J. M. Cheverud, S. F. Elena, W. Fontana, and G. Gibson. 2003. Perspective: Evolution and Detection of Genetic Robustness. *Evolution; International Journal of Organic Evolution* 57 (9): 1959–1972.

Hesiod. [ca. 700 BC] 2007. *Works and Days*, trans. H. G. Evelyn-White. Charleston, SC: Forgotten Books.

Hill, Howard C. 1919. The Americanization Movement. *American Journal of Sociology* 24 (6): 609–642.

Hill, Jason, Erik Nelson, David Tilman, Stephen Polasky, and Douglas Tiffany. 2006. Environmental, Economic, and Energetic Costs and Benefits of Biodiesel and Ethanol Biofuels. *Proceedings of the National Academy of Sciences of the United States of America* 103 (30): 11206–11210.

Hill, Lowell D. 1990. *Grain Grades and Standards: Historical Issues Shaping the Future*. Urbana: University of Illinois Press.

Hill, Lowell D., ed. 1991. *Uniformity by 2000: An International Workshop on Maize and Soybean Quality*. Urbana, IL: Scherer Publications.

Hirschman, Albert O. 1977. *The Passions and the Interests*. Princeton, NJ: Princeton University Press.

Hobbes, Thomas. [1651] 1991. *Leviathan*, ed. R. Tuck. Cambridge: Cambridge University Press.

Hobbes, Thomas. [1658/1642] 1991. *Man and Citizen (De homine and de cive)*, ed. B. Gert. Indianapolis: Hackett Publishing Co..

Hof, Robert D. 2004. Building an Idea Factory. *Business Week*, October 11, 128–132.

Hogle, Linda F. 1995. Standardization across Non-standard Domains: The Case of Organ Procurement. *Science, Technology & Human Values* 20 (4): 482–500.

Holden, Constance. 2007. Rehabilitating Pluto. *Science* 315:1643.

Hooker, Clarence. 1997. Ford's Sociology Department and the Americanization Campaign and the Manufacture of Popular Culture among Assembly Line Workers c.1910–1917. *Journal of American Culture* 20 (1): 47–53.

Hoos, Ida. 1972. *Systems Analysis in Public Policy*. Berkeley and Los Angeles: University of California Press.

Hoover, Herbert. [1924] 1937. Moral Standards in an Industrial Era. In *The Hoover Policies*, ed. R. L. Wilbur and A. M. Hyde, 300–305. New York: Charles Scribner's Sons.

Hoover, Herbert. 1952. *The Memoirs of Herbert Hoover: The Cabinet and the Presidency, 1920–1933*. New York: Macmillan.

Hoyt, Homer. 1919. Standardization and Its Relation to Industrial Concentration. *Annals of the American Academy of Political and Social Science* 82:271–277.

Hudson, Ray M. 1928. Organized Effort in Simplification. *Annals of the American Academy of Political and Social Science* 137:1–8.

Hughes, Everett C. 1963. The Professions. *Daedalus* 92 (4): 655–668.

Huizinga, Johan. 1950. *Homo ludens*. Boston: Beacon Press.

Hull, David L. 1988. *Science as a Process: An Evolutionary Account of the Social and Conceptual Development of Science*. Chicago: University of Chicago Press.

Hutchins, Edwin. 1987. *Mediation and Automatization*. San Diego: University of California, Institute for Cognitive Science.

Huxley, Aldous. [1932] 1998. *Brave New World*. New York: HarperPerennial.

Idhe, Don. 1979. *Technics and Praxis*. Dordrecht: D. Reidel.

Idhe, Don. 1983. *Existential Technics*. Albany: State University of New York Press.

Inman, R. P., and D. L. Rubinfeld. 1998. *Subsidiarity and the European Union*. Cambridge, MA: National Bureau of Economic Research.

International Accreditation Forum. 2007. *International Accreditation Forum, Inc.* http://www.iaf.nu.

International Bank for Reconstruction and Development. 2007. *Food Safety and Agricultural Health Management in CIS Countries: Completing the Transition*. Washington, DC: World Bank, Agriculture and Rural Development Department.

International Laboratory Accreditation Cooperation. 2007. Welcome to ILAC. International Laboratory Accreditation Cooperation. http://www.ilac.org/home.html.

International Organization for Standardization. 2004. *Conformity Assessment—Code of Good Practice*. Geneva: ISO.

Ithaca Hours. 2008. *Ithaca Hours Local Currency, Ithaca*. New York. http://www.ithacahours.org.

Jaffee, Steven, and Spencer Henson. 2004. *Standards and Agro-food Exports from Developing Countries: Rebalancing the Debate*. World Bank Policy Research Working Paper 3348. Washington, DC: World Bank..

Jasanoff, Sheila. 2000. Technological Risk and Cultures of Rationality. In *Incorporating Science, Economics, and Sociology in Developing Sanitary and Phytosanitary Standards in International Trade: Proceedings of a Conference*, National Academy of Sciences, 65–84. Irvine, CA: National Academy Press.

Jasanoff, Sheila. 2003. (No?) Accounting for Expertise. *Science & Public Policy* 30 (3): 157–162.

Jasanoff, Sheila, Robin Grove-White, Lawrence Busch, Brian Wynne, and David Winickoff. 2005. Adjudicating the GM Food Wars: Science, Risk, and Democracy in World Trade Law. *Yale Journal of International Law* 30 (1): 81–123.

Jensen, Karsten Klint, and Peter Sandøe. 2002. Food Safety and Ethics: The Interplay between Science and Values. *Journal of Agricultural & Environmental Ethics* 15:245–253.

Jones, C. R. [1916] 1917. Scientific management as applied to the farm, home, and manufacturing plants. Paper read at the Proceedings of the 30th Annual Convention of the Association of American Agricultural Colleges and Experiment Stations.

Josephus, Flavius. [94] 2009. Antiquities of the Jews. http://www.gutenberg.org/ebooks/2848.

Kafka, Franz. [1926] 1998. *The Castle*, trans. M. Harman. New York: Schocken Books.

Kaiser, Jocelyn. 2008. Industrial-Style Screening Meets Academic Biology. *Science* 321:764–766.

Karier, Clarence. 1973. Testing for Order and Control in the Corporate Liberal State. In *Roots of Crisis: American Education in the Twentieth Century*, ed. C. J. Karier, P. Violas, and J. Spring. Chicago: Rand McNally.

Katsanis, S. H., G. Javitt, and K. Hudson. 2008. A Case Study of Personalized Medicine. *Science* 320:53–54.

Kelly, G. P. 1979. The Relation between Colonial and Metropolitan Schools: A Structural Analysis. *Comparative Education* 15 (2): 209–215.

Kennedy, Donald. 2007. Approaching Science. *Science* 318:715.

Klaidman, Stephen. 2007. *Coronary: A True Story of Medicine Gone Awry*. New York: Simon and Schuster.

Klein, Judy L. 2000. Economics for a Client: The Case of Statistical Quality Control and Sequential Analysis. *History of Political Economy* 32 (Suppl.): 27–70.

Knorr Cetina, Karin. 1981. *The Manufacture of Knowledge*. Oxford: Pergamon Press.

Knorr Cetina, Karin. 2003. *Epistemic Cultures: How the Sciences Make Knowledge*. Cambridge, MA: Harvard University Press.

Kohler, Robert E. 1980. Warren Weaver and the Rockefeller Foundation Program in Molecular Biology: A Case Study in the Management of Science. In *Sciences in the American Context*, ed. N. Reingold, 249–293. Washington, DC: Smithsonian Institution Press.

Krimsky, Sheldon. 1984. Citizen Participation in Science Policy. In *Beyond Technology: New Roots for Citizen Involvement in Social Risk Assessment*, ed. J. C. Petersen, 43–61. Amherst: University of Massachusetts Press.

Kuang, Cliff. 2008. Tracking Air Fares: Elaborate Algorithms Predict Ticket Prices. *Wired*, July, 118.

Kuhn, Markus. 2008. *Metric Typographic Units*. Cambridge University. http://www.cl.cam.ac.uk/~mgk25/metric-typo.

Kula, Witold. 1986. *Measures and Men*, trans. R. Szreter. Princeton, NJ: Princeton University Press.

Kysar, Douglas A. 2006. It Might Have Been: Risk, Precaution, and Opportunity Costs. Cornell Law School Legal Studies Research Paper Series. Ithaca, NY: Cornell Law School.

Lafaye, Claudette, and Laurent Thévenot. 1993. Une justification écologique? Conflits dans l'aménagement de la nature. *Revue Française de Sociologie* 34 (4): 495–524.

Lakoff, George, and Mark Johnson. 1980. The Metaphorical Structure of the Human Conceptual System. *Cognitive Science* 4: 195–208.

Lampland, Martha. 2009. Classifying Laborers: Instinct, Property, and the Psychology of Productivity in Hungary (1920–1956). In *Standards and Their Stories: How Quantifying, Classifying, and Formalizing Practices Shape Everyday Life*, ed. M. Lampland and S. L. Star, 123–142. Ithaca, NY: Cornell University Press.

Lash, Scott. 2001. Technological Forms of Life. *Theory, Culture & Society* 18 (1): 105–120.

Latour, Bruno. 1987. *Science in Action: How to Follow Scientists and Engineers through Society. Milton Keynes*. England: Open University Press.

Latour, Bruno. 1993. *We Have Never Been Modern*. Cambridge, MA: Harvard University Press.

Latour, Bruno. 2005. *Reassembling the Social: An Introduction to Actor-Network-Theory*. Oxford: Oxford University Press.

Lave, Jean, and Etienne Wenger. 1991. *Situated Learning: Legitimate Peripheral Participation*. Cambridge: Cambridge University Press.

Lavin, Danielle, and Douglas W. Maynard. 2001. Standardization vs. Rapport: Respondent Laughter and Interviewer Reaction during Telephone Surveys. *American Sociological Review* 66 (3): 453–479.

Lazear, Edward P. 2000. Economic Imperialism. *Quarterly Journal of Economics* 115 (1): 99–146.

Lazzarini, Sergio G., Fabio R. Chaddad, and Michael L. Cook. 2001. Integrating Supply Chain and Network Analyses: The Study of Netchains. *Journal of Chain and Network Science* 1 (1): 7–22.

Lee, Guy A. 1937. The Historical Significance of the Chicago Grain Elevator System. *Agricultural History* 11 (1): 16–32.

Lee, Richard P. 2009. Agri-food Governance and Expertise: The Production of International Food Standards. *Sociologia Ruralis* 49 (4): 415–431.

LeMaistre, C. 1919. Summary of the Work of the British Engineering Standards Association. *Annals of the American Academy of Political and Social Science* 82:247–252.

Lemke, Thomas. 2001. The Birth of Bio-politics: Michel Foucault's Lecture at the Collège de France on Neo-liberal Governmentality. *Economy and Society* 30 (2): 190–207.

Lengwiler, Martin. 2009. Double Standards: The History of Standardizing Humans in Modern Life Insurance. In *Standards and Their Stories: How Quantifying, Classifying, and Formalizing Practices Shape Everyday Life*, ed. M. Lampland and S. L. Star, 95–113. Ithaca, NY: Cornell University Press.

Levenstein, Harvey. 1988. *Revolution at the Table: The Transformation of the American Diet*. New York: Oxford University Press.

Levinson, Marc. 2006. *The Box: How the Shipping Container Made the World Smaller and the World Economy Bigger*. Princeton, NJ: Princeton University Press.

Liebowitz, S. J., and Stephen E. Margolis. 1990. The Fable of the Keys. *Journal of Law & Economics* 33 (1): 1–25.

Lippmann, Walter. 1922. *Public Opinion*. New York: Macmillan.

Lippmann, Walter. 1938. *The Good Society*. London: George Allen and Unwin Ltd.

Lippmann, Walter. 1939. Discussion. In *Le colloque Walter Lippmann*. Travaux du centre international d'études pour la rénovation du libéralisme, cahier no. 1, 99–101. Paris: Librairie de Médicis.

Lipton, Eric, and Gardiner Harris. 2007. In Turnaround, Industries Seek U.S. Regulation. *New York Times*, September 16, Finance, 12.

Loconto, Allison, and Lawrence Busch. 2010. Standards, Techno-economic Networks, and Playing Fields: Performing the Global Market Economy. *Review of International Political Economy* 17 (3): 507–536.

Lonsdale, Chris, and Andrew Cox. 2000. The Historical Development of Outsourcing: The Latest Fad? *Industrial Management & Data Systems* 100 (9): 444–450.

Loya, Thomas A., and John Boli. 1999. Standardization in the World Polity: Technical Rationality over Power. In *Constructing World Culture: International Nongovernmental Organizations since 1875*, ed. J. Boli and G. M. Thomas, 169–197. Stanford: Stanford Unversity Press.

MacIntyre, Alasdair. 1984. *After Virtue*, 2nd ed. Notre Dame, IN: University of Notre Dame Press.

Marres, Noortje. 2007. The Issues Deserve More Credit: Pragmatist Contributions to the Study of Public Involvement in Controversy. *Social Studies of Science* 37 (5): 759–780.

Martin, Glen. 2006. Farms May Cut Habitat Renewal over E. coli Fears. *San Francisco Chronicle*, December 19, A1.

Marx, Karl. 1906. *Capital: A Critique of Political Economy* ed. F. Engels. New York: Modern Library.

Marx, Karl. 1977. *Karl Marx: Selected Writings*, ed. D. McLellan. Oxford: Oxford University Press.

Marx, Karl. [1844] 2009. Introduction to a Contribution to the Critique of Hegel's Philosophy of Right. http://www.marxists.org/archive/marx/works/1843/critique-hpr/intro.htm.

Mead, George Herbert. [1934] 1962. *Mind, Self, and Society from the Standpoint of a Social Behaviorist*. Chicago: University of Chicago Press.

Megay, Edward N. 1970. Anti-pluralist Liberalism: The German Neoliberals. *Political Science Quarterly* 85 (3): 422–442.

Melman, Seymour. 1981. Alternative Criteria for the Design of the Means of Production. *Theory and Society* 10 (May): 325–336.

Mercuro, Nicholas, and Steven G. Medema. 2006. *Economics and the Law: From Posner to Post-modernism and Beyond*, 2nd ed. Princeton, NJ: Princeton University Press.

Meyer, John W., John Boli, George M. Thomas, and Francisco O. Ramirez. 1997. World Society and the Nation-State. *American Journal of Sociology* 103 (1): 144–181.

Michaels, William Benn. 2006. *The Trouble with Diversity: How We Learned to Love Identity and Ignore Inequality*. New York: Metropolitan Books.

Michels, Robert. [1915] 2001. *Political Parties: A Sociological Study of the Oligarchical Tendencies of Modern Democracy*, trans. E. Paul and C. Paul. Kitchener: Batoche Books.

Miele, Mara. 2008. Città Slow: Producing Slowness against the Fast Life. *Space and Polity* 12 (1): 135–156.

Miller, David L. 1965. On Ordering. *Ethics* 75 (2): 112–116.

Millerand, Florence, and Geoffrey C. Bowker. 2009. Metadata Standards: Trajectories and Enactment in the Life of an Ontology. In *Standards and Their Stories: How Quantifying, Classifying, and Formalizing Practices shape everyday life*, ed. M. Lampland and S. L. Star, 149–165. Ithaca, NY: Cornell University Press.

Mills, C. Wright. [1976] 1956. *The Sociological Imagination*. New York: Oxford University Press.

Milton, J. R. 2001. Locke, Medicine and the Mechanical Philosophy. *British Journal for the History of Philosophy* 9 (2): 221–243.

Miniwatts Marketing Group. 2007. *Internet World Stats: Usage and Population Statistics*. Miniwatts Marketing Group. http://www.internetworldstats.com/stats.htm.

Mirowski, P., and D. Plehwe, eds. 2009. *The Road from Mont Pèlerin: The Making of the Neoliberal Thought Collective*. Cambridge, MA: Harvard University Press.

Mitman, Gregg, and Anne Fausto-Sterling. 1992. Whatever Happened to Planaria? C. M. Child and the Physiology of Inheritance. In *The Right Tools for the Job: At Work in Twentieth-Century Life Sciences*, ed. A. E. Clarke and J. H. Fujimura, 172–197. Princeton, NJ: Princeton University Press.

Mol, Annemarie. 2002. *The Body Multiple*. Durham, NC: Duke University Press.

Mont Pèlerin Society. 2009. *The Mont Pelerin Society*. https://www.montpelerin.org/montpelerin/index.html.

Moore, Sally Falk. 1992. Treating Law as Knowledge: Telling Colonial Officers What to Say to Africans about Running "Their Own" Native Courts. *Law & Society Review* 26 (1): 11–46.

More, Sir Thomas. n.d. [1535]. *Utopia*. Pleasanton, CA: PDF Nitro.

Mumford, Lewis. 1967. *The Myth of the Machine: Technics and Human Development*. New York: Harcourt, Brace, Jovanovich.

Nast, Thomas. 1871. Who Stole the People's Money? *Harper's Weekly*, August 19, 764.

National Center for Education Statistics. 2010. Digest of Education Statistics, 2009. Publication NCES 2010-013. Washington, DC: National Center for Education Statistics.. (accessed August 27, 20xx). http://nces.ed.gov/Programs/digest.

National Guideline Clearinghouse. 2007. Ngc. http://www.guideline.gov.

National Industrial Conference Board. 1929. Industrial Standardization. New York: National Industrial Conference Board.

National Institute of Standards and Technology. 2007. Standard Reference Materials® Catalog. Washington, DC: U.S. Government Printing Office.

National Institute of Standards and Technology. 2007. A Walk through Time. National Institute of Standards and Technology. http://www.nist.gov/pml/general/time/index.cfm.

National Quality Measures Clearinghouse. 2007. NQMC. http://www.qualitymeasures.ahrq.gov.

Nee, Victor, and Paul Ingram. 1998. Embeddedness and Beyond: Institutions, Exchange, and Social Structure. In The New Institutionalism in Sociology, ed. M. C. Brinton and V. Nee, 19–44. New York: Russell Sage Foundation.

New York Times. 1906. Butchers and Grocers Accused of Cheating. New York Times, January 12, 5.

New York Times. 1910. City Finds Weights and Measures Bad. New York Times, March 17, 7.

Noddings, Nel. 1997. Thinking about Standards. Phi Beta Kappan 78:184–189.

Nordstrom, Pam. 2006. San Patricio Battalion. http://www.tshaonline.org/handbook/online/articles/qis01.

O'Connell, Joseph. 1993. The Creation of Universality by the Circulation of Particulars. Social Studies of Science 23 (February): 129–173.

Oberg, Jan. 1965. Serlon de Wilton: Poèmes latins. Stockholm: Almquist and Wiksell.

Office of Management and Budget. 2008. Circular no. A-119, revised. White House 1998. http://www.whitehouse.gov/omb/circulars_a119.

Office of Technology Assessment. 1992. Global Standards: Building Blocks for the Future. Washington, DC: U.S. Government Printing Office.

Olshan, Marc A. 1993. Standards-Making Organizations and the Rationalization of American Life. Sociological Quarterly 34 (2): 319–335.

Organisation for Economic Co-operation and Development. 2007. OECD Guidelines for Quality Assurance in Molecular Genetic Testing. Paris: OECD.

Ormerod, Paul. 1994. The Death of Economics. London: Faber and Faber.

Orwell, George. [1949] 1990. *1984*. New York: Signet Classic.

Parfomak, Paul W. 2004. *Guarding America: Security Guards and U.S. Critical Infrastructure Protection*. Washington, DC: Congressional Research Service.

Pargman, Daniel, and Jacob Palme. 2009. ASCII Imperialism. In *Standards and Their Stories: How Quantifying, Classifying, and Formalizing Practices Shape Everyday Life*, ed. M. Lampland and S. L. Star, 177–199. Ithaca, NY: Cornell University Press.

Park, Robert E., and Ernest W. Burgess. 1921. *Introduction to the Science of Sociology*. Chicago: University of Chicago Press.

Parnas, David L., A. John van Schouwen, and Shu Po Kwan. 1990. Evaluation of Safety-Critical Software. *Communications of the ACM* 33 (6): 636–648.

Pauly, Philip J. 1987. *Controlling Life: Jacques Loeb and the Engineering Ideal in Biology*. New York: Oxford University Press.

Paynter, Ben. 2008. Feeding the Masses: Data in, Crop Predictions out. *Wired*, July, 110.

Pearson, E. S. 1935. *The Application of Statistical Methods to Industrial Standardisation and Quality Control*. London: British Standards Institution.

Pelkmans, Jacques. 2001. The GSM Standard: Explaining a Success Story. *Journal of European Public Policy* 8 (3): 432–453.

Perrow, Charles. 1984. *Normal Accidents*. New York: Basic Books.

Perry, John. 1955. *The Story of Standards*. New York: Funk and Wagnalls.

Peters, Tom. 1997. The Brand Called You. *FastCompany*, August, 83.

Pimentel, David, and Tad W. Patzek. 2005. Ethanol Production Using Corn, Switchgrass, and Wood; Biodiesel Production Using Soybean and Sunflower. *Natural Resources Research* 14 (1): 65–76.

Plato. [360 BC] 1994. *The Republic*, trans. B. Jowett. New York: Vintage Books.

Pollack, Andrew. 2009. Crop Scientists Say Biotechnology Seed Companies Are Thwarting Research." *New York Times*, February 20, B3.

Porter, Theodore M. 1995. *Trust in Numbers: The Pursuit of Objectivity in Science and Public Life*. Princeton, NJ: Princeton University Press.

Posner, Richard A. 1987. What Am I? A Potted Plant? *New Republic* 23:23–25.

Power, Michael. 1997. *The Audit Society: Rituals of Verification*. Oxford: Oxford University Press.

Power, Michael. 2003. Evaluating the Audit Explosion. *Law & Policy* 25 (3): 185–202.

Price, Derek J. 1957. The Manufacture of Scientific Instruments from c 1500 to c 1700. In *A History of Technology*. vol. 3, ed. C. Singer, E. J. Holmyard, and A. R. Hall, 620–647. Oxford: Oxford University Press.

Prokofyev, V. V. 1933. *Industrial and Technical Intelligentsia in the U.S.S.R.* Moscow: Cooperative Publishing Society of Foreign Workers in the USSR.

Puffert, Douglas J. 2002. Path Dependence in Spatial Networks: The Standardization of Railway Track Gauge. *Explorations in Economic History* 39:282–314.

Rasmussen, David M. 1973. Between Autonomy and Sociality. *Cultural Hermeneutics* 1 (1): 3–45.

Rawls, John. 1971. *A Theory of Justice*. Cambridge, MA: Belknap Press of Harvard University Press.

Rehm, Norman F. 1910. *Track Standards*. Chicago: Railway List Company.

Reich, Leonard S. 1985. *The Making of American Industrial Research: Science and Business at GE and Bell, 1876–1926*. Cambridge: Cambridge University Press.

Reich, Robert. 2007. *Supercapitalism: The Transformation of Business, Democracy, and Everyday Life*. New York: Alfred A. Knopf.

Reiter, Ester. 1991. *Making Fast Food: From the Frying Pan into the Fryer*. Montreal: McGill-Queens University Press.

Rice, William B. 1946. Statistical Quality Control: What It Is and What It Does. *Industrial Quality Control* 2 (4): 6–10.

Richard of Hexham. [1138] 1988. The Acts of King Stephen and the Battle of the Standard. In *Contemporary Chronicles of the Middle Ages: Sources of Twelfth-Century History*, ed. J. Stephenson, 53–76. Dyfed: Llanerch.

Rissik, H. 1942. Quality Control (letter to the editor). *Engineer* 173:389.

Ritzer, George. 1993. *The McDonaldization of Society*. Thousand Oaks, CA: Pine Forge Press.

Roberts, Paul. 2004. *The End of Oil: On the Edge of a Perilous New World*. New York: Houghton Mifflin.

Robson, William A. 1926. *Socialism and the Standardised Life*. London: Fabian Society.

Roeder, Philip G. 1991. Soviet Federalism and Ethnic Mobilization. *World Politics* 43 (2): 196–232.

Rogers, Adam. 2008. Tracking the News: A Smarter Way to Predict Riots and Wars. *Wired* 16 (7): 113.

Romanyshyn, Robert D. 1989. *Technology as Symptom and Dream*. New York: Routledge.

Rones, Phillip L., Jennifer M. Gardner, and Randy E. Ilg. 1997. Trends in Hours of Work since the Mid-1970s. *Monthly Labor Review* 120 (4): 3–14.

Rosa, Eugene A. 1998. Metatheoretical Foundations of Post-Normal Risk. *Journal of Risk Research* 1 (1): 15–44.

Rose, Nikolas. 1996. *Inventing Our Selves: Psychology, Power, and Personhood.* Cambridge: Cambridge University Press.

Rougier, Louis. 1939. Allocution du professor Louis Rougier. In *Le colloque Walter Lippmann.* Travaux du centre international d'études pour la rénovation du libéralisme, cahier no. 1, 13–20. Paris: Librairie de Médicis.

Rougier, Louis. 1939. Avant-propos. In *Le colloque Walter Lippmann.* Travaux du centre international d'études pour la rénovation du libéralisme, cahier no. 1, 7–8. Paris: Librairie de Médicis.

Russell, Andrew L. 2006. Industrial Legislatures: The American System of Standardization. In *International Standardization as a Strategic Tool,* 71–79. Geneva: International Electrotechnical Commission.

Russell, Maud. 1966. *Men along the Shore.* New York: Brussel & Brussel.

Rüstow, Alexander. 1939. Discussion. In *Le colloque Walter Lippmann.* Travaux du centre international d'études pour la rénovation du libéralisme, cahier no. 1, 77–83. Paris: Librairie de Médicis.

Sagoff, Mark. 1988. *The Economy of the Earth: Philosophy, Law and the Environment.* Cambridge: Cambridge University Press.

Sahal, Devendra. 1981. The Farm Tractor and the Nature of Technological Innovation. *Research Policy* 10 (4): 368–402.

Salter, Liora. 1988. *Mandated Science: Science and Scientists in the Making of Standards.* Dordrecht: Kluwer.

Samuelson, Paul Anthony, and William D. Nordhaus. 1995. *Economics,* 15th ed. New York: McGraw-Hill.

Sandrieu, Dominique. 1988. *Cinq cents lettres pour tous les jours.* Paris: Librairie Larousse.

Sarewitz, Daniel. 1996. *Frontiers of Illusion: Science, Technology, and the Politics of Progress.* Philadelphia: Temple University Press.

Schivelbusch, Wolfgang. 1978. Railroad Space and Railroad Time. *New German Critique* 14:31–40.

Schlosser, Eric. 2002. *Fast Food Nation.* Boston: Houghton Mifflin.

Schmandt-Besserat, Denise. 1980. The Envelopes That Bear the First Writing. *Technology and Culture* 21 (3): 357–385.

Schmidt, Susanne K., and Raymund Werle. 1998. *Coordinating Technology: Studies in the International Standardization of Telecommunications*. Cambridge, MA: MIT Press.

Schutz, Alfred. [1932] 1967. *The Phenomenology of the Social World*. Evanston, IL: Northwestern University Press.

Schutz, Alfred, and Thomas Luckmann. 1973. *The Structures of the Life-World*. Evanston, IL: Northwestern University Press.

Science. 1929. Industrial Standardization. *Science* 70:60.

Scott, James C. 1998. *Seeing Like a State: How Certain Schemes to Improve the Human Condition Have Failed*. New Haven, CT: Yale University Press.

Self Storage Association. 2006. *Self Storage Association: Official Not-for-Profit Organization of the Self Storage Industry*. Self Storage Association. http://www.selfstorage.org.

Shaevich, A. B. 2001. Standard Reference Materials in the Former Soviet Union and Russia: Historical Aspects. *Journal of Analytical Chemistry* 56 (7): 689–695.

Shapin, Steven, and Simon Schaffer. 1985. *Leviathan and the Air-Pump: Hobbes, Boyle, and the Experimental Life*. Princeton, NJ: Princeton University Press.

Shapiro, Carl. 2000. *Navigating the Patent Thicket: Cross Licenses, Patent Pools, and Standard-Setting*. Berkeley: University of California, Competition Policy Center.

Shewhart, W. A. 1931. *Economic Control of Quality of Manufactured Product*. Princeton, NJ: D. Van Nostrand Co.

Shewhart, W. A., and W. Edwards Deming. 1939. *Statistical Method from the Viewpoint of Quality Control*. Washington, DC: U.S. Department of Agriculture, Graduate School.

Ship-Technology. 2008. *Mol encore—Panamax Containership*. SPG Media Ltd. http://www.ship-technology.com/projects/mol_encore.

Silbey, Susan S., and Patricia Ewick. 2003. The Architecture of Authority: The Place of Law in the Space of Science. In *The Place of Law*, ed. A. Sarat, L. Douglas, and M. Umphrey, 75–108. Ann Arbor: University of Michigan Press.

Silver, Brian. 1974. Social Mobilization and the Russification of Soviet Nationalities. *American Political Science Review* 68 (1): 45–66.

Simons, Henry C. [1934] 1948. A Positive Program for Laissez Faire: Some Proposals for a Liberal Economic Policy. In *Economic Policy for a Free Society*, 40–77. Chicago: University of Chicago Press.

Slaton, Amy E. 2001. *Reinforced Concrete and the Modernization of American Building, 1900–1930*. Baltimore: Johns Hopkins University Press.

Smith, Adam. [1759] 1982. *The Theory of Moral Sentiments*. Indianapolis: Liberty Fund.

Smith, Adam. [1776] 1994. *An Inquiry into the Nature and Causes of the Wealth of Nations.* New York: Modern Library.

Smith, David C. 1964. Wood Pulp and Newspapers, 1867-1900. *Business History Review* 38 (3): 328–345.

Sobel, Dava. 1995. *Longitude: The True Story of a Lone Genius Who Solved the Greatest Scientific Problem of His Time.* New York: Penguin.

Sohn-Rethel, Alfred. 1978. *Intellectual and Manual Labor.* Atlantic Highlands, NJ: Humanities Press.

Sorokin, Pitirim A. 1937–41. *Social and Cultural Dynamics.* 4 vols. New York: American Book Co.

de Soto, Hernando. 2000. *The Mystery of Capital: Why Capitalism Triumphs in the West and Fails Everywhere Else.* New York: Basic Books.

Sprague, Paul E. 1981. The Origin of Balloon Framing. *Journal of the Society of Architectural Historians* 40 (4): 311–319.

Spring, Joel. 1972. *Education and the Rise of the Corporate State.* Boston: Beacon Press.

Stalin, Joseph. 1932. *On Technology.* Moscow: Cooperative Publishing Society of Foreign Workers in the USSR.

Stanziani, Alessandro. 2005. *Histoire de la qualité alimentaire.* Paris: Seuil.

Star, Susan Leigh, and James R. Griesemer. 1989. Institutional Ecology, "Translations" and Boundary Objects: Amateurs and Professionals in Berkeley's Museum of Vertebrate Zoology, 1907–39. *Social Studies of Science* 19 (3): 387–420.

Star, Susan Leigh, and Martha Lampland. 2009. Reckoning with Standards. In *Standards and Their Stories: How Quantifying, Classifying, and Formalizing Practices Shape Everyday Life,* ed. M. Lampland and S. L. Star, 3–24. Ithaca, NY: Cornell University Press.

Star, Susan Leigh, and Karen Ruhleder. 1996. Steps toward an Ecology of Infrastructure: Design and Access for Large Information Spaces. *Information Systems Research* 7 (1): 111–134.

Stier, Richard F., and Nancy E. Nagle. 2001. Growers Beware: Adopt GAPs or Else. *Food Safety Magazine* 7 (5): 26–28, 30, 32.

Strathern, Marilyn. 2006. A Community of Critics? Thoughts on New Knowledge. *Journal of the Royal Anthropological Institute* 12:191–209.

Sugden, Robert. 1989. Spontaneous Order. *Journal of Economic Perspectives* 3 (4): 85–97.

Swift, Jonathan. [1726] 1947. *Gulliver's Travels.* New York: Grosset and Dunlap.

Sykes, Mark V. 2008. Planetary Science: The Planet Debate Continues. *Science* 319:1765.

Talwalker, Clare. 2005. Colonial Dreaming: Textbooks in the Mythology of "Primitive Accumulation." *Dialectical Anthropology* 29:1–34.

Tannenbaum, Andrew S. 2003. *Computer Networks*, 4th edition. Upper Saddle River, NJ: Pearson Education.

Tassey, Gregory. 2000. Standardization in Technology-Based Markets. *Research Policy* 29:587–602.

Taylor, Chris F., Dawn Field, Susanna-Assunta Sansone, Jan Aerts, Rolf Apweiler, Michael Ashburner, Catherine A. Ball, Pierre-Alain Binz, Molly Bogue, Tim Booth, Alvis Brazma, et al. 2008. Promoting Coherent Minimum Reporting Guidelines for Biological and Biomedical Investigations: The MIBBI Project. *Nature Biotechnology* 26 (8): 889–896.

Taylor, Frederick Winslow. 1911. *The Principles of Scientific Management*. New York: Harper.

Taylor, Paul W. 1962. Can We Grade without Criteria? *American Journal of Philosophy* 40 (2): 187–203.

Thaler, Richard H., and Cass R. Sunstein. 2008. *Nudge: Improving Decisions about Health, Wealth and Happiness*. London: Penguin.

Thévenot, Laurent. 2006. *L'action au pluriel: Sociologie des régimes d'engagement*. Paris: La découverte.

Thompson, Paul B., and Wesley Dean. 1996. Competing Conceptions of Risk. *Risk: Health, Safety and Environment* 7 (Fall): 361–384.

Timmermans, Stefan, and Marc Berg. 2003. *The Gold Standard: The Challenge of Evidence-Based Medicine and Standardization in Health Care*. Philadelphia: Temple University Press.

Transition Town Lewes. 2008. *The Lewes Pound*. Transition Town Lewes. http://www.thelewespound.org.

Travaux du centre international d'études pour la rénovation du libéralisme. 1939. *Le colloque Walter Lippmann*, cahier no. 1. Paris: Librairie de Médicis.

Travers, Max. 2007. *The New Bureaucracy*. Bristol: Policy Press.

Trevelyan, Charles E. 1838. *On the Education of the People of India*. London: Longman Orme Brown Green & Longmans.

Turnbull, David. 1993. The ad hoc Collective Work of Building Gothic Cathedrals with Templates, String, and Geometry. *Science, Technology & Human Values* 18 (3): 315–340.

University of Phoenix. 2007. *University of Phoenix*.http://www.phoenix.edu.

Ure, Andrew. 1835. *The Philosophy of Manufactures*. London: Charles Knight.

Urmson, J. O. 1950. On Grading. *Mind*, New Series 59 (234): 145–169.

U.S. Citizenship and Immigration Services. 2006. *Naturalization*. U.S. Citizenship and Immigration Services. http://www.uscis.gov/portal/site/uscis.

U.S. Department of Agriculture. 1997. *Grain Inspection Handbook*. Washington, DC: Grain Inspection, Packers and Stockyards Administration, Federal Grain Inspection Service.

van Gennep, Arnold. 1960. *The Rites of Passage*. Chicago: University of Chicago Press.

Veblen, Thorstein. 1904. *The Theory of Business Enterprise*. New York: Charles Scribner's Sons.

Veblen, Thorstein. 1921. *The Engineers and the Price System*. New York: B. H. Huebsch.

Venkatachalam, K. V. 1970. Strategy for the Change-over. In *Metric Change in India*, ed. L. C. Verman and J. Kaul, 128–144. New Delhi: Indian Standards Institution.

Verman, Lal C., and Jainath Kaul. 1970. History of Metric Debate in India. In *Metric Change in India*, ed. L. C. Verman and J. Kaul, 15–56. New Delhi: Indian Standards Institution.

Viner, Jacob. 1940. The Short View and the Long in Economic Policy. *American Economic Review* 30 (1, Part 1): 1–15.

von Mises, Ludwig. 1939. Discussion. In *Le colloque Walter Lippmann*. Travaux du centre international d'études pour la rénovation du libéralisme, cahier no. 1, 41. Paris: Librairie de Médicis.

von Mises, Ludwig. [1927] 1985. *Liberalism in the Classical Tradition*, 3rd ed. San Francisco: Cobden Press.

Walzer, Michael. 1983. *Spheres of Justice: A Defense of Pluralism and Equality*. New York: Basic Books.

Walzer, Michael. 1994. *Thick and Thin: Moral Argument at Home and Abroad*. Notre Dame, IN: University of Notre Dame Press.

Waterman, Michael S. 2000. *Introduction to Computational Biology: Maps, Sequences and Genomes*. Boca Raton, FL: CRC Press.

Weber, Max. [1905] 1958. *The Protestant Ethic and the Spirit of Capitalism*. New York: Charles Scribner's Sons.

Weber, Max. [1922] 1978. *Economy and Society*. Berkeley and Los Angeles: University of California Press.

Weinberger, Sharon. 2008. Spotting the Hot Zones: Now We Can Monitor Epidemics Hour by Hour. *Wired*, July, 114.

Weller, Toni, and David Bawden. 2005. The Social and Technological Origins of the Information Society: An Analysis of the Crisis of Control in England, 1830–1900. *Journal of Documentation* 61 (6): 777–802.

Whipple, M. 2005. The Dewey-Lippmann Debate Today: Communication Distortions, Reflective Agency, and Participatory Democracy. *Sociological Theory* 23 (2): 156–178.

White, Lynn Jr. 1962. *Medieval Technology and Social Change*. New York: Oxford University Press.

Whitney, Albert W., and Herbert Hoover. 1924. *The Place of Standardization in Modern Life*. Washington, DC: Inter American High Commission.

Wiese, Heike. 2007. The Co-evolution of Number Concepts and Counting Words. *Lingua* 117:758–772.

Williams, Raymond. 1983. *Keywords: A Vocabulary of Culture and Society*, 2nd ed. New York: Oxford University Press.

Williamson, Oliver E. 1975. *Markets and Hierarchies*. New York: Free Press.

Williamson, Oliver E. 1994. Transaction Cost Economics and Organizational Theory. In *The Handbook of Economic Sociology*, ed. N. J. Smelser and R. Swedberg, 77–107. Princeton, NJ: Princeton University Press.

Willis, J. W. 1938. *One Variety Cotton Community Organization*. Mississippi State, MS: Extension Department of Mississippi State College.

Wilson, Edward O. 1998. *Consilience: The Unity of Knowledge*. New York: Alfred A. Knopf.

Wimsatt, William C. 2007. *Re-engineering Philosophy for Limited Beings*. Cambridge, MA: Harvard University Press.

Winner, Langdon. 1986. *The Whale and the Reactor*. Chicago: University of Chicago Press.

Wolf, Oliver, Dolores Ibarreta, and Per Sørup. 2004. *Science in Trade Disputes Related to Potential Risks: Comparative Case Studies*. Seville: European Commission, Joint Research Centre, Institute for Prospective Technological Studies.

World Commission on Dams. 2000. *Dams and Development: A New Framework*. London: Earthscan Publications.

World Health Organization. 2007. *International Classification of Diseases*. World Health Organization. http://www.who.int/classifications/icd/en.

World Standards Services Network. 2009. *WSSN: World Standards Services Network.* http://www.wssn.net/WSSN/index.html.

Wuthnow, Robert. 1987. *Meaning and Moral Order.* Berkeley and Los Angeles: University Of California Press.

Wynne, Brian. 1989. Sheepfarming after Chernobyl. *Environment* 31 (2): 10–15, 33–39.

Wynne, Brian, Michel Callon, Maria Eduarda Gonçalves, Sheila Jasanoff, Maria Jepsen, Pierre-Benoît Joly, Zdenek Konopasek, et al. 2007. *Taking European Knowledge Society Seriously.* Brussels: Directorate-General for Research.

Zamyatin, Yevgeny. [1924] 1993. *We.* New York: Penguin.

Zelizer, Viviana A. 1998. The Proliferation of Social Currencies. In *The Laws of the Markets,* ed. M. Callon, 58–68. Oxford: Basil Blackwell.

Zerubavel, Eviatar. 1982. The Standardization of Time: A Sociohistorical Perspective. *American Journal of Sociology* 88 (1): 1–23.

Zimmerman, Ann S. 2008. New Knowledge from Old Data: The Role of Standards in the Sharing and Reuse of Ecological Data. *Science, Technology & Human Values* 33 (5): 631–652.

Index

Printed in the United States
by Baker & Taylor Publisher Services